Till / a Symposium

TILL

a Symposium

Edited by Richard P. Goldthwait

Assisted by:
Jane L. Forsyth
David L. Gross
Fred Pessl, Jr.

OHIO STATE UNIVERSITY PRESS

Copyright © 1971 by the Ohio State University Press
All Rights Reserved
Library of Congress Catalogue Card Number 70-153422
Standard Book Number 8142-0148-2
Manufactured in the United States of America

Dedication

To George W. White,
at the time of his retirement,
to honor his devotion, inspiration,
and teaching on the subject of till.

Contents

Preface

Symposia can be challenging, or deadly. This subject, till, excited the two hundred participants so that they urged that all the papers be gathered in one volume. These papers challenged a few participants to get out and solve some problems in the field, and some differences were argued on the floor. Subsequent solutions to many of these problems are incorporated in the articles published here. The symposium may guide many in the future to engage in more discerning or more useful investigations of glacial till.

The idea of glacial till as the subject of a symposium was originally suggested by Professor George W. White of Illinois, and enthusiastically endorsed by this editor, and by Professor Dreimanis and the local committee. It was a subject which was to attract a large group of geologists to the regional (North-Central) meetings of the Geological Society of America in Columbus, Ohio, in 1969. The oral papers were invited by open letter in November, 1968. The subject was so timely that it elicited 24 papers (plus three "by title"), which were formed into three half-day sessions on May 15 and 16, 1969. Twenty of these oral presentations are given here as full-length papers in which are incorporated additions and emendations in response to the lively floor discussion.

To the original twenty papers are added two new complimentary studies: (1) a general subject and literature review, with an appraisal of the status of till studies, written by this editor, to relate the papers and to set them in context; and (2) one of the very few recent detailed studies of till in and on an active glacier by G. S. Boulton. As with all spontaneous papers, these do not cover every facet of the subject; soils and weathering aspects and applied and economic studies are conspicuously absent. The subjects do conveniently fall, however, within certain areas of current interest: genesis, structure, stratigraphic correlation, composition, and fabric. The papers are designed on a professional level for the co-worker and scientist in related fields. This volume should serve as a reference volume for all interested in this common subject.

There are a few related papers presented at that Columbus meeting which are not here, either because they were submitted elsewhere or were not completed in publishable form. These papers are listed below, and most of their abstracts are published in "Abstracts with Programs for 1969, Part 6, North Central Section, The Geological Society of America":

Brace, B. R., "Till identification methods recently used in Butler and Preble Counties, Ohio."

Favéde, F., "Differentiation of ice deposits of Wisconsin age in Portage County, Ohio."

Flemal, R. C., Hinckley, K., and Hessler, J. L., "Fossil pingo field in north-central Illinois."

Harris, S. A., "The meaning of till fabrics" (title listed without abstract).

Jacobs, A. M., "Sand-silt-clay ratios of two Illinoian tills."

Karrow, P. F., "Character and variations in multiple-till sequences in the Waterloo interlobate area of Ontario."

Lindsay, J. F., Summerson, C. H., and Barrett, P. J., "A long-axis clast fabric comparison of the Squantum 'Tillite,' Massachusetts, and the Gowganda Formation, Ontario."

Schafer, J. P., "Structural relationships of tills in western Connecticut."

Sitler, R. F., "Weathering in glacial till."

Tipton, M. J., and Leap, D. I., "Pre-Wisconsin glaciation as the builder of the Coteau des Prairies."

Myriads of earlier studies of glacial till in English are referenced at the end of each paper here, especially the first. Still other older sources are listed in Charlesworth, J. K., "The Quaternary Era", v. 1, Chapters XVI to XX, p. 359-414. One symposium on till was held (1966) in Poznan, Poland, and is soon to be published in languages other than English. This volume is the first group of papers in English on till.

Not the least of the indirect contributions to quality in this volume are the high standards and careful critique performed promptly and eagerly by the assistant editors. To them and their organizations (Bowling Green State University, Illinois State Geological Survey, and United States Geological Survey) is due gratitude for the many "free" hours, and three trips to Columbus for editorial meetings. The cochairman of the original oral sessions, R. F. Black, R. H. Durrell, W. R. Farrand, J. L. Forsyth, A. Gooding, J. L. Rau, A. F. Schneider, and W. J. Wayne, carefully steered and conducted the discussions. This is, then, an integrated published effort of a large group of experts.

Richard P. Goldthwait

Introduction 1

Till / a Symposium

Introduction to Till, Today

R. P. Goldthwait

WHY TILL?

Till is the only sediment stemming directly and solely from glacial ice. It and its rock equivalent, tillite, are used to identify the former presence and extent of glaciers, and to define the sequence of climatic events leading to repeated glaciations extending far into temperature latitudes. The unconsolidated deposit, variously referred to as hardpan, boulder clay (Charlesworth, 1957), unsorted glacial drift, Geschiebelehm (German), keileem (Dutch), or just "moraine," forms patchy to continuous cover over 30 percent of the continents. Another 10 percent or more of the land may be underlain by consolidated tillites from former glaciers hundreds of millions of years ago (principally Permocarboniferous and Huronian).

In addition to its universality, till holds the secrets to the action of ice at the bottom of a glacier. Tunnels and stream openings under glaciers penetrate only the outer few tens or hundreds of feet of ice, and involve an artificial boundary condition. Actual drilling in deep ice is just beginning to supply some answers. Another approach to basal action in ice comes from theoretical physics (Nye, 1952) and has grown rapidly since 1960 (e.g., Shumskii, 1964; Lliboutry, 1964; Pounder, 1965; Oura, 1969), but in most ways it is still untested. One paper here (Nobles and Weertman, 1971; this volume, p. 117) is concerned entirely with this approach. So we rely on the product itself, till, for many answers. It has been studied far too little; indeed, this is the first compendium in English on the subject.

Positive identification of the glacial origin of till is important, both as to interpretation of the extent and frequency of drastic prehistoric climatic events, and as to the tracing and locating of outcropping mineral resources, overrun and buried by the glacial drift. The latter is an art long developed in Finland and rapidly developing in Canada. Till is also the foundation material with which almost every engineer and builder in latitudes above

40° must deal. Till is often thin (White, 1971; this volume, p. 149), but it is the top material upon which roads, houses, factories, and power lines must be based. Loose sandy till may contain water, or if compact it may form a cover of low permeability which holds in the water in most areas of thick drift. Thin, loose till acts as a sponge to bring the precipitation to joints in the underlying rock. Any waste disposal in shallow wells or sanitary landfill in glacial drift requires a knowledge of the continuity and permeability of tills. This volume is not concerned with engineering properties *per se*, though they can be a key to till identification because the chemical, lithologic, and mineralogic compositions determine its physical properties.

WHAT IS TILL?

Till was defined originally (Geikie, 1863, p. 185) as "a stiff clay full of stones varying in size up to boulders produced by abrasion carried on by the ice sheet as it moved over the land." Till is a clastic sediment, not a landform (Harland and others, 1966). The word *moraine*, on the other hand, commonly and confusingly implies a hummocky landform. Perhaps at the start it was considered to be a material, for de Saussure (1803, p. 253) called it "this mass of sand and clays which are deposited at the edge of a glacier moraine." But common use for a century has made moraine a drift form, deposited chiefly by direct glacial action, and having constructional topography. Although most moraines in the United States are shown to be composed of till, and the term *ground moraine* is used synonymously in Europe (Woldstedt, 1961, p. 90), some of the most famous end moraines, like the Kettle Moraine in Wisconsin or Salpausselka-Central Swedish-Ra Moraines in Scandinavia, are composed of sorted material, in beds. Thus moraine is not a good synonym for till.

Till is one of the two kinds of material implied by the term *glacial drift* (Lyell, 1839). It is the non-sorted variety, in contrast to *sorted* glacial drift, which includes the sand and gravel of high-energy glacial streams (glaciofluvial), or the sand, silt, and clay of most glaciolacustrine (low-energy) environments, or fine sand, silt, and coarse clay of most glacioeolian blankets. All of these may represent facies resulting from but one glacial event, but, unlike the till, they involve clear-cut transportation and deposition by an agent other than ice.

Till has more variations than any other sediment with a single name. The most-used hallmarks of identification are: (1) a lack of complete sorting, which usually means the presence of some pebbles or boulders much larger than the dominant clay, silt, or sand; (2) a homogeneous mix lacking any smooth lamination or regular graded bedding, combined with (3) a mixture of mineral and rock types, some from far away. There are a number of additional identifying features frequently found associated with till: (4) at least a small proportion of striated stones and microstriated grains; (5) common orientation of the long axes of elongated grains and pebbles; (6) compactness or close packing as contrasted to neighboring sediments, due to the variety of grain sizes available and to the pressures involved; (7) a striated surface on the rock or sediment basement beneath it; and (8) subangularity in clasts of all sizes, due to frequent breakage in transport coupled with partial smoothing by abrasion.

When is till not a till? The conservatives say that, as soon as the clastic material is handled by any agent other than glacier ice, such as being moved in a landslide or settling through sea water or being rolled fifty feet in rivulets, it bears another stamp, so it should carry another name. Liberals say that, as long as it has the imprint of glacial ice (striae, lack of sorting, mixed lithology including material of distant transport), the word *till* should be used. One case in point is the "till balls" (Pettyjohn and Lemke, 1971; this volume, p. 383) which are unaltered internally, but changed in orientation. Here the obvious solution is to use that dual name "till ball." Another is flowtill (used by Hester and DuMontelle, 1971; this volume, p. 367), which forms by mudflow off the ice (Hartshorn, 1958). Less clear is "marine-till," which settles through salt water beneath floating shelf ice, because it may be confused with material which the glacier scraped up off the sea floor as it gathered sediment (e.g., "shelly-till" in Ireland: Synge and Stephens, 1960). At least, everyone agrees that the word till *does* mean glacial handling.

More difficult to name is the marginal material which has one or two characteristics of till, but which may have a completely nonglacial origin. Certainly there are mudflows in tropical mountains, solifluction materials on Alaskan-Siberian thawed slopes, colluvial soils in most wooded temperate areas, laha deposits far across valleys of some active volcanic regions, and continental-slope sediments under a hundred or more fathoms of sea water, which do have a mechanical composition identical to till, and which often have a few scratches with some mixing of lithologies. The word *diamicton* has been proposed (Flint, Sanders, and Rogers, 1960, p. 1) and it enjoys a limited use for "essentially non-sorted, non-calcareous, terrigenous deposits composed of sand and/or larger particles in a muddy matrix." The rock is *diamictite*, and tillite would be one specific subclass. Certainly it is to be preferred over "till-like" or "tilloid" (Pettijohn, 1957; Harland and others, 1966). All classifications agree on glacial involvement in "till," and only an astute and extensive appraisal of all the criteria on a regional basis will show whether a given diamicton is till or not (e.g., Hester and DuMontelle, 1971; this volume, p. 367).

What proportion of an exposed mass of glacial drift must be non-sorted and non-bedded to call it till? Nearly every drill hole or large cut in till exposes some inclusion of well-sorted clay or silt, or bedded sand, or a lens of gravel. Practical experience suggests that, where the relatively non-sorted material makes up more than 50 percent of the entire mass or stratum, the whole group of subunits may still be generalized as till. And clearly, with such variation in sorting, till is just one end member in a graded series which at some point, say at Trask coefficient 4.5 (which compares two quartile points on the cumulative curve $\sqrt{Q_3/Q_1}$), becomes sorted drift. Better measures of grain-size deviation today are phi standard deviation measures, which represent the spread of the grain-size curve about the median size (Inman, 1952, $\dfrac{P_{84} - P_{16}}{2}$), but it is impossible to set a practical specific limit. More important is the homogeneity of the layer and the relative spread of sizes in contrast to adjacent sediments of similar lithology.

WHERE IS TILL FOUND?

The seasoned glacial geologist tends to forget some simple empirical truths about till distribution, which stand out in the hundreds of areal studies. Some possible implications of these are drawn below.

1. The major volume of all drift, including till, is within the *outer* third of the ice-covered area of any one glacial episode. From all knowledge of glaciology today, we infer that it is generally a one-way transport.

2. Till becomes generally thinner, sometimes to the extent of being just a few stones and boulders, as one proceeds geographically from the ice edge to the center and source of ice movement, or topographically from the valley floor to high prominent bedrock knobs covered by ice.

3. Tills must necessarily postdate the markings on the ledges or layers beneath them even though both may stem from a single glacial episode. In other words, at any spot, till is deposited *after* erosion is largely completed. This probably means most tills are emplaced late in the glacial cycle, and all covering till layers come later; this is the old law of stratigraphic superposition.

4. Multiple till layers are common in the outer peripheral part of every large glaciated area. Some represent changing fluxes in a single glacial episode, so two tills generally do not demonstrate two or more separate glaciations. Lengthy ice withdrawal is shown in some sequences by interglacial sediments, organic deposits, weathering

horizons, or buried periglacial features, but one cannot develop a safe glacial chronology even here by just counting the layers.

5. Till is a uniform mixture of most or many things which the basal ice passed over, including soil, organic matter, bedrock of any age, and preglacial regolith. And the lowest part of the till at any spot is *most* like the material just beneath it.

6. It is remarkable that former (Pleistocene) glaciers in repeated episodes so often started from the same general central area and ended within a few miles of the same outer boundary. The physical conditions for repetition of glaciation (cycling climate and ice dynamics) seem to have remained the same many times over.

Correlation of till sheets and masses from exposure to exposure, or from drill hole to drill hole and outcrop to outcrop, is the big problem and objective. Most of the following papers and one special section (p. 165-233) are concerned with this. Where unusually long exposed sections have been available, as in strip mines at Danville, Illinois (Johnson and others, 1971; this volume, p. 184), or in long interstate road cuts (White, 1971; this volume, p. 149), the individual till layers are seen to be discontinuous. One wonders how, in till studies in northern Vermont (Stewart and MacClintock, 1971; this volume, p. 106), the *outstanding* criterion of correlation is pebble fabric, and in northeastern Ohio it is granulomatric analysis (Shepps, 1953), whereas in southern Ontario heavy-mineral and carbonate characteristics were mostly relied on (Dreimanis and others, 1957; Dreimanis, 1961); or in western Ohio it is chemical analyses and clays (Steiger and Holowaychuck, 1971, this volume, p. 275; and Wilding and others, 1971; this volume, p. 290) or in central-southwestern Ohio it is pebble lithologies (Goldthwait, 1969), while in southern New England it is color (Pessl, 1971; this volume, p. 92), and in Illinois tills are "tied down" by clay mineralogy (Lineback, 1971; this volume, p. 328, or Willman, Glass, and Frye, 1963) and by color (Kempton and others, 1971; this volume, p. 217). The approach to till is still empirical. Interpretation and physical justification follow analysis of field data. Certainly each of these workers has used and generated data based on several criteria and used several of these for his decision. But while to some extent these single criteria may indeed identify valid regional and stratigraphic differences between tills, certainly they reflect also the care and persistence of each investigator with the techniques which he found *best* in his earlier work. As judged from the cross-continent survey of till laboratories (Dreimanis, 1971; this volume, p. 27), it may be essential to use several carefully documented criteria and to establish their validity in each area (as is done by Christiansen, 1971; this volume, p. 167; or Westgate and Dreimanis, 1967). It is to be hoped that techniques can be streamlined to the essential, quick, and diagnostic parameters, through studies like those in this volume, without thousands of hours wasted in calculating needless statistics for the correlation of tills.

In making areal correlations by such local and variable parameters, head-on arguments are bound to appear. One here concerns the thin tills (White, 1971; this volume, p. 149), for Totten's (1969) recent deduction that the Powell and Broadway Moraines in central Ohio are buried topographic forms inherited from Altonian time (50,000 years ago) and covered by three thin tills only one to a few tens of feet thick flatly denies other studies just to the west in Delaware County. These soils studies involve calcium/magnesium ratios and silt/clay ratios of till four to six feet down (Steiger and Holowaychuck, 1971; this volume, p. 275), or mechanical analyses and stone counts (Goldthwait, 1969) which indicate a clear surficial break at these moraines and support a late Wisconsin (16,000 B.P.) readvance of ice. It is hardly likely that one section of a continuous belt of end moraine is 50,000 years old while the next part in the same county is 15,000 years old.

HOW DOES TILL FORM?

Entrainment

Evidence for the entrainment of till into the glacier ice comes primarily from glaciological studies. The bulldozing of ridges or piles of already-loosened rock debris or soil subsequently incorporated in the ice and carried off is a popular concept which is almost never seen. Only in certain end moraines at the former glacier periphery (in Alberta, Martha's Vineyard, Denmark, Germany, and Poland) are there large masses of preglacial rock piled up by marginal ice, and these are not overridden far or carried onward. Perhaps this is because most glacier advance is in the long autumn and winter season; either permafrost or temperate terrain is frozen hard then. One such study of a slow advance in Greenland (Goldthwait, 1960, 1971) concluded that the ice front was cliffed, and calved off as the ice tended to roll over into winter snowdrifts.

Dirt is brought forward from within the advancing ice sheet, and is found in closely stacked basal layers. Debris may include large lumps of frozen outwash, or lenses of solid till or boulders, but layers generally become thicker and debris more massive nearer the bottom of the ice at its outer edge. In spite of a very dirty surface appearance, random sampling of the debris-ice mixture at several sites in Greenland yielded a permanent solid-debris content of only 0.03 to 3 percent by weight and to 50 percent locally in dark bands of Spitsbergen ice sheet (Boulton, 1971; this volume, p. 41), or 64 percent in the dirtiest part of the Casement mountain glacier (G.

McKenzie, personal communication, 1970). The observed dirty-layered ice seen on dozens of glaciers, some advancing but most retreating, extends generally only three to ten feet above any marginal rock hill. Rock debris is abundant and extends 20 to 100 feet above any valley floor. These common observations point to the bottom of the glacier as the source of dirt, and imply that more debris comes down valleys, and that it is raised highest above the valley floor.

In many places above the sharp debris zone with mixed clasts, there is a zone of brown or yellow ice called "amber ice" half as thick as the dirty ice beneath (Goldthwait, 1960). Out of it one can melt only a few grains of very fine sand, some silt, and much clay, all of which are disseminated uniformly throughout ice crystals. These grains are very sparse indeed (under 100 ppm, 1 cc/m²). Presumably they deploy by upward diffusion as a result of vertical movement of the lower ice. This has been worked out theoretically by Nye (1957) and Weertman (1968a) and applied to amber ice and salts in a cold (−18°C) slow-moving Antarctic glacier (Holdsworth, 1969).

Arguments have been developed over 20 years concerning just how coarse dirt is raised up into the ice. Goldthwait (1951) measured the dips of dirty *shear planes* in the Barnes Ice Cap (10° to 36°) and calculated that these curve back into the ice so as to intersect the ground (a potential debris source) 1,600 feet back and 250 feet under clean white ice. Long tunnels into basal Greenland ice did indeed show nearly horizontal dirt bands as far back as 1,100 feet (Swinzow, 1962). A shorter tunnel under nearby Red Rock Ice Cliff showed such bands only one to three feet above the bottom 60 to 100 feet from the margin (Goldthwait, 1960). Rarely has the rate of shearing been measured, but in each case, some peg or a grid showed sharp distortion, so the only question is, how sharp or discrete are the shear planes. If local shearing near the edge were true in all glaciers, till sheets would have very local sources and would be built seriatim as the ice margin retreated. Significant but small percentages of far-travelled (300-500 miles) pebbles, such as the two to 25 percent crystalline pebbles in till of Ohio, Indiana, and Illinois, suggest that this marginal shearing is not the only mechanism of transport.

Material is frozen into basal ice by *regelation* wherever ice slides (Weertman, 1961b). Actual and theoretical studies (Kamb and LaChapelle, 1964; Peterson, 1969; Holdsworth, 1969) of clear regelation layers trapping air bubbles or dirt film, stemming from small bedrock or drift knobs, do confirm that rock material becomes frozen and encased by thin layers of ice wherever pressure-melting occurs. Theoretically this should be prevalent at the base in the large central parts of an ice sheet, where earth and frictional heat melt the ice (Weertman, 1961a; Gow and others, 1968), or under the ice margins in a temperature climate.

Delicate structures in soils, plants, and laminations are sometimes preserved in thicker masses of debris within bottom ice, so some other additional mechanism captures masses of frozen ground. Perhaps winter-frozen layers of wet ground or ground above cryostatic permafrost pressure are vulnerable. These freeze-on mechanisms may well involve colder basal ice ($-1°$ to $-5°C$) nearer the margin. In several documented cases of thrust-stacking of large till sheets, hydrostatic pressure under the melting basal ice has reactivated earlier-deposited till (Ramsden and Westgate, 1971; this volume, p. 325; Moran, 1971; this volume, p. 127; or MacClintock and Dreimanis, 1964).

One more obvious source of till matrix is the glacial *abrasion* of rock material beneath the base. Physically this is possible wherever the rocks are massive and firm enough to be engraved and polished in place, and the ice **does carry** some rock fragments. Every field glacial geologist knows that some rock materials, like hard felsites, vein quartz, or fine limestone, take striae and polish better than do many other types of rock. This must create the fine powder (Glen and Lewis, 1961) that is called glacial rock flour in till, so it is no surprise that 60 to 90 percent of the silts in many tills have shown freshly ground quartz, calcite, or feldspar predominating, rather than weathered clay minerals (L. P. Wilding, personal communication, 1970). However, since Krynine (1937) could find secondary weathered iron minerals making up nearly 25 percent of the mineral grains in one western till, there may be a liberal mixture of preglacial soils as abrading material in the glacial ice (Feininger, 1970). In spite of spectacular corrugated grooves and ridges, each abraded by thousands of cutting grains to depths of up to six feet (Carney, 1909; Smith, 1948), there is local evidence, through studies of joint spacing on one hill of granite (Jahns, 1943) that lee-end plucking went ten times as deep and therefore ten times as fast as stoss-side abrasion. On lee slopes plucking may well outrun abrasion as a source for clasts.

The basal till in some deposits gives evidence that fissile rocks like shale *stream off* in a train of chips down-glacier. The sources described are usually a cliffed, rough slope facing away from the oncoming ice. Similarly, in the lee of coarse granitic ledges there have been found large angular and rectangular pieces which fit back exactly into niches left in the bedrock. Since the yield stress of granite is on the order of 100 to 200 times the common stresses in a glacier base (Glen and Lewis, 1961), it is argued that the granite must have been already jointed and greatly weakened by prior strain. Pressure changes under thick ice, resulting from kinematic, surge, or even accumulation waves, can repeatedly alter the local pressure-melting point the fraction of a degree necessary to force water into joints and then expand the water to ice (Holmes, 1944). Measured velocities on the stoss slope of a rôche moutonnée under an alpine glacier were sufficient to produce water which might seep into joints on the lee side and expand to ice (Carol, 1947).

Debris is acquired by *mass movement* from slopes along a valley or around a nunatak. Where the ice edge moves along the contour of a slope, the additional processes of atmospheric freeze-thaw are involved. Probably this creates movable joint-pieces, a process described originally as sapping (Johnson, 1904) and later shown (Battle and Lewis, 1951) to be limited to near-surface situations. Certainly it involves nivation (Bowman, 1916), the extra-rapid comminution of debris adjacent to the ice. Mass movement (wash, rockfall, debris avalanche, mudflow, etc.) and revitalization of lateral dead ice will introduce some of this material onto the moving ice. At every glacier junction and every termination of a protruding knob, the stream of rock particles from high valley-side sources becomes a *medial moraine.* Thus it introduces rock debris well up in the ice mass (Sharp, 1948). It has been proposed that the rougher the terrain inundated by a glacier, the higher and thicker is the debris train in the ice. Certainly striae "wrap around" higher mountains, but they cross the top directly, so a debris stream must trail off to the lee as is observed at every nunatak near the margin of continental ice.

Transport

Four things may happen to the larger clasts as they travel in the ice from point of pickup to point of deposition: (1) they may become worn, striated, and smoothed by glacial abrasion; (2) they may become broken, returning to angular shapes in small sizes by surface shattering and/or ice-bottom conflicts; (3) they may become oriented preferentially by shear; and (4) they may become displaced along a path related to the flow lines within the ice.

Where does the debris ride? It is certain that some clasts do change from a basal to englacial position, but the material still moves in the lower 3 to 300 feet of ice, for no debris is seen over most of the surface of a continental ice sheet. Much of this same material will go from englacial to superglacial position by ablation just on the marginal slope. Of course, in mountain glaciers there are broad lateral moraines exposed below the snow line, and, where these become medial moraines or heavily crevassed, some debris may fall in and become englacial again.

No maximum size is commonly recognized for the size of rock fragments which ice can carry. Meneley (1964) has calculated, from the physical properties of ice (yield stress 0.3 to 1.5 bars), that more or less equidimensional boulders up to 15 or 75 feet in diameter can be stably supported by the shear stresses acting on the sides and base of the boulder within the moving ice. Tabular slabs 5 to 25 feet thick can only be as much as 1,000 feet long if they are to be transported by ice. These are approximately the sizes of the common largest erratics (Flint, 1957, p. 130). Every ice movement must mix all sizes of clasts, for this mixing is characteristic of the till it leaves. Apparently there is no settling or sorting mechanism within glacier ice, unless it be the raising of the finest clasts into the amber ice.

The abrasion of clasts en route was described two centuries ago, but only within the last decade has quantitative shape analysis been made. Holmes (1960) studied shapes of over 3,000 limestone, sandstone, and shale pebbles and cobbles from known sources in upstate New York to demonstrate that sedimentary pebbles and cobbles increased in ovoid to discoid shape and in degree of rounding with distance. Drake (1968) showed that initial shapes of 1,852 clasts from New Hampshire depend upon the breaking characteristics of the 16 igneous and metamorphic lithologies, but rounding increased greatly (from Krumbein numbers 0.1 to 0.5) either in the first mile (1.1 miles) of artificial tumbling or in the first mile from the bedrock outcrop. Some shapes, such as rods, are preferentially destroyed. Others, such as the dolostone studied by Vagners (1966), dominantly tend to be crushed, so that rounding of only 0.2±.05 is maintained. The half-rounded, average material is maintained at a steady state of rounding, in spite of new breakages, until all clasts of the given size range (pebble) are gone. Less than 0.1 percent of any lithology of clasts survives beyond 21 miles. Drake (1971; this volume, p. 73) shows how this is associated with basal transport only.

Striations engraving the soled and faceted surfaces of glacially transported clasts are the clearest evidence of abrasion. Flint summarizes the numerous observations: "Usually no more than 5% to 10% of the stones in a till of varied lithology have striations" (1957, p. 117). Among thousands of pebbles, Drake (1968) finds 2.9 percent of the crystalline rocks (schists, gneiss, granite) to be striated in "hard" basal New England tills and only 0.1 percent in "soft" tills, whereas Holmes (1960) finds 27.9 percent of the glacially transported fragments of sedimentary rocks (sandstone, limestone, shale) in New York to be striated. Soled or "rocker-shaped" facets with slightly criss-crossing striae are ascribed to rotational forces tipping the stone forward into the place of greatest shear as it moves differently over englacial debris below, or as it slides against till and bedrock beneath a wet, melting base. Perfectly flat facets with parallel striae indicate the more fixed grip of a rock held rigidly in compact till and abraded by the dragging of rock tools over it. A mixture of these two types of surfaces in some tills indicates that pieces may be stopped and shaped at the bottom below the ice (boulder pavement) and then picked up again.

Attrition to smaller and smaller pieces during transport is well established. Several early glacial geologists (e.g., Shaler, 1896) showed this roughly with figures on boulder size from a unique, identifiable small source.

The pebble-boulder ratios in upstate New York indicate that limestones are more crushed en route, and sandstones have superior resistance (Holmes, 1960). Apparently most of the reduction is through breakage. Drake (1968) found that the average time between crushings is sufficient for an average pebble to reattain a roundness of 0.5 and for "most pebbles to go through a minimum of several cycles in the first mile." Peterson (1969) actually observed one cobble emerging over a rock knob just as it was crushed. Dreimanis and Vagners (1971; this volume, p. 237) show that there is a terminal grade size for each lithology: fine sands are the final stage from granitic and metamorphic rocks, silts from carbonate rocks, and clays from shales. Such attrition emphasizes the particle interference and conflicts actually seen under temperate (near $0°C$) glaciers by Carol (1947) and by Peterson (1969).

Many have observed, on medial and ablation moraines, that boulders of one or more distinctive lithologies are reduced to a heap of angular pebbles and gruss (Sharp, 1949). Statistical association of angularity with superglacial features (Drake, 1971; this volume, p. 73) argues that superglacial materials suffer far more breakage (angularity) than do basal clasts, and that many end up angular with little or no abrasion. Salt solutions penetrate into fine cracks and expand into salt crystals, and filaments of interstitial water expanded into ice, so either or both are blamed qualitatively for rapid physical breakdown of blocks in ablation moraine. Ablation moraine on the low part of a glacier does have a microclimate where humidity varies greatly and temperatures pass the freezing line very frequently.

Chemical weathering of clasts is now believed by most authors to have been almost exclusively a preglacial process amplified only during superglacial transport. Statistically, the compact basal tills have half as many (5 percent) rotted crushable pieces among 29 lithologies as do loose ablation tills (Drake, 1971; this volume, p. 73). Water-worn sorted deposits had still fewer (3.4 percent) rotten pebbles. Stained oxidation rims and a crumbling shell are common on a quarter of all pebbles in most tills. Since these occur right next to unstained solid pebbles of the same lithology, the chemical alteration is interpreted by many (e.g., Goldthwait and Kruger, 1938) to have taken place prior to glacial transport. The destruction of clasts by weathering of a till *after* deposition is well recognized in soil studies (Wilding and others, 1971; this volume, p. 290).

Orientation of pebbles in ice during transport was recognized in glaciers over 30 years ago (Perutz and Seligman, 1939; Richter, 1932), but only recently has the interest in deposited till-fabrics, arising out of Holmes's study (1941), led to a flood of till investigations. Fabric in till was noted over a century ago (Elson, 1965). Theoretical studies (Glen, Donner, and West, 1957) based on Jeffrey's fluid-flow equations show that elongate pebbles should have a "flip-flop" rotation where shear is involved, but elongate and flattened pieces spend more residence time in a near-horizontal attitude. Well-oriented elongate pebbles have been noted in basal parts of several glaciers (G. D. McKenzie, personal communication, 1970), but there remains some question as to whether this is the set retained in deposited till fabric (Evenson, 1971, this volume, p. 345; Young, 1969).

Direction of transport has been interpreted from the positive identification of indicator rocks and minerals from well-mapped source outcrops in literally hundreds of studies in many countries. Abundant tracers are used to follow rare or valuable materials (diamonds: Gunn, 1968). Peculiar heavy-mineral grains, such as differently colored garnet grains, are used to identify sources of lobes (Dreimanis and others, 1957). Adams (1960) has studied sources from which lobate motion may spread materials 2° to 10° sidewise. However, the spreading fan of distribution from small oval sources (Shaler, 1893; J. W. Goldthwait, 1925; R. P. Goldthwait, 1968) indicates a greater lateral dispersal of from 20° to 60°. The exponential density distribution of pieces outward from the axis of a boulder train and down the center line of some known fans (Krumbein, 1937; Dreimanis, 1956) denies simple variation in general ice-flow direction, but it suggests some basal mechanism of lateral dispersion. Recent physical theory calculating regelation around small basal obstructions (Weertman, 1961a) and divergent accelerated flow around larger obstructions (Carol, 1947; Lliboutry, 1959) certainly can explain intricate lateral dispersal beneath even unidirectional upper ice. Partial divergence of striae obliquely around a rôche moutonnée is common, and complete deflection into a larger oblique valley may deviate 45° either side of the regional mean.

Distance of transport is equally well studied in myriads of places (especially Gross and Moran, 1971; this volume, p. 251, and Drake, 1971; this volume, p. 73), but two very opposed interpretations arise. All studies agree that some clasts are carried many hundreds of miles. From Ohio to Iowa these are the Canadian crystalline rock types, because these are the only coarsely jointed types of rock which quarry out in large pieces. Shales produce a constant quantity of small fragments in till immediately above the outcrop but disappear within a mile of the source, and rarely produce boulders because they grind to silt and clay so readily (they are prolific in sizes below 1 cm.). In studies of the pebble lithology of several lobes, Anderson (1957) concludes that local bedrock did not make significant contributions to the till. Harrison (1960) made one of the most detailed studies of the sources of all sizes of material at five adjacent cuts in Indiana, in contrast to many of the articles here which use certain size fractions of the till. The three or four till sheets involved were so similar as to be indistinguishable. Assuming a flow path from the final ice center, parallel to existing striae, the weight percent of material from each source rock mapped (e.g., 33 percent shale, 18 percent limestone, 29 percent crystalline) is equal to the total area of source rock. In other words, bedrock 200 miles away contributed just as much debris as bedrock two miles away. All till was carried from its pick-up point to its destination and then deposited en masse after motion. But some assumption as to the identity of clasts in the Indiana study can be argued.

The great number of authors with detailed studies of certain clasts have arrived at the opposite conclusion: "more than 90 percent of the exposed drift of this region (New York) was transported less than 50 miles" (Holmes, 1952). To be sure, there is a great difference between grade sizes introduced from each lithology, and important percentage effects result from the dilution provided by one prolific bedrock (e.g., Gillberg, 1965). These are discussed by Gross and Moran (1971; this volume, p. 251) and by Dreimanis and Vagners (1971; this volume, p. 237). Careful studies of matrix fines would probably show fairly large quantities of pulverized distant material at terminal size of grinding, but the middle-sized material is local in inverse proportion to its resistance to crushing-weathering and modified by dilution from other sources. Large clasts come from as far away as one must go to find coarse-jointed bedrock.

Deposition

Debris entrained and carried by moving ice must be emplaced either underneath the ice, where the basal ice-debris mix last moved, or at the surface of the glacier where the ice melts out from underneath. All englacial debris thus becomes subglacial by bottom melting or superglacial by surface ablation, even if for only a minute, as it is being deposited.

Nearly a century ago (Torrell, 1877; Upham, 1891), tills were divided into a "lower" compact *basal till*, pressed down hard by the ice and an "upper" loose *ablation till*, sloughed off a melting ice surface. This is still a principal theme of one whole section of these papers (Drake, 1971; this volume, p. 73; Pessl, 1971; this volume, p. 92; and Stewart and MacClintock, 1971; this volume, p. 106). The primary evidence of the origin of till lies in structures found in the deposited till, whereas most of the secondary evidence, such as clast shape, assumes that debris carried near the base (striated pieces, etc.) does indeed become deposited at the base and that debris reaching the surface (broken, angular, and lacking fines) is let down from that surface. Recent studies on Spitsbergen ice sheets (Boulton, 1969; and 1971; this volume, p. 41) question the "criteria which have previously been accepted" and subdivide the basal tills further into those smeared on at the bottom versus those which just melt out at the bottom, and the supraglacial tills, which slide or flow off the ice surface, versus those which just accumulate on the surface (Boulton, 1971; this volume, p. 41).

For till deposited at the bottom of the glacier, several mechanisms have been proposed, but none has been demonstrated and none excluded by actual observations. (1) Particles might collect one by one through frictional interference and catch on an obstacle already present on the floor; certainly the ice at tunnel floors bears delicate groovings. (2) Sheets of till or ice-debris mix could be left behind or "sheared" over wherever the friction on the bottom becomes so great that the basal velocity is reduced to zero. Later any entrained ice would melt from above or below. (3) Basal melting may simply bring slow-moving bands of englacial debris close to the floor and abandon them as the ice matrix melts from beneath (Nobles and Weertman, 1971; this volume, p. 117). Some shear and thrust structures are interpreted to show this (Pessl, 1971; this volume, p. 92). Basal melting is more permanent than the temporary regelation around obstructions (p. 9). It is shown to exist beneath central parts of cold arctic ice sheets ($-5°$ to $-20°C$), because thermal gradients trend toward $0°C$ in deep holes in all large ice sheets today (Weertman, 1968b; Gow and others, 1968). Calculations of the average earth's heat flow from below indicate that geothermal heat must melt nearly one centimeter of ice per year. Motion measurements indicate that an additional one to four centimeters must melt by the heat of shear friction, if the ice is moving (Robin, 1955; Weertman, 1964).

Perhaps *lodgement* is the right word for these combined processes. T. C. Chamberlin defined "lodge moraine" (1894, p. 28): "A glacier may fail to carry forward to its actual extremity the material which it is pushing at its base, and this may lodge under the margin, forming a submarginal accumulation." Flint (1957, p. 120) refines this to lodgement till "deposited from drift in transport in the base — specifically under the surface — of a glacier." Flint further deduces that "slow pressure melting of the flowing ice frees the particles and allows them to be plastered, one by one and under pressure, on the subglacial floor. . . ."

Two other hypotheses for final late movement arise from observations of fluidity of till to knee depth next to receding temperate glaciers, coupled with recent discoveries about bottom melting. (4) MacClintock and Dreimanis (1964) and now Evenson (1971; this volume, p. 345) show that reorientation of fabric in saturated till under deep ice may occur to great depths (35 feet). (5) An old idea of wave-induced ice erosion (von Engeln, 1937) was shown to be doubtful in that application (Demorest, 1939), but it is about to be revived for ice flowing over completely saturated till of lower viscosity. Under conditions of convective (non-laminar) and compressional ice flow, this would produce alternate bands of till squeezed out from under wave depressions and till injected from wave crests into folds of moving ice above. (6) Alternately the till does flow just under the ice margin into corrugations in the ice base. This is well demonstrated by grooved tills (Ramsden and Westgate, 1971; this volume, p. 335). Many forms of minor moraine ridges (Elson, 1969) are supposed to involve late secondary movements which would alter the fabric of the till.

All these hypothetical mechanisms involve an ice sheet sliding at its base. Numerous measurements in thin portions of glacier show that this motion commonly amounts to 50 percent of surface velocity. However, there are some arctic glaciers which are frozen to their base and are not sliding at all (Goldthwait, 1960). There are at least three studies of till, already mentioned (Harrison, 1960; Anderson, 1957; Dreimanis and Vagners, 1969), which show, on a lithologic basis, that the volume of debris transported from great distances is equal to or nearly as great as that from local sources. If a significant part of the drift comes from great distances, this poses the final (7) depositional possibility, that much till is gathered throughout glaciation, but it is not put down until the ice virtually stops flowing, wastes away, and bottom melting brings the low dirty ice layers vertically down to earth.

Fabric of aligned pebbles has been attributed to basal motion since its discovery by Richter (1932) and elaboration by Holmes (1941). A whole section is devoted to it in this volume (p. 318-64) and other papers (Stewart and MacClintock, 1971; this volume, p. 106) deal with it in some detail. A similar microfabric has been shown to exist for clay, silt, and sand clasts (Harrison, 1957; and Evenson, 1971; this volume, p. 345). Hopefully some rapid laboratory or field machine-method to determine this microfabric on oriented samples will soon replace the laborious search for elongate pebbles and measurement of each pebble (Dreimanis, 1959). Certainly methods need to be standardized (Andrews, 1971; this volume, p. 321).

Whether these orientations are inherited from the orientations of pebbles in the ice discussed earlier (p. 13–14), or by shearing action during deposition, or by reorientation under overriding ice while saturated is not clear. If it is inherited from ice, as the well-known Glen, Donner, and West treatment (1957) deduces, why is the up-glacier dip of the long axis preserved in a basal-melting process? Certainly secondary orientations or complete reorientation may be superposed in the surface few feet of till when it is in saturated juicy plastic condition under varying ice pressures (Ramsden and Westgate, 1971; this volume, p. 335). Some even question now whether the orientation does express direction of ice motion at all (Kauranne, 1960; Harris, 1969; Young, 1969), but so many strong fabrics do nearly parallel known striae and indicator paths (Drake, 1971; this volume, p. 73; Lineback, 1971; this volume, p. 328) that the relation can hardly be fortuitous.

Ablation till actually seen on ice may be as rich in clay and silt as basal till, and as heterogeneous through mixing from frost action. Most samples from ice surfaces, however, are more sandy, and the closely observed product has all kinds of inclusions of basal till, large angular clasts, and half-sorted layers. Frequently these are contorted, wrinkled, overturned, or on end. Very little study has been made of the actual origin of these structures, but articles in this volume imply that these may form both by sliding down underlying melting-ice slopes and by shove from winter refreezing of meltwaters (Embleton and King, 1968). Boulton (1968, and 1971; this volume, p. 41) has observed great masses of thin washed silty sands interlayered with flowed till-sheets. These become increasingly tipped and contorted toward the margin, as the ice core melts out. They are genetically similar to flowtill originally described by Hartshorn (1958). A more detailed study is presented in this volume by Hester and DuMontelle (p. 367). Clearly the cover masses accumulating on stagnant ice may be one to 20 feet thick, as is upper "ablation" till described in this volume.

Fabric is sometimes good, but often is very weak in known ablation moraine on ice. In final deposits it is variable at best (Drake, 1968). Perhaps fabric will serve to show which element of the superglacial till flowed into place, and the alteration of various ice-surface slopes (Boulton, 1971; this volume, p. 41).

All these deductions concerning the multiple origin of tills focus on an old "two-till" argument waged in the field in Massachusetts by earlier geologists (Bryan, Denny, Chute, Currier, Mather, Moss, J. W. Goldthwait) from 1938 to 1948. For some reason not explained here, many midcontinent areas (Wisconsin: Evenson, 1971; this volume, p. 345; Ohio: White, 1971; this volume, p. 149; Alberta: Ramsden and Westgate, 1971; this volume, p. 335) do not present significant masses of loose upper drift below the topsoil horizons to command explanation and argument. Did two entirely different glaciers produce the New England tills at two different times? Or do they show two kinds of transportation and deposition from just one glacier?

Now one group (Pessl, 1971; this volume, p. 92; Pessl, 1967; Pessl and Schafer, 1968) interprets compact lower tills from Connecticut and southern New Hampshire as originating from an earlier glaciation because of their olive-gray-brown (weathered) color to a depth of 20 or 30 feet, and dark iron-staining deep along joints truncated at the top (see also Flint, 1961). Furthermore both pebble fabric and streamlined forms within the lower till show northwest orientation, in contrast to less consistent directional data from the upper till. The loose upper till is interpreted to be related to a separate, late glaciation because it is often gray (unweathered), and because it rests on a sharp erosional contact with structures thrusting in one direction and the loose till contains blocky inclusions of the lower till. Only the joints inside the inclusions are limonite coated, so they were weathered before detachment.

The second group (Drake, 1971; this volume, p. 73; Goldthwait, 1968) emphatically denies this interpretation, based on statistical data from central New Hampshire. It claims that, despite slight differences in lithology, both tills relate to one glaciation. The loose upper till is superglacial because large clasts are so much more angular and drift is sandier, while the lower compact till is gray and hard, and contains more rounded and striated pebbles, indicating its subglacial origin. The upper till is stained yellowish, by frequent water passage, but is disoriented, or shows variable fabric orientation, due to gravity flow. Hard-till inclusions and shear structures in the upper till are accounted for by late basal shearing as ablation moraine suffered collapse or by local ice shove.

Happily, a week-long field conference (L. D. Drake, R. P. Goldthwait, K. Koteff, F. Pessl, P. Schafer) subsequent to the symposium has indicated a possible compromise and justification of both sets of observations and interpretation. There are indeed some few localities in southern New England where a hard gray "basal" till lies below the upper loose sandy till; there are indeed a few localities in New Hampshire where a hard till, olive-colored to a depth of 20 to 30 feet with iron-stained joints, lies below both hard and soft tills. No interglacial soil horizons have been located yet. Oxidation in the loose, generally permeable, upper till has produced all colors from yellow gray to bright red in streaky fashion, but only the iron-rich boulders are oxide encrusted. Thrust structures may have several origins (ablation-collapse or basal-ice shear); only regional studies of directional relations, where there are abundant exposures, will discern which is true in any one exposure. And thus a two-till problem raised again by this symposium is amenable to a three-till solution.

SUMMARY

Till is the ubiquitous glacial deposit by which every former glacier is most surely traced. It is a homogeneous sediment of mixed sizes (clay to boulders) and assorted kinds of rock-mineral grains. Some grains and stones are striated, but subangularity (partial-rounding) is common, and very elongate clasts often have predominant orientations. Masses of sorted, bedded drift are common within a till body, especially in the upper or loose till. Frequently, till lies on a bedrock pavement with striations which just predate the till. The total glacial drift including till tends to be thickest in the outer part of each glaciated area so multiple-till layers which register successive earlier glacial advances are most commonly preserved in the peripheral areas. The focus of studies thus far is the correlation of successive and discontinuous till sheets, often thin, from exposure to exposure. This has been done empirically and perhaps accurately by mechanical analysis (sand-silt-clay), by lithology (heavy minerals, pebble types, clay minerals), by color, chemical ratios, and sometimes by fabric.

The exact origin of till is very imperfectly known. First the material must be picked up or entrained in glacier ice. The debris represents some of every material which the glacier has passed over, but particularly former regolith and soil. Fine layers stream off small roughnesses of the bottom as they are trapped by formation of regelation ice which occurs at small obstacles. Abrasion of stoss slopes on large rock masses creates a belt of fine materials and rock flour. To it are added coarse clasts which are excavated even more rapidly (ten times) on the lee slopes by plucking. Transportation is accomplished within 100 to 200 feet of the bottom. Very fine materials are dispersed upward in the ice sheet from 50 to 200 feet, depending on time. Coarser silt, sand, and pebbles travel in bands which are concentrated near the ground and intersect it at places.

Shape and roundness statistics show that granules and pebbles are both abraded en route to become more round, but they are broken from time to time to become angular again at the bottom. Lithology determines the frequency of breaking and the shape of the broken product, as well as the terminal size of comminution. Longer pebbles become oriented by ice sheer, but they become disoriented and are rebroken to be very angular if they reach the surface near the terminal as ablation moraine. Materials are dispersed laterally by topographic deflections of bottom ice in a 20° to 60° fan either side of an axis of abundance. Larger grade sizes of most types are reduced to under one-tenth of a percent in five to fifty miles. The matrix, being terminal size for grinding, appears to come from farthest, although it contains bits of all bedrocks traversed.

Deposition must be either at the bottom of the ice or off the top by ablation. At the bottom either individual grains may stop by catching, or whole sheets of till may become stacked by frictional drag. In most cases bottom melting of one to five centimeters (two inches) per year must occur. Till thus saturated by water beneath the ice pretty surely deforms by flow. Under thin ice it is shown to flow into corrugated basal ice and crevasses. This may be the origin of much fabric rather than the orientation inherited from shear in flowing ice. Ablation debris on the steep slope of the terminal area has weaker or less-regionally consistent fabric through slumping or flow (flow till). Often ablation debris is more sandy and loose from partial wash, and pebbles are significantly more angular or less striated than in basal till.

The great variation in these properties of till leads to extremes of permeability vital to groundwater movements or waste-disposal or extremes of compaction critical to building excavation and to use as road fill. The mixing of lithologies and dispersal from sources is critical to finding ores. In gravel roads and concrete the lithologies of pebbles determine strength and durability.

REFERENCES

Anderson, R. C., 1957, Pebble and sand lithology of the major Wisconsin glacial lobes of the Central Lowland: Geol. Soc. America Bull., v. 68, no. 11, p. 1415-50.

Andrews, J. T., 1971, Methods in the analysis of till fabrics: *in* this volume.

Battle, W. P. V., and Lewis, W. V., 1951, Temperature observations in bergschrunds and their relationship to cirque erosion: Jour. Geology, v. 59, no. 6, p. 537-45.

Boulton, G. S., 1968, Flow tills and related deposits on some Vestspitsbergen glaciers: Jour. Glaciology, v. 7, no. 51, p. 391-412.

————, 1969, The criteria by which tills of different origin may be distinguished (Abstr.): VIII Congress INQUA (Paris, 1969) Résumés des Communications, 269 p.

————, 1971, Till genesis and fabric in Svalbard, Spitsbergen: *in* this volume.

Bowman, I., 1916, The Andes of southern Peru: Amer. Geographical Soc., Spec. Pub. 2, p. 285-313.

Carney, F., 1909, Glacial erosion on Kelley's Island, Ohio: Geol. Soc. America Bull., v. 20, December, p. 640-44.

Carol, H., 1947, The formation of rôches moutonées: Jour. Glaciology, v. 1, no. 2, p. 57-62.

Chamberlin, T. C., 1894, Proposed genetic classification of Pleistocene glacial formations: Jour. Geology, v. 2, no. 5, p. 517-38.

Charlesworth, J. K., 1957, Boulder Clay, *in* The Quaternary Era (Chap. 18), v. 1, London, Arnold, p. 376-88.

Christiansen, E. A., 1971, Tills in southern Saskatchewan, Canada: *in* this volume

Demorest, M., 1939, Glacial movement and erosion: a criticism: Amer. Jour. Sci., v. 237, no. 8, p. 594-605.

Drake, L. D., 1968, Till Studies in New Hampshire: Ph.D. dissertation. Ohio State University, 112 p.

————, 1971, Evidence for ablation and basil till in New Hampshire: *in* this volume.

Dreimanis, A., 1956, Steep Rock iron ore boulder train: Proc. Geological Assoc. Canada, v. 8, pt. I., p. 28-70.

Dreimanis, A., Reavely, G. H., Cook, R. J. B., Knox, K. S., and Moretti, F. J., 1957, Heavy mineral studies in tills of Ontario and adjacent areas: Jour. Sed. Petrology, v. 27, no. 2, p. 148-61.

Dreimanis, A., 1959, Rapid macroscopic fabric studies in drill-cores and hand specimens of till and tillite: Jour. Sed. Petrology, v. 29, p. 459-63.

————, 1961, Tills of southern Ontario: *in* Soils in Canada, R. F. Legget (ed.), Royal Society of Canada Pub. No. 3, p. 80-96.

Dreimanis, A. and Vagners, U. J., 1969, Lithologic relation of till to bedrock: *in* Quaternary Geology and Climate, H. E. Wright, Jr. (ed.) Proc. VII Congress INQUA, Nat. Acad. Sci., v. 16, p. 93-98.

Dreimanis, A., 1971, Procedures of till investigations in North America: a general review: *in* this volume.

Dreimanis, A., and Vagners, U. J., 1971, Bimodal distribution of rock and mineral fragments in basal till: *in* this volume.

Elson, J. A., 1960, Geology of glacial tills: 14th Canadian Soils Mechanics Conference (preprint), Niagara Falls, 30 p.

———, 1965, Early discoveries XXIII, Till stone orientation: Henry Youle Hind (1823-1908): Jour. Glaciology, v. 6, no. 44, p. 303-6.

———, 1969, Washboard moraines and other minor moraine types: *in* Encyclopedia of Geomorphology, R. W. Fairbridge (ed.), v. 3, New York, Rheinold, p. 1213-19.

Embleton, C., and King, C. A. M., 1968, Glacial and periglacial geomorphology: London, E. Arnold, p. 300-321.

von Engeln, O. D., 1937, Rock sculpture by glaciers: a review: Geog. Review, v. 27, no. 3, p. 478-82.

Evenson, E. B., 1971, The relationship of macro- and microfabrics of till and the genesis of glacial landforms in Jefferson County, Wisconsin: *in* this volume.

Feininger, T., 1970, Chemical weathering and glacial erosion of crystalline rocks and the origin of till: (in prep.).

Flint, R. F., 1957, Glacial drift I. Till; Moraines: *in* Glacial and Pleistocene geology: New York, John Wiley, p. 108-35.

Flint, R. F., Sanders, J. E., and Rogers, J., 1960, Diamictite: a name for unsorted terrigenous sedimentary rocks that contain a wide range of particle sizes: Geol. Soc. America Bull., v. 71, no. 4, p. 507-9.

Flint, R. F., 1961, Two tills in southern Connecticut: Geol. Soc. America Bull., v. 72, no. 11, 1687-92.

Geikie, A., 1863, On the phenomena of the glacial drift of Scotland: Geological Soc. Glasgow Trans., v. 1, pt. 2, p. 190.

Gillberg, G., 1965, Till distribution and ice movements on the northern slope of the south Swedish highlands: Geologiska, Forenningensi Stockholm, Forhandlingar, v. 86, no. 519, p. 433-84.

Glen, J. W., Donner, J. J., and West, R. G., 1957, On the mechanism by which stones in till become oriented: Amer. Jour. Sci., v. 255, no. 3, p. 194-205.

Glen, J. W., and Lewis, W. V., 1961, Measurements of side-slip at Austerdalsbrun, 1959: Jour. Glaciology, v. 3, no. 30, p. 1109-22.

Goldthwait, J. W., 1925, The geology of New Hampshire: N. H. Academy of Science, Handbook No. 1, p. 16-25.

Goldthwait, J. W., and Kruger, F. C., 1938, Weathered rock in and under the drift in New Hampshire: Geol. Soc. America Bull., v. 49, no. 8, p. 1183-98.

Goldthwait, R. P., 1951, Development of end moraines in east central Baffin Island: Jour. Geology, v. 59, no. 6, p. 567-77.

———, 1960, Study of an ice cliff in Nunatarassuaq, Greenland: U. S. Army, Snow Ice Permafrost Research Estb., Tech. Rept. 39, 108 p.

———, 1968, Surficial geology of the Wolfeboro and Winnipesaukee Quadrangles, New Hampshire: N. H. State Dept. of Resources and Economic Development. 60 p.

Goldthwait, R. P., and Rosengreen, T., 1969, Till stratigraphy from Columbus southwest to Highland County, Ohio: Field Trip no. 2 *in* Field Trip Guidebook, North Central Section, Geological Soc. of America, p. 2-1 to 2-17.

Goldthwait, R. P., 1971, Restudy of Red Rock Ice Cliff in Nunatarssuaq, Greenland: U. S. Army Terrestrial Sciences Lab. Tech. Rept. 224 (in press).

Gow, A. J., Ueda, H. T., and Garfield, D. E., 1968, Antarctic ice sheet: preliminary results of first core hole to bedrock: Science, v. 161, no. 3845, p. 1011-13.

Gross, D. L., and Moran, S. R., 1971, Grain-size and mineralogical gradations within tills of the Allegheny Plateau: *in* this volume.

Gunn, C., 1968, Provenance of diamonds in the glacial drift of the Great Lakes region, North America (Abstr.): Bull. Canadian Petroleum Geology, v. 16, no. 3, p. 418.

Harland, W. B., Herod, K. N., and Krinsley, D. H., 1966, The definition and identification of tills and tillites: Earth-Science Reviews, Elsevier, v. 2, p. 225-56.

Harris, S. A., 1969, The meaning of till fabrics: Canadian Geographer, v. 13, no. 4, p. 317-37.

Harrison, W., 1957, A clay-till fabric: its character and origin: Jour. Geology, v. 65, no. 3, p. 275-308.

———, 1960, Original bedrock composition of Wisconsin till in central Indiana: Jour. Sed. Petrology, v. 30, no. 3, p. 432-46.

Hartshorn, J. H., 1958, Flowtill in southeast Massachusetts: Geol. Soc. America Bull., v. 69, no. 4, p. 477-82.

Hester, N. C., and DuMontelle, P. B., 1971, A Pleistocene mudflow along the shelbyville Moraine front, Macon County, Illinois: *in* this volume.

Holdsworth, G., 1969, Mode of flow of Meserve Glacier, Wright Valley, Antarctica: Ph.D. dissertation, Ohio State University, 342 p.

Holmes, C. D., 1941, Till fabric: Geol. Soc. America Bull., v. 52, no. 9, p. 1299-1354.

———, 1944, Hypothesis of subglacial erosion: Jour. Geology, v. 52, no. 3, p. 184-90.

———, 1952, Drift dispersion in west central New York: Geol. Soc. America Bull., v. 63, no. 10, p. 995-1010.

———, 1960, Evolution of till-stone shapes, central New York: Geol. Soc. America Bull., v. 71, no. 11, p. 1645-60.

Inman, D. L., 1952, Measures for describing the size distribution of sediments: Jour. Sed. Petrology, v. 22, no. 3, p. 125-45.

Jahns, R. H., 1943, Sheet structures in granites: Its origin and use as a measure of glacial erosion in New England: Jour. Geology, v. 51, no. 2, p. 71-98.

Johnson, W. D., 1904, The profile of maturity in alpine glacial erosion: Jour. Geology, v. 12, no. 7, p. 569-78.

Johnson, W. H., Gross, D. L., and Moran, S. R., 1971, Till stratigraphy of the Danville region, east-central Illinois: *in* this volume.

Kamb, W. B., and LaChapelle, E., 1964, Direct observation on the mechanism of glacier sliding over bedrock: Jour. Glaciology, v. 5, no. 38, p. 159-72.

Kauranne, L. K., 1960, A statistical study of stone orientation data in glacial till: Finland Comm. Geology Bull., v. 32, no. 188, p. 87-97.

Kempton, J. P., DuMontelle, P. B., and Glass, H. D., 1971, Subsurface stratigraphy of the Woodfordian tills in the McLean County region, Illinois: *in* this volume.

Krumbein, W. C., 1937, Sediments and exponential curves: Jour. Geology, v. 45, no. 6, p. 577-601.

Krynine, P. D., 1937, Age of till on "Palouse Soil" from Washington: Amer. Jour. Sci., v. 34, no. 203, p. 345-63.

Lineback, J. A., 1971, Pebble orientation and ice movement in south-central Illinois: *in* this volume.

Lliboutry, L., 1959, Une théorie du frottement du glacier sur son lit: Annals de Géophysique, v. 15, no. 3, p. 250-65.

———, 1964, Traité de glaciologie: Paris, Masson, 1040 p.

Lyell, C., 1839, Nouveaux éléments de géologie: Paris, Pitois-Levrault, 648 p.

MacClintock, P., and Dreimanis, A., 1964, Reorientation of till fabric by overriding glacier in the St. Lawrence Valley: Amer. Jour. Sci., v. 262, no. 1, p. 133-42.

Meneley, W. A., 1964, Geology of the Melfort Area (73-A) Saskatchewan: Ph.D., dissertation, Univ. of Illinois, p. 33-39.

Moran, S. R., 1971, Glaciotectonic structures in drift: *in* this volume.

Nobles, L. H., and Weertman, J., 1971, Influence of irregularities of the bed of an ice sheet on deposition rate of till: *in* this volume.

Nye, J. F., 1952, The mechanics of glacier flow: Jour. Glaciology, v. 2, no. 12, p. 82-93.

———, 1957, The distribution of stress and velocity in glaciers and ice sheets: Royal Society Proc., v. 239A, p. 113-33.

Oura, H. (ed.), 1969, Physics of snow and ice: Internat. Conf. on Low Temperature Science, 711 p.

Pessl, F., 1967, A two-till locality in northeastern Connecticut: U. S. Geol. Survey Prof. Paper 550D, p. D89-D93.

Pessl, F., and Schafer, J. P., 1968, Two-till problem in Naugatuck-Torrington Area, western Connecticut: Guidebook No. 2, State Geological and Natural History Survey of Connecticut, Trip B-1, 25 p.

Pessl, F., 1971, Till fabrics and till stratigraphy in western Connecticut: *in* this volume.

Perutz, M. F., and Seligman, G., 1939, A crystallographic investigation of glacier structure and the mechanism of glacier flow: Royal Soc. London, Proc. ser. A, v. 172, no. 950, p. 335-60.

Peterson, D. N., 1969, Glaciological investigations on the Casement Glacier, southeast Alaska: Ph.D. dissertation, Ohio State University, 183 p.

Pettijohn, F. J., 1957, Sedimentary Rocks: New York, Harper, 718 p.

Pettyjohn, W. A., and Lemke, R. W., 1971, Unarmored till balls in unusual abundance near Minot, North Dakota: *in* this volume.

Pounder, E. R., 1965, Physics of ice: Elmsford, N. Y., Pergamon, 151 p.

Ramsden, J., and Westgate, J. A., 1971, Evidence for reorientation of a till fabric in Alberta: *in* this volume.

Richter, K., 1932, Die Bewegungsrichtung des Inlandeises rekonstruiert aus den Kritzen und Langsachesen der Geschiebe: Zeitschr. Geschiebeforsch. Flachidgeol., v. 8, p. 62-66.

Robin, G. deQ., 1955, Ice movement and temperature distribution in glaciers and ice sheets: Jour. Glaciology, v. 2, no. 18, p. 523-32.

de Saussure, H. B., 1803, Voyages dans les Alpes, Genève: v. 2, pt. 2.

Shaler, N. S., 1896, The conditions of erosion beneath deep glaciers, based upon a study of the boulder train from Iron Hill, Cumberland, R. I.: Bull. Museum Comparative Zoology (Harvard Col.), v. 16, no. 11, p. 185-267.

Sharp, R. P., 1948, The constitution of valley glaciers: Jour. Glaciology, v. 1, no. 4, p. 182-89.

———, 1949, Studies of superglacial debris in valley glaciers: Amer. Jour. Sci., v. 247, no. 5, p. 289-315.

Shepps, V. C., 1953, Correlation of tills of northeastern Ohio by size analysis: Jour. Sed Petrology, v. 23, no. 1, p. 34-48.

Shumskii, P. A., 1964, Principles of structural glaciology: David Kraus, (transl.), New York, Dover, 497 p.

Smith, H. T. U., 1948, Giant grooves in northwest Canada: Amer. Jour. Sci., v. 246, p. 503-514.

Steiger, J. R., and Holowaychuck, N., 1971, Particle-size and carbonate analysis of glacial till and lacustrine deposits in western Ohio: *in* this volume.

Stewart, D. P., and MacClintock, P., 1971, Ablation till in northeastern Vermont: *in* this volume.

Swinzow, G. K., 1962, Investigation of shear zones in the ice sheet margin, Thule Area, Greenland: Jour. Glaciology, v. 4, no. 32, p. 215-29.

Stankowski, A., and Stankowski, W., 1966, Prôba rozposimowania glin zwaowych Polski Zachodniej w swietle analiz mineralogicznych i chemicznych.

Torrell, O., 1877, On the glacial phenomena of North America: Amer. Jour. Sci., ser. 3, v. 13, p. 76-79.

Totten, S. M., 1969, Overridden recessional moraines of north-central Ohio: Geol. Soc. America Bull., v. 80, no. 10, p. 1931-46.

Upham, W., 1891, Distribution of englacial drift: Geol. Soc. America Bull., v. 3, no. 1, p. 134-48.

Vagners, U. J., 1966, Lithologic relationship of till to carbonate bedrock in southern Ontario: M.Sc. thesis, Geology Dept., Univ. Western Ontario, 154 p.

Weertman, J., 1961, Equilibrium profile of ice caps: U. S. Army, Snow Ice Permafrost Research Estb., Research Rept. 84, 12 p.

———, 1964, Profile and heat balance at the bottom surface of an ice sheet fringed by mountain ranges: U. S. Army, Cold Regions Research and Engr. Lab., Research Rept. 134, 7 p.

———, 1968a, Diffusion law for the dispersion of hard particles in an ice matrix that undergoes simple shear deformation: Jour. Glaciology, v. 7, no. 50, p. 161-66.

————, 1968b, Comparison between measured and theoretical temperature profiles of the Camp Century, Greenland, borehole: Jour. Geophysical Research, v. 73, no. 8, p. 2691-2700.

Westgate, J. A., and Dreimanis, A., 1967, The Pleistocene sequence at Zorra, southwestern Ontario: Canadian Jour. Earth Science, v. 4, no. 6, p. 1127-43.

White, G. W., 1971, Thickness of Wisconsinan tills in Grand River and Killbuck lobes: *in* this volume.

Wilding, L. P., Drees, L. R., Smeck, N. E., and Hall, G. F., 1971, Mineral and elemental composition of Wisconsin-age till deposits in west-central Ohio: *in* this volume.

Willman, H. B., Glass, H. D., and Frye, J. C., 1963, Mineralogy of glacial tills and their weathering profiles in Illinois, Part 1, Glacial tills: Illinois State Geological Surv., Circ. 347, 55 p.

Woldstedt, P., 1961, Das Eiszeitalter: v. 1, F. Enke, 374 p.

Young, J.A.T., 1969, Variations in till microfabric over very short distances: Geol. Soc. America Bull., v. 80, no. 11, p. 2343-52.

Till/a Symposium

Procedures of Till Investigations in North America: A General Review

A. Dreimanis

ABSTRACT

As a part of a world-wide survey on procedures of investigations of tills and till classifications, conducted by the INQUA Commission of Genesis and Lithology of Quarternary Deposits, a questionnaire was distributed among North American geologists, pedologists, soils engineers, and geographers. Ninety-nine of these questionnaires were returned, representing 545 persons who have studied or described tills during the past five years.

Results of the questionnaire are summarized in the following generalizations. Tills are classified (in decreasing importance) according to texture, color, degree of compaction and weathering, mode of deposition, and lithologic composition. Sampling is primarily from exposures, with lesser amounts from augering or drilling. Procedures used for till investigation include: visual textural description, color description (mostly verbal, some by Munsell scale), granulometric analyses (usually matrix only, some including pebbles), quantitative analyses of lithology (pebbles most commonly studied, followed by boulders, clay fraction, and entire till matrix, in that order).

Pedologists pay much attention to the carbonate and organic content of till matrix, and determine the porosity or moisture content. Engineers commonly test moisture content, plasticity, shear strength, and bulk density. Most commonly used upper boundaries of the particle-size fractions are $2\,\mu$ for clay, $62\,\mu$ for silt, 2 mm for sand, 4 mm for granules, and 64 mm for pebbles. Equipment most commonly used includes: for sampling — common field tools, hand augers, and motor-driven drills; for granulometric analyses — sieves, shakers, hydrometer, mechanical stirrers, and pipettes; for lithologic investigations — microscopes, heavy-mineral separators, X-ray diffractometer, and gasometric apparatus for carbonate analyses.

Studies of tills are conducted at least at 30 institutions: 17 in Canada, 13 in the United States, most of them (22) in universities.

INTRODUCTION

This project was initiated in the fall of 1967 at the request of B. Krygowski, co-president of the INQUA Commission on Genesis and Lithology of Quaternary Deposits, and was addressed to A. Dreimanis. Because of the size of the area from which the information had to be gathered (Canada and United States), the following people helped to prepare the questionnaire and gather the results, and to critically read this paper: L. A. Bayrock, Research Council of Alberta, for western Canada; J. A. Elson, McGill University, for eastern Canada; S. C. Porter, University of Washington (Seattle), for western United States; J. R. Kempton, Illinois State Geological Survey, for the central United States; and R. P. Goldthwait, Ohio State University, for eastern United States. R. W. May, of the University of Western Ontario, also helped. The services of all these people are gratefully acknowledged.

A questionnaire was prepared by correspondence, and its final version was distributed in the summer of 1968 to all institutions or persons who were known to have recently dealt with investigations of till. Questions were asked on the following subjects:

1. Number of persons engaged permanently or temporarily in investigation of tills during the last five years.

2. Criteria (11 suggested) for description and classification of tills used by the above workers (stratigraphic classifications excluded).

3. Procedures (34 suggested) employed for till investigations.

4. Particle-size boundaries commonly used.

5. Equipment used in investigation of tills: for granulometric analyses (10 major items listed), lithologic and chemical analyses (13), engineering tests (5), and field work (4).

6. Current studies on genesis and classification of tills.

Blank spaces were provided for additional entries in each group of questions, and 16 items, not listed in the questionnaire, were added by the participants.

Ninety-nine questionnaires were returned, 67 from Canada and 32 from United States, representing 545 investigators, 62 percent of them living in

TABLE 1

DISTRIBUTION, BY PROFESSION AND MAIN EMPLOYMENT, OF 545 TILL INVESTIGATORS
WHO RESPONDED TO QUESTIONNAIRE

Profession	Total Number	Percentage of Total	Number of till investigators employed by:		
			Universities	Non-academic Government Institutions	Private Companies
Geologists and hydro-geologists	355	65	210	143	0(?)
Engineering geologists and soils engineers	108	20	23	63	22
Pedologists	61	12	13	48	0
Geographers	21	3	16	5	0
Total	545	100	262	259	22

Canada and 38 percent in the United States. About half of the investigators are associated with universities, the remaining half with various nonacademic government agencies, particularly geological surveys and private companies (Table 1). Most of the investigators are geologists and ground-water geologists (in further discussions all of these are called geologists), followed by soils engineers and engineering geologists (these subsequently abbreviated to engineers), pedologists, and geographers. Their numbers are approximated in Table 1, as the division by professions was guided mainly by the specialities of the institutions, unless the employees were known personally.

The questionnaire was arranged so that most of the answers could be processed by computer, a task carried out by R. W. May. A. Dreimanis was responsible for final compilation and evaluation of the data.

Questions about the use of till classifications, procedures employed, and equipment used were answered by placing a checkmark under one of the following column heads: Used: Commonly ($>50\%$), Occasionally (10-50%), Rarely ($<10\%$), Never. Absence of a checkmark answer for a given question was considered as implying "Never."

The first step in the evaluation of the answers was to determine the frequency of response, among the 545 workers, for each use classification, procedure, and item of equipment. For instance, of all 545 investigators, the genetic classification of tills has been used: commonly, by 30 percent; occasionally, by 28 percent; rarely, by 21 percent; never, by 21 percent. Similar average percentages were determined for each profession and its subgroups, as listed along the right sides of Figures 1 through 6.

To facilitate a more compact presentation of the frequency percentages, an application index was computed in the following manner: the percentage of the common usage was multiplied by 1.0, the occasional usage by 0.5,

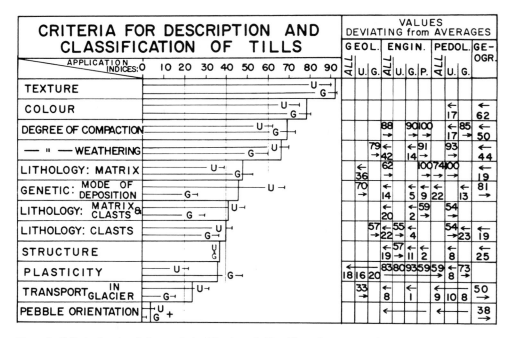

Figure 1. Criteria for description and classification of tills. All numbers are application indices. Abbreviations: U. = Universities, G. = Nonacademic government institutions, P. = Private companies. The horizontal bars and lines are general averages from all answers; the numbers in the right-side columns refer to the indices of separate groups listed by their headings. Only those with notable deviations are given (See text for further explanation.)

and the rare usage by 0.1, and the products were then added. For example, in the case of the data relating to the use of the genetic classification of tills above, the application index would be $30 \times 1.0 + 28 \times 0.5 + 21 \times 0.1 = 46.1$. The maximum possible index figure is 100, representing 100 percent in the column "commonly used."

The application indices are presented in Figures 1 through 6, both graphically as horizontal bars with identifying lines through the middles of the bars, and as numbers in the right-hand block. In the right-hand block, under "Values deviating from averages," are listed the values of those indices which deviate furthest from the average. From this it can be determined if any profession responded far above or below the average. If the index figure for the subgroup does not deviate significantly from the general average, the space is left blank. Considering the previous example of genetic classification (line 6 in Fig. 1), the blank space in the right-hand-block column representing all 355 geologists indicates that this group does not deviate very much from the average index of 46, while "70" in the next column (geologists at universities) shows that their index is 70, being significantly above the general average. For faster orientation, arrows are added to the index numbers; an arrow pointing to the right indicates that the index number for the subgroup is higher than average, whereas one pointing to the left indicates a lower average index number.

DESCRIPTION AND CLASSIFICATION OF TILLS

Before discussing the criteria of description and classification of tills, two comments quoted from the returned questionnaires explain some of the approaches used in selecting certain criteria. A geologist from a large government institution writes:

No uniform classification or recipe for description is used by [name of the institution]. Each individual is left to his own preferences. Basically, the tills are described, and the nomenclature reflects the one or two parameters that are judged most significant and/or prominent. This in turn commonly depends on what aspect of till geology most concerns the investigator.

A pedologist writes:

In soil survey work, although we make no attempt to classify glacial till *per se*, we do attempt to describe it in the field in some detail as part of our soil profile description.

For most soils engineers, who are more interested in properties than classification, till is merely one kind of soil.

Nearly all investigators describe or classify tills by their texture (Fig. 1). Among other criteria, color, degree of compaction, and degree of weathering (in decreasing order) are used in most cases, except by some subgroups of engineers and pedologists.

It seems surprising that the lithology of the till matrix and its clasts is considered in less than a half of all till investigations, though it is of equal significance to texture. Engineers in private companies and pedologists at universities appear to be most aware of the importance of the lithologic composition of till matrix.

Among other descriptive criteria, structure and plasticity are determined almost as commonly as the lithology. However, considerable difference in interest exists among various professions and their subdivisions. For instance, all engineers and most pedologists consider plasticity a very important criterion, whereas most geologists and university pedologists give it little attention. Pebble orientation in till actually is used more widely than shown in Figure 1, as indicated by the relatively higher index (about 25) for till-fabric determinations among the answers on procedures in Figure 2.

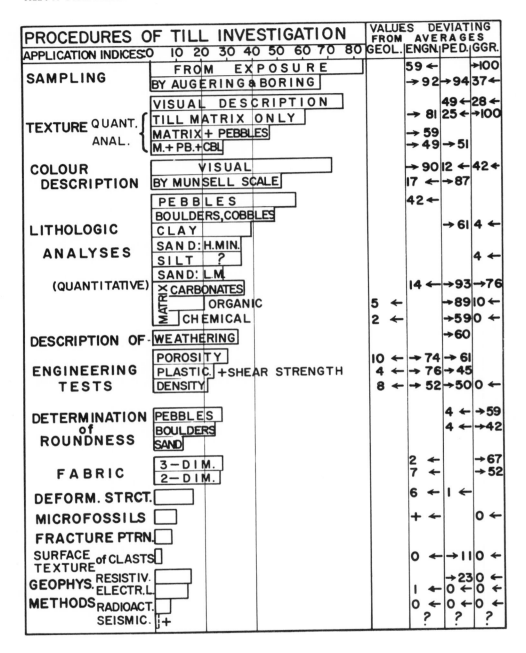

Figure 2. Procedures of till investigation. See Figure 1 for explanation.

This apparent discrepancy can be explained by the inadvertent omission of pebble orientation from the questionnaire, although this criterion was added in some returns.

Among the interpretative classifications, mode of desposition is inferred more often than is the probable mode of transport in glacier prior to the deposition of till. Both criteria are used, particularly by the theoretically minded geographers and geologists at universities.

PROCEDURES EMPLOYED FOR TILL INVESTIGATION

These procedures (Fig. 2) duplicate the classification and description criteria to some extent, but they give a more thorough insight into how the till investigators have arrived at their classification. Most investigators take till samples directly from exposures; engineers are the exception, for they generally use drilling equipment. This difference is due largely to the timing and extent of sampling, for most engineering investigations start with test drilling as one of the routine procedures, whereas geological work usually begins with regional mapping and examination of natural exposures. Geophysical methods, which may either replace sampling or supplement it, are very seldom used. The seismic method is probably more popular than indicated in Figure 2, as it was unintentionally omitted from the questionnaire.

Differences in the use of various procedures have been influenced by specific professional practices and interests. For instance, texture is determined in most cases mainly by visual examination during geological field mapping, while, in engineering investigations, emphasis is placed upon more exact quantitative laboratory analyses. Pedologists, because they pay more attention to the weathering of till, prefer the Munsell scale for color determination, while many geologists and particularly soils engineers feel satisfied with visual descriptions of color.

When doing quantitative lithologic investigations, geologists place their primary emphasis on clasts (pebbles, cobbles, and boulders). Pedologists, on the other hand, pay special attention to the till matrix. However, it is possible that most of those who indicated that the silt fraction was a subject of quantitative analyses actually meant the entire till matrix rich in silt, for many of them did not indicate that they were separating clay from silt. Chemical analyses of tills are very seldom done, except by pedologists.

The engineering tests on porosity, plasticity, shear strength, and density are performed both by engineers and pedologists, but are commonly neglected by most geologists. The fabric and shape of clasts, on the other hand, have been studied particularly by geologists and geographers, but elicit little attention from engineers. Certain till characteristics, such as deformational structures, fracture patterns, and incorporated micro-fossils, are investigated only in special cases.

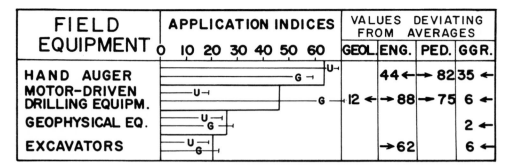

Figure 3. Field equipment used for till sampling. See Figure 1 for explanation.

PARTICLE SIZE BOUNDARIES IN GRANULOMETRIC ANALYSES

The Wentworth scale is most commonly used to define upper boundaries of the particle size fractions: 2 μ for clay, 62 μ for silt, 2 mm for sand, 4 mm for granules, and 64 mm for pebbles. The 4μ upper boundary for clay is nearly as popular as the 2μ, particularly among geologists.

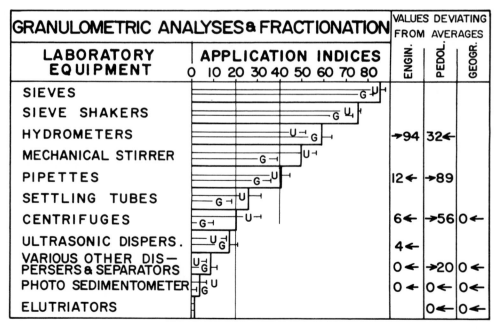

Figure 4. Laboratory equipment used for granulometric analyses and fractionation of tills. See Figure 1 for explanation. The application indices of geologists differ little from the general averages; therefore they are not listed under "Values deviating from averages."

EQUIPMENT USED FOR INVESTIGATING TILLS

Field Equipment

Besides the pick and shovel, items not included in the questionnaire, hand augers are the most frequently used item of equipment among geologists (Fig. 3). Motor-driven drilling equipment is used mostly by engineers and pedologists, particularly those employed by government institutions and by private companies.

Granulometric Analyses and Fractionation

Sieves and shakers are the most commonly used granulometric equipment (Fig. 4). For study of fine-particle fractions, hydrometers are more popular than pipettes, except among pedologists. Dispersion prior to granulometric analyses is done mainly by mechanical stirrers, though ultrasonic dispersers are being used in several laboratories. Fractionation, judging from the listing of the equipment, apparently is carried out only infrequently. Pedologists are the main group using centrifuges.

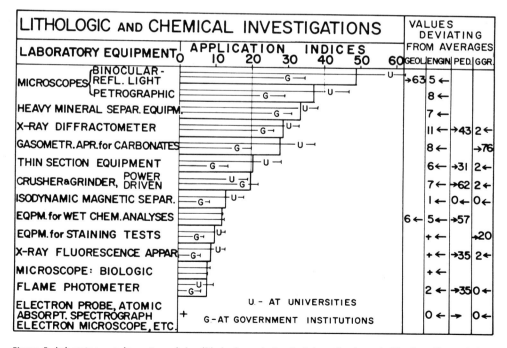

Figure 5. Laboratory equipment used for lithologic and chemical investigation of tills. See Figure 1 for explanation.

Equipment for Lithologic and Chemical Investigations

This equipment (Fig. 5) is used more at universities than at nonacademic government institutions, and is used mainly by pedologists and geologists. The most common items are microscopes, followed, in order by equipment for separating heavy minerals, X-ray diffractometers, and gasometric apparatus for carbonate determination. Pedologists also commonly use power-driven crushers and grinders, equipment for wet chemical analyses, X-ray-fluorescence apparatus, and flame photometers. Among the more sophisticated apparatus are the electron probe, the atomic absorption spectograph, and the electron microscope, each listed by several laboratories. As a group, engineers appear least equipped for lithologic and chemical analyses.

Equipment for Engineering Tests

Plasticity tests appear to be a routine procedure by all engineers involved in till investigations (Fig. 6), but other equipment is used by the private companies. Considering all the professional groups together, the nonacademic government agencies appear to be better equipped with instruments for engineering tests than are universities.

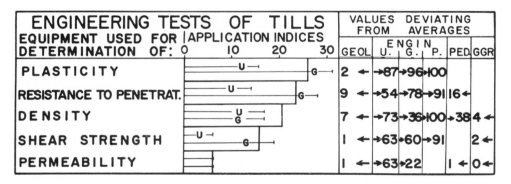

ENGINEERING TESTS OF TILLS EQUIPMENT USED FOR DETERMINATION OF: (APPLICATION INDICES 0 10 20 30)	VALUES DEVIATING FROM AVERAGES					
	GEOL	ENGIN U.	ENGIN G.	ENGIN P.	PED	GGR
PLASTICITY	2	87	96	100		
RESISTANCE TO PENETRAT.	9	54	78	91	16	
DENSITY	7	73	36	100	38	4
SHEAR STRENGTH	1	63	60	91		2
PERMEABILITY	1	63	22		1	0

Figure 6. Equipment used for engineering tests of tills. See Figure 1 for explanation.

CURRENT STUDIES ON GENESIS AND CLASSIFICATION OF TILLS

Studies of tills are being conducted at some 30 institutions: 17 in Canada and 13 in the United States. The total number may be somewhat higher, as some questionnaires were not returned. Among the above thirty, 21 are at universities representing 16 geology, three geography, one pedology, and one soil mechanics departments. The remaining nine are nonacademic government institutions, five of them dealing mainly with the geologic aspects of till, particularly geological surveys, one with pedology, and one with soil mechanics. Eighteen institutions (or 21, if papers presented by title are included) were represented in the till symposium in Columbus, Ohio

The following aspects of tills, not reported at this till symposium, are currently being investigated: engineering properties and permeability of tills, homogeneity or heterogeneity of a till layer, indicator tracing by lithologic studies of tills, water-laid tills, transport of drift and deposition of till by active glaciers, and energy represented by till.

Three classifications of tills were suggested in the answers to the questionnaires: one using texture, and two based on genesis.

ACKNOWLEDGMENTS

The authors are very grateful to all those who answered the questionnaires, especially to those who added comments which will be used for a more detailed further evaluation of the classification of tills and procedures of till investigation. Thanks are due also to the secretarial staff of the institutions listed with the authors for typing and mailing the questionnaires, and to Mrs. R. Ringsman for drafting the figures.

2

Genesis

Till Genesis and Fabric in Svalbard,
Spitsbergen

Geoffrey S. Boulton

ABSTRACT

Svalbard glaciers are characterized by their transport of considerable volumes of englacial debris, which are released from the ice to form three main types of till. *Flowtill* is released as a fluid mass from the englacial-debris load when this is exposed by down-wasting of the glacier surface; *melt-out till* is deposited by slow melting out of the top surface of masses of dead ice covered by a stable overburden; *subglacial lodgement till* is material released from the basal ice either by pressure melting against bedrock obstructions, or by melting of debris-rich ice masses which have become stagnant beneath the moving glacier sole.

Englaciál stone orientations are described, but these are rarely preserved, except, in part, in melt-out tills and in certain subglacial tills. Flowtills often except, in part, in melt-out tills ad in certain subglacial tills. Flowtills often show an upper, more fluid element in which fabrics both transverse and parallel to flow develop, while the lower part shows a parallel fabric with up-slope, *a*-axis imbrication. Melt-out tills have fabrics in which a/b planes tend to lie in the plane of deposition, but which reflect azimuthally the former englacial fabrics. Subglacial tills show variable fabrics dependent upon the shape of the underlying bedrock surface, although where this is planar, parallel fabrics are dominant. All three main types of till can show systematic regional fabric patterns, although fabrics at one site can rarely be taken to represent the direction of glacier movement.

INTRODUCTION

Till fabric analysis is one of the most commonly used techniques in glacial geology, although very little is known about the actual process by which tills acquire these fabrics. This difficulty is shared by the field of glaciology as a whole, for there are indeed very few observations of the ways in which tills are deposited by modern glaciers. Over the years, however, a considerable body of hypothesis has been built up regarding the mode of origin of tills and of the sequences in which they occur, especially in the lowland areas of Europe and North America. Much of this hypothesis is entirely speculative and little based on studies of those modern glaciers which are now producing similar sequences. Thus has been built up the hypothetical model that compact, fine-grained tills are, almost by definition, subglacial in origin, while observations of thin moraines on the surfaces of valley glaciers (Sharp, 1949) have added an upper element, the washed, ablation till of supraglacial origin (Flint, 1957).

The number of observations on tills of known origin are very few, and there has been a certain reluctance to use even these few results. Perhaps this reluctance is based on the fact that relatively easily accessible alpine valley glaciers in both Europe and North America provide rather poor depositional models for Pleistocene ice caps on lowland plains.

The ice caps and valley glaciers of Svalbard have long been known to deposit tills of very considerable thickness over wide areas, and to produce multi-till sequences similar to those of Pleistocene age in lowland areas of temperate latitudes (Garwood and Gregory, 1898; Lamplugh, 1911; Gripp, 1929). In view of this similarity, a project was undertaken in 1964 to identify, as far as possible, the modes of origin of these tills and the sequences in which they lie, and to investigate the relation between till genesis and character. It is possible to be very specific about the origin of tills within the Svalbard sequences, simply because they are often so intimately related to the ice from which they are derived.

Svalbard glaciers tend to transport considerable volumes of debris in an englacial position, and at their margins, where there is often a strong upward component of flow, this debris is released at the glacier surface as it wastes down, forming a cover of supraglacial till (Boulton, 1968). As the thickness of this till builds up, ablation of the underlying ice is increasingly inhibited, resulting, if the glacier is retreating, in very large areas of supraglacial till, which tend to acquire a hummocky surface as a result of differential ablation. Fluvial or lacustrine activity in this marginal zone also produces thick supraglacial accumulations of stratified sediments between ice-cored hummocks and ridges, which, when sedimentation ceases, are covered by thick deposits of till continuously moving off the ice-cored ridges (*flowtill*) (Fig. 1).

Masses of buried stagnant ice, with its high debris-content, are also an important source of another type of till. This is till which melts out slowly from beneath overburden, either at the top surface or bottom surface of buried stagnant ice, and which retains much of its original englacial struc-

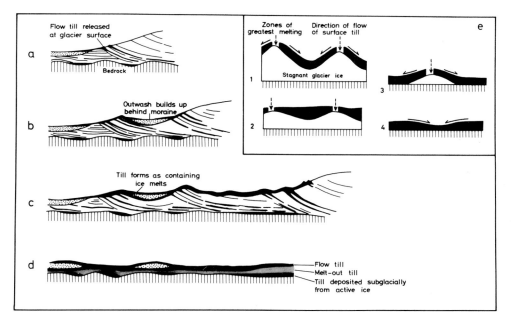

Figure 1. A schematic diagram showing the typical development of till sequences in Svalbard (a-c). d is a prediction of the resulting sequence if all buried ice were to melt, while e shows the changing pattern of differential ablation and flow which produces flow till sheets of low relief and allows buried ice to melt. Absence of further flow beyond stage e (2) would lead to a hummocky till surface.

ture (*melt-out till*). A third type of till is that which is deposited beneath active ice (*lodgement till*). A series of schematic diagrams (Fig. 1a-1c) contrasts the ways in which these three different types of till develop, and Figure 1d illustrates the probable result if all the buried ice were to melt; the left-hand side of Figure 1d shows the type of sequence which would develop where supraglacial fluvial and lacustrine deposits were extensive, and the right-hand side shows the type of sequence which would develop where they were not.

This classification of tills is presented in Table 1, arranged so as to show their relationship with the existing terminology. This classification has been chosen because the three kinds of tills—flowtill, melt-out till, and lodgement till—have fundamentally different stratigraphic implications, and because they are relatively easy to distinguish.

TABLE 1

CLASSSIFICATION OF TILLS

Classification Adopted Here	Origin	Existing Classification
Flowtill	} Supraglacial	Ablation Till
Melt-out Till		
Lodgement Till	Subglacial	Lodgement Till

The intention in what follows is to describe the stone orientation fabrics of tills whose origin is known, in order that this data can be applied to tills of unknown origin. In some cases the process of deposition is so easily studied that fabric variation is readily predicted; in other cases, where the mode of deposition cannot be so closely specified, considerable sampling problems are met with.

Collection and Treatment of Data

It has not been possible at any one site to standardize the number of measurements made, the methods of sampling, the accuracy of these measurements, or the types of stones sampled. For instance, some subglacial till pockets contain very few stones. Stone shapes and dimensions vary with varying bedrock. In some cases it was possible to collect oriented till samples and to measure stone orientations in the laboratory, but this was impossible with englacial stones; and in all cases the number of measurements, and probably accuracy, varied with temperature.

Much of the three-dimensional directional data is presented by Schmitt equal-area projection of the lower hemisphere. This is then contoured by a method suggested by Kamb (1959), which has the dual advantages of presenting the data in a visual form and giving an indication of the statistical significance of apparent orientation peaks. In this method the area of the counter used in the conventional contouring procedure is chosen so that data-point densities can be contoured at the desired levels of significance. I have adopted contour intervals of 2σ and a significance level of 3σ. Significant maxima have then been analyzed individually using Fisher's (1953) function for probability density on a sphere (Steinmetz, 1962), which gives orientation of the resultant vector ($A =$ angle of dip, $D =$ direction of dip), its magnitude (R) in relation to the total number of observations (N), a radius of circle of confidence (Θ), and a precision parameter (K). This overcomes some, although not all, of the difficulties of attempting to analyze spherical distributions, which are often non-normal, by methods which are only suitable for the analysis of spherical normal distributions (Green, 1962).

Several conventions are necessary. The direction normal to the glacier front, parallel to the general axis of advance of the glacier, is termed the "parallel" direction, while the "transverse" direction is at right-angles to this. In the local deformational system, which may be unrelated to the parallel and transverse directions, the direction of major tectonic transport is denoted A, with B and C as intermediate and minor axes, respectively. These terms are used simply because till-stone orientations have often been assumed to be due entirely to direct glacier-generated stresses related to the parallel and transverse directions, and thus it is useful to compare the local stress situation responsible for a till fabric with these two directions. The long, intermediate, and short axes of till clasts are referred to as the a, b, and c axes, respectively; prolate particles are those in which $b \simeq c$ and $a/b \geqq 2$; blades are those in which $a: b: c \geqq 3: 2: 1$; and plates are those in which $a \simeq b$ and $a/c \geqq 2$.

ENGLACIAL TRANSPORT

By far the largest part of the materials which make up the three main types of till are transported englacially to their point of deposition. The original clast-orientation fabrics of englacial debris are partly preserved in melt-out tills, and may, in special cases, influence the final fabrics of supraglacial flowtill and subglacial lodgement till. Englacial-debris fabrics are thus of some importance.

Figure 2. Basal ice in the compressive zone on Makarovbreen. Thrust planes cut the ice foliation planes and boulders are rotated so that a-axes lie parallel to the direction of movement along the thrust planes. Ice movement is from left to right and boulders are generally of ellipsoidal shape.

The greatest amount of englacially transported debris is derived from the glacier bed and can be divided into two main types: masses of unlithified subglacial sediment which have been incorporated *en bloc* into the glacier, in which the included ice is only interstitial, and, by far the more important, particulate suspensions in glacier ice. These particles, which vary in size throughout the range of sizes normally found in till, may be merely aggregations of smaller particles, or they may be individual grains. The concentrations of these suspensions vary from below 10 percent to above 50 percent by volume in the parallel-to-foliation bands in which they tend to occur.

Observations of the orientations of englacial clasts at ten sites in areas of planar, sub-horizontal, or dipping foliation show that the *a/b* planes of blade- and plate-shaped particles or the *a*-axes of prolate particles tend to lie within the foliation plane (Fig. 3). The orientation of *a*-axes within this plane appears to depend upon the local stress situation. In the snout area of the glacier Makarovbreen in Ny Friesland, there is a transition from a zone of extending flow to a zone of compressive flow (Boulton, 1970a). In the zone of extending flow, the ice foliation is sub-horizontal, parallel to the glacier substratum, and *a/b* planes of stones in the basal debris-rich layers lie parallel to this foliation, while long axes lie parallel to the direction of flow. In the terminal compressive zone, the foliation has a steeper, up-glacier dip, and although the *a/b* planes of blade- and plate-shaped stones still lie in the foliation planes, the *a*-axes of prolate and blade-shaped stones now tend to lie transverse to glacier movement (Fig. 3). These observations

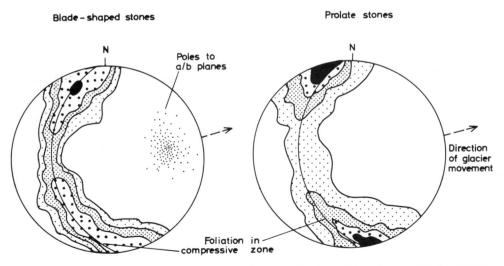

Figure 3. Transverse *a*-axis orientations in a zone of compressive flow, Makarovbreen. Note the stronger orientation of prolate stones, Contours at 2σ intervals. For blade-shaped stones; A = 4°, D = 353°, N = 100, R = 67.8, Θ = 9°40′, K = 3.1. For prolate stones; A = 3°, D = 342°, N = 100, R = 76.6, Θ = 7°46′, K = 4.2.

suggest the possibility that stone orientations are strongly affected by the change from a tensile- to a compressive-stress system, and that the direction of maximum extension of the triaxial strain ellipsoid has had a controlling influence on *a*-axis orientation.

Further evidence for this thesis was provided by observations at the northern margin of Aavatsmarkbreen in Oscar II Land where the measured geometry of the strain ellipsoid was compared with the orientation of englacial-debris fragments and other structures. This glacier flows rapidly into the sea in a westerly direction, but at its northwestern boundary, it flows up-slope (Fig. 4), as a result of which the glacier foliation planes have a steep up-glacier dip (of as much as 60°) and are truncated by even more steeply dipping thrust planes. Structures within the ice indicate considerable parallel compression and transverse extension, an assessment in accord with the measured orientation of the strain ellipsoid (Nye, 1959). Up-thrust slices of debris-rich basal ice are responsible for the outcrops of numerous debris bands on the glacier foreslope, a number of which were

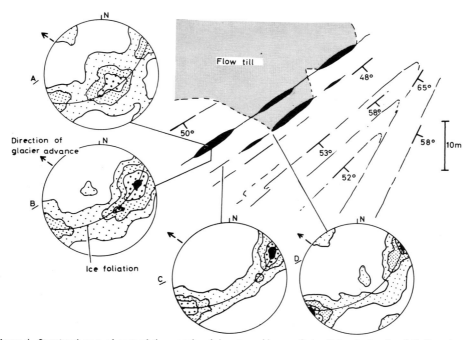

Figure 4. Structural map of part of the margin of Aavatsmarkbreen, Oscar II Land, showing foliation planes in debris-rich ice. The principle strain-rate component at the glacier surface was oriented at 42.5° east of north. A and B show the orientations of the *a*-axes of blades and prolate particles in the middle and at the margin of a lense of debris which contains only interstitial ice. C and D show long axis orientations of blades and prolate particles of debris suspended ice. Contours at 2σ intervals. (A); $A = 58°$, $D = 123°$, $N = 70$, $R = 33.3$, $\theta = 18°12'$, $K = 1.88$. (B); $A = 15°$, $D = 68°$, $N = 70$, $R = 37.8$, $\theta = 16°14'$, $K = 2.14$. (C); $A = 7°$, $D = 59°$, $N = 70$, $R = 40.8$, $\theta = 14°45'$, $K = 2.36$. (D); $A = 0°$, $D = 51°$, $N = 70$, $R = 39.7$, $\theta = 15°12'$, $K = 2.28$.

investigated. Two bands containing englacial debris as particulate suspensions yielded clast-fabric analyses showing transverse peaks of *a*-axis orientation of varying strengths, and *a/b* planes of blade- and plate-shaped stones lying in the ice-foliation planes, which dip at angles varying from 48° to 65° (Fig. 4c-4d). In addition, a number of debris bands consisting of lenses of till which contain merely interstitial ice, crop out on the glacier surface. In places several of these lenses lie along the line of a single foliation plane. Orientation measurements made on particles lying along the plane of contact between glacier ice and debris and within the body of the debris band gave rather different results. In the first case, *a*-axes of prolate particles were oriented with a mean vector within 20° of the long-axis of the strain ellipsoid, with a slight down-dip component (Fig. 4b), while in the second case, there was a diffused down-dip orientation maximum and a weak secondary peak at right angles to this direction (Fig. 4a).

The discontinuous lenticular nature of some debris banks in which the ice content is merely interstitial in nature has been noted before (Boulton, 1968). Dort (1967) described englacial bands from Antarctica consisting of sand which had originally been deposited by wind on the upper part of the glacier; where these were exposed near the glacier snout, the bands formed lenticular rods with their axes parallel to glacier movement, and it was suggested that they had originally formed parts of more extensive sheets which had broken up by transverse elongation. The debris bands referred to on Aavatsmarkbreen, comparable to *boudins* in metamorphic rocks, could have reacted to stress in a relatively brittle fashion, being pulled apart by lateral extension of the more plastic ice matrix. In this case, the dominantly transverse fabric at the debris-ice interface would result from relative movement of ice over the upper and lower surfaces of the bands, while, as long as the materials forming the bands were relatively rigid, there would be little or no re-orientation of particles during transport within this zone. The parallel fabrics could either be derived from the time before the debris was picked up *en bloc* from the glacier bed, or from some earlier englacial stress system.

These observations suggest that lateral extension in a zone of longitudinal compression can have a considerable effect on englacial-stone fabrics. The effect of this extension, however, will depend upon the magnitude of the simple shear strain in a vertical plane within the ice. Glen, Donner, and West (1957) based their theoretical analysis of the orientations of stones within glacier ice upon the assumption that parallel-to-foliation shear would be dominant. If this were true, and the glacier were to flow down a smooth channel of uniform cross-section, then it would seem most plausible that the stones would take up orientations such as those predicted by Glen, Donner, and West (1957). If, however, lateral extension occurs within the glacier, stone orientations would be influenced by both major components of flow. Another possibility is that, especially in the snout area of the glacier, shear could be restricted to narrow zones. On Makarovbreen, relict thrust planes are completely obliterated in the basal two to three meters of the glacier, suggesting the presence of intense shear at this level, whereas at higher levels such structures survive considerable transport without being deformed. This suggests that shearing parallel to the foliation planes is relatively unimportant and therefore that clast-rotation is also unimportant at these higher levels. The distinction between zones of shear and zones of triaxial compression is well-marked on Makarovbreen. Between high-angle thrust planes in the ice, a/b planes of blades and plates lies along the ice-foliation planes. In contrast, those stones which are intersected by thrust planes (which also cut and displace the ice foliation) have been rotated so that their a/b planes lie within the thrust plane, at a high angle to the ice foliation, while the a-axes of these stones tend to be parallel to the direction of movement along the thrust plane. A similar phenomenon was reported from Dunerbreen in Ny Friesland (Boulton, 1968), where stones lying in a zone of shear had a preferred a-axis orientation parallel to movement along the shear planes, whereas away from the shear planes, the a-axes of blade-shaped stones merely lay in the foliation plane but had no significant preferred orientation within this plane. Thus it seems likely that, along shear planes, stone orientations are controlled by the movement of one ice surface across another, which tends to produce a-axis orientation parallel to movement.

The effect of shape on a-axis orientation is seen in Figure 3, where a-axes of prolate particles are much more strongly grouped about the axis of maximum strain than are the a-axes of blades. At three localities, where there was neither marked parallel extension (no crevasses at surface) nor compression (horizontal foliation present, with no obvious thrust planes), there was no significant preferred orientation of a-axes of blades, and only a very weak preferred orientation of a-axes of prolate particles.

Other observations of englacial-clast fabrics show variations similar to those reported here. Donner and West (1957) reported measurements on steeply dipping bands of englacial till from the margin of Vestfonna and the ice cap on Nordaustlandet, Svalbard. The fabric maxima varied greatly; some were parallel to the glacier advance, some were parallel with a secondary transverse peak, and some were transverse with a secondary parallel peak. Unfortunately, structural relationships were difficult to determine, and there was no indication whether the englacial "till" contained merely interstitial ice or whether it consisted of debris suspensions in ice. Harrison (1957) also reported fabrics determined on a block of englacial till from the ice-cap margin at Thule, Greenland. Particles within the till showed both parallel and transverse preferred orientation, and the poles to the *a/b* planes of blades and plates were normal to ice-foliation planes. Measurements by Richter (1936) and by Galloway (1956a) also recorded parallel englacial fabrics.

FABRICS ACQUIRED DURING SUPRAGLACIAL DEPOSITION

Flow Tills

Supraglacial tills develop on those glaciers in which there is a strong upward component of movement in the terminal area, and in which there is a thick englacial-debris sequence. Ablation of the glacier surface in the terminal zones exposes the englacial debris at the surface, where, on melting out, it produces a highly fluid till. Movement of this till down the glacier surface can give rise to a supraglacial cover over wide areas. When the thickness of this till cover increases to more than about three centimeters, the rate of ablation of the underlying ice decreases. The thickness of this supraglacial till may be increased from two sources: by further flows from up-slope positions, and by accretion of till from the underlying melting ice. Thus a supraglacial till may be composed of two elements: an upper allochthonous element, which has been subjected to sub-aerial processes, and a lower autochthonous or parautochthonous element, which has not been exposed at the surface. If the thickness of till accumulating on the glacier surface increases to more than two to three meters, the depth of summer melting is equalled and no further accumulation of autochthonous till takes place at the ice/till interface, although allochthonous till flows may still cause accumulation of till at the surface. Thus very large areas of sediment-covered glacier ice may survive glacier retreat.

There are three basic flow processes which affect supraglacial tills such as those described above: mobile, liquid flow; semi-plastic flow; and down-slope creep, the nature of the flow depending on the nature of the slopes and the water content of the till. These three processes are considered separately below.

Mobile, Liquid Flow

The first type of flow affects those tills which have just been released sub-aerially directly onto glacier ice, or where an excessive amount of water occurs on the surface of an existing till. The water content of these tills is very high, above the liquid limit, and very mobile downslope flows are produced. These flows are rarely more than 20 cm thick, and boulders and stones tend to settle through them, producing a distinction between a lower, stony, relatively slowly moving element, and an upper, stone-free, rapidly moving element, with velocities in this upper element exceeding one meter per hour. The frequent release of water at the surface of these flows produces normal water sorting, and thus they may acquire a stratification. The flows tend to be elongate, with a lobate front. In the body of the flows, stones tend to move with their a/b planes parallel to the underlying surface and a-axes parallel to the direction of transport, whereas, in the nose of the flow, a-axis orientations develop that are transverse to the flow direction.

Semi-plastic Flow

The second type of movement occurs on more stable slopes, where changes in surface loading or pore-pressure cause the shear strength of the till to be exceeded, and failure takes place along an arcuate slip-plane, down

Figure 5. The surface of a small lobate till flow, showing compression at its front with transverse-to-flow stone orientations. In the body of the flow, parallel-to-flow orientations dominate. When sectioned it has a structure similar to that shown in Figure 7B.

which the till moves as a semi-plastic mass. The slip-plane is usually the junction between unfrozen and frozen till, or unfrozen till and glacier ice. The till moves downslope as a lobate flow, and the slip plane becomes an arcuate scar (Boulton, 1968, Fig. 11), the back wall of which is oversteep-ened and unstable, and therefore continues to feed the flow and to enlarge itself, often to considerable size. These failures are important in early sum-mer, when the ground is thawing and there are rapid increases of pore-pressure.

Observations on the surfaces of flows show that stone orientations vary systematically in different parts of the flow. A very small-scale flow which illustrates this is shown in Figure 5. The variation in orientations of blade-shaped stones in an ideal flow is shown in Figure 7a in a schematic diagram (made up from many observations), which is very similar to the observa-tions of Lundqvist (1949). In the body of the flow, rapid longitudinal exten-sion and vertical settling produce a preferred orientation of *a*-axes in the direction of flow and *a/b* planes which either are parallel to the flow surface or dip toward or away from the direction of flow, depending on the shape of the underlying bed. In the nose of the flow, longitudinal compression tends to produce *a*-axes oriented transverse to flow and *a/b* planes that dip up-flow, whereas, at the lateral margins, *a/b* planes dip toward the axis of the flow with *a*-axes oriented parallel to the flow direction. If the water content is low, dips are higher and vice versa, except for flows in con-stricted channels which tend to show high-angle *a/b* planes throughout.

Internally, most of these flows show little sign of stratification as a result of washing by excess water, although streaking out of inhomogeneities does sometimes produce some pseudo-stratification. However, on the surfaces of some flows, especially soon after the initial movement (failure), excess water does occur, resulting in some liquid flow that tends to produce strati-fication (Fig. 6). In such a flow, the stratification, though it may be limited, is very valuable in that it will reflect any subsequent deformation of the flow. Figure 7b is one flow of a series in vertical sequence which has ac-quired some stratified horizons as a result of washing, and whose structure is thus revealed. The flow structure is essentially that of a flat-lying isoclinal fold, with a well-marked basal shearplane. Stone-orientation fabrics were determined at four points within the fold. In order to sample only a small area, oriented samples of 300-500 cc were collected, and the small stones (between one and five centimeters) were measured in the laboratory. Measurements of *a*-axis orientation produced A-direction peaks of varying strength throughout the flow (near the top surface, in the middle, and at the bottom surface of the flow), whereas in the fold nose, a strong B-direction peak was produced. Considerable variation occurred, within the flow in the orientations of blades and plates, which were imbricated up-

Figure 6. A flow till which carries evidence of surface washing. Some of the laminae are isoclinally folded.

flow at the base, formed a transverse stereographic girdle in the middle, had a down-flow imbrication at the top, and in the nose had a high-angle,

A.

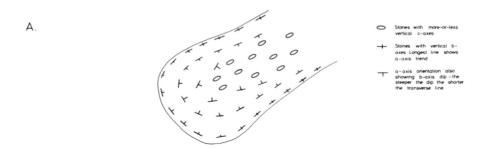

Figure 7A. Idealized lobate till flow showing the pattern of orientations which would be taken up by blade-shaped particles.

B.

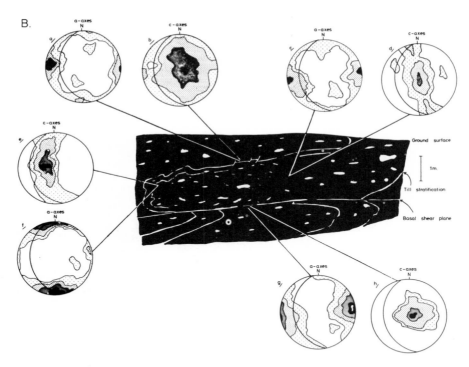

Figure 7B. The pattern of a-axis (prolates and blades) and c-axis (blades and plates) fabrics from a section in an old lobate till flow in which washed horizons reveal the structure. Contours at 2σ intervals. (a); Major peak: $A = 7°$, $D = 227°$, $N = 55$, $R = 36.91$, $\Theta = 20°56'$, $K = 2.98$. Minor peak: $A = 4°$, $D = 189$, $\Theta = 26°14'$, $K = 2.6$. (b); $A = 75°$, $D = 112°$, $N = 70$, $R = 30.8$, $\Theta = 19°42'$, $K = 2.4$. (c); $A = 2°$, $D = 268°$ $N = 80$, $R = 47.2$, $\Theta = 13°18'$, $K = 2.4$. (d); $A = 86°$, $D = 67°$ $N = 70$, $R = 36.9$, $\Theta = 16°31'$, $K = 2.1$. (e); $A = 66°$, $D = 271°$, $N = 70$, $R = 38.2$, $\Theta = 15°54'$, $K = 2.2$. (f); Major peak: $A = 4°$, $D = 181°$, $N = 64$, $R = 47.5$, $\Theta = 10°30'$, $K = 3.8$. Minor peak: $A = 3°$, $D = 77°$, $N = 13$, $R = 9.2$, $\Theta = 18°42'$, $K = 3.2$. (g); $A = 7°$, $D = 82°$, $N = 80$, $R = 56.3$, $\Theta = 10°21'$, $K = 33.3$. (h); $A = 77°$, $D = 102°$, $N = 70$, $R = 50.5$, $\Theta = 10°51'$, $K = 3.5$.

up-slope dip, though slightly oblique to the direction of movement. A very similar fabric distribution has been reported in flat-lying isoclinal folds in Norfolk tills (Banham, 1966).

Downslope Creep

On slopes where the supraglacial-till cover is stable, tills are composed of two elements: an upper allochthonous element, which has flowed into place by a combination of both liquid and plastic flow, and a lower autochthonous or parautochthonous element derived from the underlying ice, which has not been exposed to sub-aerial processes. There are many exposures in which this distinction cannot be recognized, but in others the two elements are clear: (1) an upper allochthonous element, which shows signs of washing (loss of silt-clay as in the conventional ablation till, or stratification) over (2) a lower element which tends to be compact and unstratified (Fig. 8).

Figure 8. The lower, parauthochtonous part of a flow till, which has not been subjected to sub-aerial processes. Flow from right to left, note imbrication.

The majority of such compact tills, although they appear to be quite stable, are in fact undergoing slow downslope creep under stresses less than the shear strength of the material, and with relatively low water contents. Where the underlying ice surface is melting, creep velocities are relatively high, on the order of one to five centimeters per month, whereas, when melting does not penetrate to the glacier surface, velocities are much less, undetectable except over several years' observations, although shear planes at the freezing surface and deformation of segregation-ice layers indicate that creep is taking place.

The parautochthonous lower element of a flow till has a characteristic

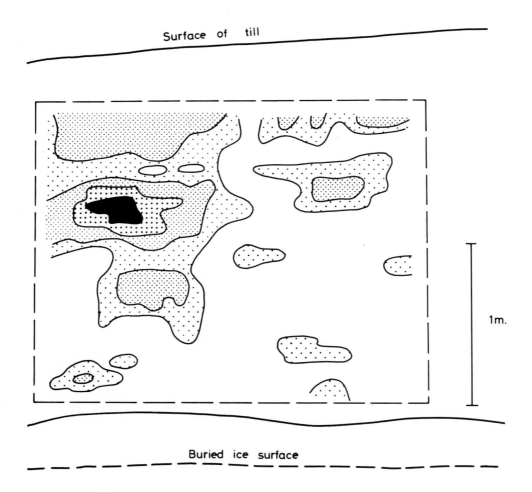

Figure 9. The frequency of occurrence of transverse-to-flow a-axes in a supraglacial flow till in which a-axes are predominantly parallel to flow. Stones selected on the basis of a 10 cm. grid, and contoured at 20, 30, 40, and 50 cm intervals.

fabric in which the *a*-axes of prolate stones tend to lie parallel to the direction of slope and to be imbricated in an up-slope direction. The *a/b* planes of blade-shaped stones tend to lie horizontally and, if the *a/b* ratio is high, the *a*-axis also tends to be oriented parallel to the direction of slope, but if the *a/b* ratio is low, *a*-axis orientation is not so well controlled. Thus fabric diagrams made from two samples taken from a flow till might show significantly different *a*-axis fabric patterns, with respect to the direction of flow, if one contained a higher percentage of blade-shaped stones than of the other shape. Indeed, flow till fabric diagrams made up of the *a*-axes of blades often show a pronounced partial horizontal girdle, whereas a fabric diagram made from measurements on prolate particles shows a single parallel peak.

One of the controls on the fabric of the parautochthonous element of a flow till is the original englacial fabric. This is largely preserved in a till which has just melted out, but not yet moved, though such fabric changes relatively rapidly during downslope creep. An exposure at the margin of Valhallfonna, an ice cap in Ny Friesland, shows glacier ice beneath flow till and intersecting the junction of the two, a distinctive englacial band of red-brown debris dipping up-glacier at 52°, which is easily distinguished

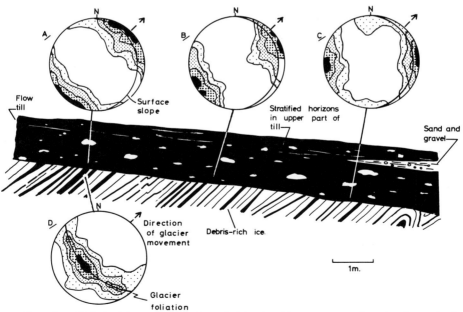

Figure 10. A flow till derived from debris in the underlying ice. The stereograms show the change in clast fabrics resulting from melting of the containing ice, and those which develop subsequently as a result of flow. Contours at 2σ intervals. (A); $A = 3°$, $D = 44°$, $N = 80$, $R = 50.2$, $\Theta = 12°15'$, $K = 2.65$. (B); $A = 1°$, $D = 241$, $N = 80$, $R = 56.1$, $\Theta = 10°59'$, $K = 3.3$, (C); $A = 2°$, $D = 268°$, $N = 80$, $R = 60.9$, $\Theta = 8°54'$, $K = 41.1$. (D); $A = 51°$, $D = 237°$, $N = 80$, $R = 62.7$, $\Theta = 8°22'$, $K = 4.6$.

from the other debris with its gray-black color. This red-brown debris can also be traced in the overlying till for a distance of 10 meters downslope from the point at which the debris band intersects the till-ice junction (Fig. 10), presumably indicating 10 meters of downslope creep since the debris band started to contribute to the base of the till. Stereograms show the fabric in the red-brown debris-band (Fig. 10d), in the till immediately above this band (Fig. 10a), and after three meters and six meters of downslope flow (Figs. 10b, 10c). Initially there is a clear azimuthal similarity in preferred orientation between stones in the till and in the englacial band, but subsequent downslope flow changes this orientation (Θ) by 40°, and the further the till flows, the greater the strength (R) of orientation becomes.

Spatial Variation in Flowtill Fabrics

Variation in flowtill fabric tends to be greater in the vertical direction than in the horizontal. A flowtill occurring above ice or above bedded sediments may show a lower parautochthonous element with a strong parallel fabric acquired during slow creep, and an upper allochthonous element showing a much more variable fabric, acquired initially as the result of accumulation of a series of relatively rapidly flowing tongues, and perhaps modified subsequently by creep. Alternatively, the flowtill may consist entirely of the upper element, or may even interfinger with stratified sediments, as a result of interplay between fluvial and lacustrine deposition and mudflow deposition (Boulton 1968, p. 408).

An example of vertical variation was seen in an exposure of 2.5 m of compact, fine-grained till resting on dead ice at the margin of Aavatsmarkbreen in Oscar II Land. From its position, this was obviously a flowtill, although if the ice were to melt and the till were to be deposited on bedrock, it would show none of those structures, such as washed horizons or stratified lenses, which occasionally occur in flowtills. The till contained a large number of small prolate stones in a fine-grained matrix. By means of a net-sampling system applied over an area of two and a half meters by two meters, the a-axis orientations of 286 prolate ($a/b \geq 2$) particles with a-axes of between two and five centimeters were measured and their positions plotted on a plan of the site. The resultant fabric showed two clear peaks of a-axis orientation, one oriented normal to the glacier front, and one parallel to it. Each stone was then correlated with either the parallel or transverse peak and the densities of stones in each mode were then contoured using a 2σ contour interval (Fig. 9). The result showed that significant transverse peaks tended to lie in certain well-defined domains, rather than being randomly distributed, and that there were few significant transverse peaks in the fabric diagram for the lower part of the till. This diagram does *not* show the orientation strengths, nor does it imply that a fabric analysis of a small part of the lower till would not show a significant transverse element. It simply shows unstratified variation for this minimal sampling density. The most probable explanation of this variation is that the fabrics in the lowest meter of the till have developed as a result of creep, while those in the uppermost meter and a half represent the effects of the accumulation of a series of either thin flows with different transport directions or lobate flows whose fronts are represented by the transverse peaks.

In some sections, however, there is no clear differentiation into obvious parallel and transverse peaks; an orientation at one horizon may be consistent laterally, but in a vertical direction there may be considerable azimuthal variation. This may be brought about by a succession of mudflows, which in some cases may be separated by thin fluvial or lacustrine beds. Flowtill fabrics also show considerable areal variations, where an important factor is the change in both direction and inclination of the slope down which the tills move, because of slow melting of buried ice. Tills initially form on the glacier foreslope, but as the active glacier margin retreats, that slope is changed by ablation and tends to develop a slope toward the glacier on its proximal side (Fig. 1a-1b). In its subsequent development and enlargement, such ice-cored moraine may be "controlled" with a ridge system parallel to the glacier front, or "uncontrolled," a random arrangement of ridges and hillocks. But no matter which, the configuration of the till surface is constantly changing in response to patterns of differential ablation and till flow (Fig. 1e). During this process of adjustment, the upper part of the till deforms most easily and thus changes its fabric most readily, while the lower part may well retain a fabric derived from its initial flow down the glacier foreslope. Indeed, if the prediction from Jeffery's (1922) model is correct, that the a/b planes of particles in laminar flow are horizontal, then it should be possible to identify a flow which formed on a slope of different inclination from its present slope.

The term *controlled* used above in reference to a moraine system is used in the sense defined by Gravenor and Kupsch (1959), as a feature which reflects in plan the pattern of some glacial structure from which it is derived. Most "controlled" ice-cored moraine ridges tend to lie parallel to the glacier margin probably because of the fact that they are derived from series of debris bands in the underlying ice which themselves lie parallel to the glacier margin (for further discussion, see Boulton, 1968). In such a moraine, much of the flow in the supraglacial till will be either directly toward or directly away from the glacier. Thus, as flowtill fabrics tend to show peaks either parallel or transverse to flow, so in an area of controlled ice-cored moraine, flowtill peaks will tend to lie parallel or transverse to glacier movement. Areal variations in the fabrics of tills formed supraglacially in a zone of "uncontrolled" ridges would in contrast tend to be random.

Other observations on solifluction and mudflow fabrics show results comparable to those reported above. Glen, Donner, and West (1957) determined long-axis fabric orientations at five sites at the margin of Vestfonna, Nordaustlandet, in what they termed "ground moraine," although this was in fact supraglacial flow till and fabrics were produced by downslope flow. At all sites there was a dominant parallel peak, two of which showed secondary transverse peaks. Rudberg (1958), measuring long-axis orientations on solifluction slopes, determined a downslope peak of *a*-axis orientation with up-slope imbrication, a result similar to that of Caine (1968), who measured orientations of boulders in a periglacial blockfield (although he also found a secondary transverse peak).

Melt-out Tills

The entirely autochthonous element which occurs at the base of some stable flow till sheets is different from the overlying till in that it retains part of the original englacial fabric. Most melt-out tills which initially lie between a thin flowtill cover and glacier ice are likely to be deformed later by creep and thus to become flowtills (Fig. 10), but much of the debris contained within dead-ice masses overlain by thick overburden could well be released as melt-out till on final disappearance of the ice. Whether or not this process would take place would depend upon whether the overburden was sufficiently thick to inhibit flow and sufficiently permeable to allow water to escape from the melting ice, so that effective consolidation of the melt-out till could take place. If water was not able to escape, high pore pressures within the till and differential loading under an overburden of variable thickness (see Fig. 1) could produce flow or diapirism within the till. The effect of the diapirism in tills is well seen in the tills of north-east Norfolk, England (West and Banham, 1968), where it is recognized by the involution of the till stratification (in an unstratified till, recognition would be very difficult).

Melt-out tills may form either at the base of stagnant ice, or at its surface, by slow melting *in situ*. In Svalbard at present, only surface melting occurs. However, although the depth of seasonal thawing is only two to three meters on the average, melting of buried ice surfaces can occur at considerable depth beneath supraglacial streams or lakes, and thus melt-out tills can occur beneath thick stratified fluvial and lacustrine sequences. Their stone-orientation fabrics are essentially similar, in the plane of deposition, to the fabrics of the englacial debris from which they are derived, although they show somewhat weaker maxima, presumably because of disturbance and interactions between stones during deposition (Boulton, 1970b, Fig. 2; and Fig. 10 above). However, the angles of dip of long axes tend to be reduced during deposition by amounts which depend upon the englacial concentration of the debris from which they are derived. If this concentration is high, the original three-dimensional orientation will tend to be preserved; on the other hand, if this concentration of debris is low, vertical distances will be considerably reduced and only the projections of these

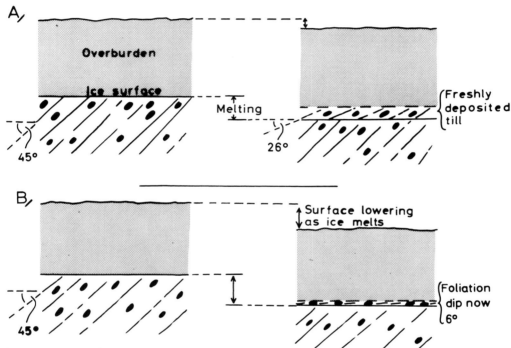

Figure 11. A. schematic diagram showing melting beneath an overburden of a mass of debris-rich ice with a debris content of 50 percent by volume and a foliation dip of 45°. The resultant melt-out till shows an a/b plane dip of 26°.
B. The production of melt-out till from ice with a debris content of 10 percent by volume. a/b planes now dip at 6°.

orientations onto the plane of deposition will be preserved (Fig. 10 and Fig. 11). Most melt-out tills develop from ice in which the foliation has a high up-glacier dip; however, because most debris bands have a relatively low ice content, this up-glacier foliation will be reflected only rarely in melt-out tills. This contrasts with the conclusion drawn by Harrison (1957). Similarly, in such a zone, a high proportion of stones show transverse orientations, and thus melt-out tills could be expected to show flat-lying a-axes and a/b planes, and commonly a transverse component of a-axis orientation.

FABRICS ACQUIRED DURING SUBGLACIAL DEPOSITION

Subglacial Lodgement Till

The thin marginal zones of many Svalbard glaciers which terminate on land are probably frozen to their substratum (Palosuo and Schytt, 1960), but some more rapidly flowing glaciers, which terminate in or near the sea, actively move over their beds in the terminal zone, and often give opportunity for inspection of the moving glacier sole. The lowest elements of the englacial load are plastically held within the ice and protrude through this sole; and, coming into contact with underlying bedrock, they are probably responsible for much of the striation and abrasion on bedrock surfaces (Carol, 1947). Where the glacier sole loses contact with the bedrock surface, above steep-sided cavities, the glacier sole itself can be seen to be heavily striated, and the *a*-axes of particles embedded within it lie parallel to these striations. Observations of the glacier sole were made in tunnels, in ice caves, and at the bases of ice cliffs at 17 localities. Subglacial till was present at less than half of these; at the others, the glacier rested upon smoothed and striated bedrock or deformed unlithified fluvial and lacustrine sediments. Most subglacial tills here have a drumlinoid or fluted form. Only at a limited number of localities was there any direct evidence for the processes of deposition. Two basic processes of till deposition from debris-rich ice were recognized from these exposures, although some tills may have originated by other processes for which no evidence was found. These processes were: the slow melting out of till from debris-rich ice which had become stagnant beneath the active glacier, and deposition from active ice as a result of pressure melting of the glacier sole as it passed over bedrock obstructions. The fabrics of these tills appeared to develop both from the process of deposition and from subsequent subglacial deformation. Although subglacial observations were restricted to snout areas, many glaciers have retreated considerable distances throughout recent centuries so that certain of what are now marginal subglacial deposits may have formed under a much thicker ice cover.

Lodgement Tills Released from Ice by Pressure Melting

Two localities showed definite evidence of lodgement tills actively being released as a result of pressure melting against an obstruction, and this process may well account for other subglacial till accumulations. At one locality, 25 m beneath the margin of Aavatsmarkbreen in Oscar II Land, till was actively accumulating on both sides of a bedrock knob (Fig. 12). On the up-glacier side, the debris-rich glacier sole was moving over both till and bedrock, and melting was active at this sole. The result was to

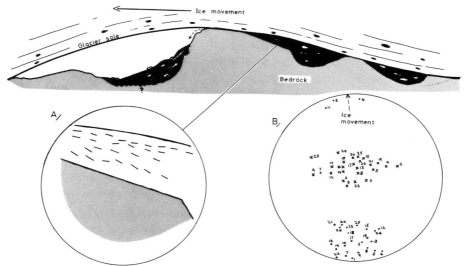

Figure 12. Section showing subglacial lodgement till built up around the flanks of a rôche moutonée. Compiled from observations made in several intersecting subglacial tunnels. A shows the a-axis dips of the magnetic susceptibility ellipsoids for 27 specimens collected from the subglacial till. The specimens are numbered from left to right, and downwards through four rows. B is a stereographic plot of the a-axes (dots) and c-axes (crosses) of the susceptibility ellipsoids for the specimens, numbered as described above.

release from the ice particles which had hitherto been partially or entirely englacial. Some of this material, a thin layer beneath 0.2 and 0.5 cm thick, seemed to be actively moved along by shearing beneath the glacier sole at about 0.7 cm/day (determined by markers). On the down-glacier side of the bedrock knob, the glacier lost contact with bedrock and thin till was oozing out like toothpaste from the ice/bedrock contact, to slump and slide down the lee side of the knob, at the foot of which was a considerable accumulation of such slumped material. This is slump deposit, so only a very indistinct bedding is acquired, presumably because much of the material falls as a semi-cohesive mass, although some sorting does occur as a result of water movement over the depositional slope. Pebbles on this slope tend to lie with their flattened surfaces on the slope and their long axes parallel to the direction of slope.

A better consolidated till had accumulated on the up-glacier flank of the obstruction; as there is no obvious subglacial source for this till, and as pressure melting and till accumulation are actively occurring on its surface, there is a strong possibility that the whole accumulation has built up by this same process. Irregularities in the underlying bedrock surface have been filled in, thus producing a smooth subglacial profile partly in rock and partly in till. Cavities can obviously be filled in, in much the same way as outlined above for the slump deposit on the down-glacier flank of the bedrock obstruction, but subglacial till has also been built up immediately above smoothed and striated bedrock surfaces, suggesting that only part of the till deposited by pressure melting is carried along at the glacier/ substratum interface, subsequently to be forced into cavities, and that a small residue is actively being plastered permanently onto the substratum. The plastered-on surface of this till in contact with ice is striated; clasts lying on it show *a*-axis orientations parallel to ice movement along the plane, while *a/b* planes show a partial transverse stereographic girdle with a mode normal to the till surface. (See theoretical treatment: Nobles and Weertman, 1971; this volume.)

In order to investigate the small-scale variations in the fabric in this till, small samples were collected at the intervals shown in Figure 12. For each of these samples, three measurements of the anisotrophy of magnetic suceptibility were determined, which, when computed, gave a susceptibility ellipsoid whose three axes reflect the physical orientations of magnetic grains within the samples. These orientations were then checked against the microfabrics determined by cutting mutually perpendicular thin sections from four of the till samples. A good agreement resulted, leading to the assumption that magnetic fabrics are equivalent to microfabrics (for details of the method, see Hamilton and Rees, 1965, and Rees, 1965). As can be seen from Figure 12, the magnetic fabric shows a tendency for long-axis orientation to acquire an up-glacier imbrication with depth, and for normals to *a/b* planes to give a single peak rather than to describe a transverse girdle, as was found near the till/ice contact. The reason for this, I would suggest, is that initial plastering-on produces a parallelism of *a*-axes to the depositional surface and a transverse girdle for *c*-axes, whereas the shearing effect transmitted to the till lying below this surface tends to stabilize *c*-axes and to make *a*-axes dip up-glacier. The addition of till derived from the glacier by successive increments means that the components of the lowest parts of the till have been subjected to shearing stresses for the greatest time and have thus probably been most effectively reoriented by stress. Several other determinations of the susceptibility ellipsoid, made at another locality on the same glacier where such till seemed to be forming, showed similar features, but in some cases there was a tendency for down-glacier imbrication of *a*-axes. In addition, incomplete studies of *a*-axes in the till for the same locality show strong transverse orientations, because the till is being compressed against the up-glacier flank of a bedrock obstruction inducing transverse elongation within the till. Indeed, at this locality, till fabrics appear to reflect the bedrock configuration, rather than the direction of ice movement.

Melting observed beneath the accessible parts of Svalbard glaciers appears largely to be pressure melting produced by obstructions, a process which may thus be responsible for the common occurrence of till-plastered rôches moutonées or rock-cored drumlins. If, however, basal melting were common at the margin of a glacier with a large englacial debris load, then the very processes referred to above could be responsible for the deposition of a widespread lodgement-till sheet on a planar surface.

Lodgement Tills Released from Debris-Rich Stagnant Ice beneath an Active Glacier

This origin can be ascribed with certainty to tills at three localities, and may apply to others. The occurrence and some fabrics of these tills are described in a recent publication (Boulton, 1970b).

Basal debris-rich ice, less plastic than cleaner ice at higher levels, can become essentially stagnant when its forward movement is blocked by bedrock obstructions over which the glacier is forced to rise. There is usually a well-marked plane of *décollement* separating the stagnant and active ice, and the former may well undergo further folding and deformation as a result of shear stress transmitted across this plane. Thus the fabrics normally characteristic of basal, highly sheared ice may well be changed if this same

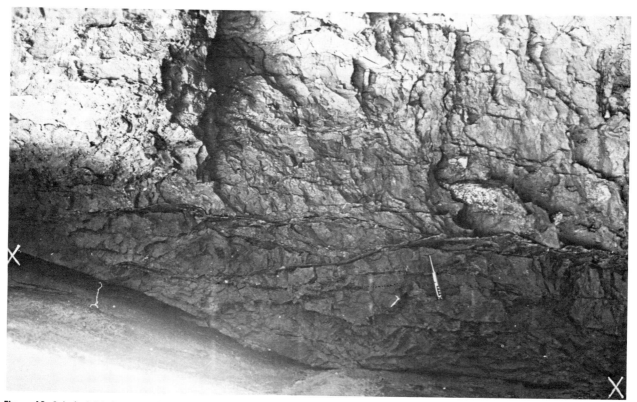

Figure 13. Subglacial lodgement till deposited as a result of melting out of a mass of debris-rich stagnant ice beneath the active sole of Nordenskioldbreen in Bunsow Land. Note the sub-horizontal sheer planes. X-X marks the glacier sole.

ice becomes an essentially subglacial stagnant mass. Such a subglacial till was described from the glacier Nordenskiöldbreen (Boulton, 1970b), in which the till fabric showed elements inherited from an englacial-fabric maximum transverse to flow and some elements induced by glacier over-riding. These latter elements were found in those parts of the till which had reacted to glacier overriding by the formation of a series of sub-horizontal shear planes, which had, where they intersected till stones, re-oriented many of them until their a-axes lay parallel to movement along the shear planes, which itself was parallel to glacier movement (Fig. 13). Thus, where the till contained few shear planes, it showed a strong transverse fabric peak with a minor parallel peak; but, where the shear-plane density was very high, the transverse peak had been almost entirely replaced by a parallel peak. Presumably, post-depositional re-orientations of the type described in the previous section are also possible in these tills.

Fluted Lodgement Till

Fluted lodgement till was investigated at two localities, and bedrock was investigated at another. Most ground-moraine flutings lie either on the lee side or on the up-glacier side of large boulders, in contrast to those observed by Hoppe and Schytt (1953), or Galloway (1956), in which till accumula-tions tended to occur only on the lee side of boulders. The flutings described by Hoppe and Schytt (1953), and also by Price (1969), tend to be low, relatively continuous ridges rarely more than 50 cm high. The ridges ob-served in Svalbard tend to form less extensive though taller ridges, up to two meters high, although this latter contrast might result from the gener-ally large size of the boulders which form their core (Fig. 14). As in most tills exposed to the atmosphere, the surface is washed clear of much fine material, and many surficial boulders tend to show orientations derived from downslope movement. Measurements on the surfaces of flutings at the points at which they emerge from beneath the glacier indicate a weak though definite a-axis orientation of particles greater than 10 cm and less than 30 cm in diameter parallel to the direction of the fluting. However, pits dug within the fluted ridges showed more complex orientations. A pit on the lee side of a large boulder yielded a fabric in which the a-axes tended to dip up-glacier at 45°; a pit on one lateral flank of the same ridge showed a-axes dipping away from the ridge crest at 25°, while a pit on the other flank revealed a-axes with a 70° dip oblique to glacier movement. Yet an-other pit, at the down-glacier extremity of the same ridge, did not yield an obvious preferred orientation. Some of these data are rather difficult to interpret. The parallel orientation at the surface is obviously produced by the drag of ice along the surface of the ridge, but the fabrics below are a problem. They could have formed by development of the mechanism suggested by Galloway (1956b), as a result of subglacial flow of till under differential overburden pressures into the cavities which formed both in front of and behind large boulders. In this case, the complex fabrics would merely reflect the complex pattern of the plastic deformation within the till as it was forced into these cavities, and, indeed, two different fabrics could both be interpreted as reflecting such flow, with a strong component upward toward the rim of the cavity. Alternatively, the ridges might have formed in a manner similar to that suggested for the tills in the two preceding sections, though the deposition of the till would be around boulder cores rather than rôches moutonnées.

These two processes are fundamentally different. In the first, the fluting is post-depositional and affects till already deposited from the ice by some other process and presumably in some other subglacial zone. This till may even have been deposited by a previous ice advance. In the second case, the flutings form as the direct result of a depositional process. Both proc-

Figure 14. Fluted lodgement till beyond the northern margin of Söre Buchananisen on Prins Karls Forland.

esses, of course, could play a part, but such a combination would be entirely coincidental.

The plausibility of the first hypothesis is elegantly demonstrated by an exposure at the margin of Aavatsmarkbreen. Bedrock at part of the margin of the glacier is represented by a series of gently dipping Tertiary clays containing thin siltstone bands. At one locality these have been deformed into a series of flutings, approximately two meters high, trending parallel to glacier movement. The internal bedding is a complex anticlinal structure, with an axial plane trending parallel to the ridge direction, and with many minor folds on its limbs displaying complex, competent/incompetent relationships between siltstone and clay. These structures have clearly formed subglacially. These are not eroded, transported masses, because their internal bedding parallels the surface; thus they are *in situ* and not erratics, representing deformed bedrock rather than materials deposited by the glacier. On this basis I would suggest that differential pressures, developed subglacially because of cavitation, had forced the clays and silts to flow into cavities which were probably present on the lee side of large boulders. Essentially similar mechanisms of flow of unfrozen till from areas of high to low confining pressures have been postulated for the formation of till ridges parallel to the ice front (Andrews and Smithson, 1966; Price, 1969).

In summary, five subglacial processes have been recognized which produce preferred orientations; drag of the glacier sole over a till surface, confined deformation by flow of till in response to the shear stress induced by overriding ice, brittle dislocations induced by the same stresses, unconfined downslope movement in subglacial cavities, and flow of till under a subglacial pressure gradient into subglacial cavities.

SUMMARY OF MAIN FABRIC TRENDS

All of the three main types of till described above show systematic areal fabric trends, which are related, directly or indirectly, to the direction of glacier movement, except perhaps for flow tills which have developed on an uncontrolled topography. Variations in flow till fabrics are basically related to the shape of the underlying ice surface, variations in melt-out till fabrics are related to the glacial stress system, and variations in subglacial lodgement till fabrics are related to the mode of deposition and the form of the surface onto which the till is deposited. All three types are likely to show considerable between-site variation, irrespective of any change in the direction of glacial movement. At individual sites, considerable vertical fabric variation can occur in all three types; in contrast, lateral variation is unlikely to be so great, except in the case of the lodgement till on an irregular substratum. The lower parts of flowtills tend to show systematic up-slope imbrication of a/b planes, whereas the a/b planes of melt-out till stones tend to lie in the horizontal plane.

The Use of Till Fabrics

Fabrics have commonly been used to give two rather different kinds of information: direction of movement of the depositing glacier, and the mode of origin of the till. In the former, that part of the till fabric which reflects the large-scale flow must be abstracted from that part of the fabric which may have been strongly influenced by local factors, such as mode of deposition, bedrock topography, local stresses within the ice, etc. In the latter case, in contrast, the purpose is to attempt both to build up a clear picture of the fabric in order to assess the mode of deformation of the till, and then to fuse this with other local details in order to infer a mode of origin, either on the basis of what is known about actual glacial deposition or (if this provides no guide) on a speculative basis.

In the first case, regional, between-site sampling is necessary in order to infer a systematic trend. If such a trend exists, one can be reasonably sure of identifying either the parallel or transverse direction. Should a till have both parallel and transverse maxima, however, extra care must be taken, for if these are all assumed to be parallel, then an essentially random pattern could be interpreted as a systematic one. The very nature of till fabrics and of the processes responsible for these fabrics imply that a single fabric taken at one locality has only a slight probability of reflecting exactly the local direction of ice movement; thus many detailed patterns of ancient ice movement inferred from the premise that a fabric lies in the parallel direction are basically unsound. In addition, it is of course axiomatic that all the fabrics should be taken from what is known to be the same till. Regional variance in till fabrics from a known till sheet is such that attempts to distinguish tills of different age by regional sampling are very suspect (c.f. West and Donner, 1956).

Unfortunately, some tills are only accessible at a small number of sites, in which case the inherent variability of till fabrics makes identification of the parallel direction extremely difficult. In some cases, such as that of fluted lodgement till, this is possible; but even if within-site fabric variation is small, identification of the parallel direction from one site may be a questionable procedure. Single sample fabrics from one site may be even less sound, as is obvious from the data from Svalbard, and as has been well demonstrated by Young (1969).

Within-site sampling also presents problems, as illustrated by several examples from Svalbard. A poorly designed sampling scheme for the flow till shown in Figure 8 could have revealed either fabric peaks parallel or transverse to flow, or a fabric with peaks both transverse and parallel to flow. The same is true for the subglacial till shown in Figure 13, in which there is a transverse *a*-axis orientation between joint planes and a parallel maximum along joint planes. Magnetic-susceptibility methods have a great potential in this context, in that they make possible the design of an adequate sampling scheme in a clay till, irrespective of the positions, sizes, and shapes of stones. Such methods also both reduce the size of the sample area so that it covers only one structural domain, and are less time-consuming than microscopic fabric methods, although they do depend upon the presence of elongate magnetic minerals within the till. The recognition of different structural fabric domains within a till and the separate analysis of these (Baird, 1962) could play a very useful role in interpretation of glacial motion.

The second main use of till fabrics, that of determination of till genesis, is obviously complex when one considers the number of variables involved. In addition, clast-orientation fabrics themselves provide only crude reflections of these variables, simply because of the relatively small number of characters included in fabric analyses, long-axis orientation, orientation strength, and influence of stone shape at any one locality. I would suggest that the best way of using fabrics for this purpose is in conjunction with other methods of till analysis, which, taken together, would probably give a much better indication of the general mode of origin. If, on the basis of these other criteria, it is possible to infer a subglacial origin, then till fabrics could be used to specify origin still further (Andrews and Smithson, 1966; Drake, 1971, this volume).

An attempt to infer genesis by till fabric alone seems to me to be extremely difficult, as so many different processes can produce similar results. Thus, although work such as that of Holmes (1941) and of Harrison (1957) is extremely valuable, their approach, using till fabric alone, has limited application. A possible approach using fabrics alone would be to undertake a thorough structural analysis of the stone orientations so as to be able to reconstruct the detailed strain pattern within the till, although this is not needed in stratified tills where the nature of strain is easily recognized. This would involve, perhaps, 200 values from each sample rather than the 50 to 100 which are normally adequate. One of the difficulties here lies in the problem of recognizing complex, three-dimensional orientation patterns, for the usual statistical analyses of till-fabric data assume a spherical normal distribution, and even the assessment of bimodal peaks by these methods involves some approximations. Possible methods which could be applied are those recently described by Kelley (1968), in which complex spatial patterns can be used as orientation models.

ACKNOWLEDGMENTS

Thanks for logistic support are due to W. B. Harland, and to R. H. Wallis and P. Maton, director and leaders of the Cambridge Spitsbergen Expedition. W. B. Horsefield, D. Smith, M. R. Rhodes, and C. Nash gave field assistance, and N. Hamilton provided valuable advice on the use of magnetic anisotropy methods.

REFERENCES

Andrews, J. T., and Smithson, B. B., 1966, Till fabrics of the cross-valley moraines of North Central Baffin Island, Northwest Territories, Canada: Geol. Soc. America Bull., v. 77, no. 3, p. 271-90.

Baird, A. K., 1962, Superposed deformation in the central Sierra Nevada foothills east of the Mother Lode: Univ. California Pub. in Geol., v. 42, p. 1-70.

Banham, P. H., 1966, The Significance of till pebble lineations and their relations to folds in two Pleistocene tills at Mundesley, Norfolk: Proc. Geologists Assoc., v. 77, p. 469-74.

Boulton, G. S., 1968, Flow tills and related deposits on some Vestspitsbergen glaciers: Jour. Glaciology, v. 7, no. 51, p. 391-412.

———, 1970a, On the origin and transport of englacial debris in Svalbard glaciers: Jour. Glaciology, v. 9, no. 56, p. 213-29.

———, 1970b, The deposition of subglacial and melt-out tills at the margins of certain Svalbard glaciers: Jour. Glaciology, v. 9, no. 56, p. 231-45.

Carol, H., 1947, The formation of rôches moutonées: Jour. Glaciology, v. 1, no. 2, p. 57-62.

Donner, J. J., and West, R.G., 1957, The Quaternary geology of Brageneset, Nordaustlandet, Spitsbergen: Norsk Polarinstitutt Skrifter, no. 109, p. 9-16.

Dort, W. J., Jr., 1967, Internal structure of Sandy Glacier, South Victoria land, Antarctica: Jour. Glaciology, v. 6, no. 46, p. 529-40.

Fisher, R. A., 1953, Dispersion on a sphere: Royal Soc. London Proc., ser. A, v. 217, p. 295-305.

Flint, R. F., 1957, Glacial and Pleistocene geology: New York, Wiley, 589 p.

Galloway, R. W., 1956a, The structure of moraines in Lyngsdalen, North Norway: Jour. Glaciology, v. 2, no. 20, p. 730-33.

———, 1956, Rides de façonnement glaciaire sur une moraine de fond dans le Lyngsdal, Norvege septentrionale; Rev. Géomorph. Dynamique, no. 11-12, p. 174-77.

Garwood, E. J., and Gregory, J. W., 1898, Contributions to the glacial geology of Spitsbergen: Quart. Jour. Geol. Soc. London, v. 54, p. 197-227.

Glen, J. W., Donner, J. J., and West, R. G., 1957, On the mechanism by which stones in till become oriented: Amer. Jour. Sci., v. 255, no. 3, p. 194-205.

Gravenor, C. P., and Kupsch, W. O., 1959, Ice-disintegration features in western Canada: Jour. Glaciology, v. 67, no. 1, p. 48-64.

Green, R., 1962, Available methods for the analysis of vectorial data: Jour. Sed. Petrology, v. 34, no. 2, p. 440-41.

Gripp, K., 1929, Glaciologische und Geologische Ergebnisse der Hamburgischen Spitzbergen-Expedition 1927. Naturwissenschaftlicher Verein in Hamburg: Abhandlungen aus dem Gebiet der Naturwissenschaften, Bd. 22, Ht. 2-4, p. 146-249.

Hamilton, N., and Rees, A. I., 1965, The anisotrophy of magnetic susceptibility of rocks by the torque method: Jour. Geophy. Research, v. 67, p. 1565-72.

Harrison, W., 1957, A clay-till fabric: its character and origin: Jour. Geology, v. 65, no. 3, p. 275-308.

Holmes, C. D., 1941, Till fabric: Geol. Soc. America Bull., v. 52, no. 9, p. 1299-1354.

Hoppe, G., and Schytt, V., 1953, Some observations on fluted moraine surfaces: Geog. Annaler, ser. A, v. 35, no. 2, p. 105-15.

Jeffery, G. B., 1922, The motion of ellipsoid particles immersed in a viscous fluid: Royal Soc. London Proc., ser. A, v. 102, no. 715, p. 161-79.

Kamb, W. B., 1959, Ice petrofabric observations from Blue Glacier, Washington, in relation to theory and experiment: Jour. Geophys. Research, v. 64, no. 11, p. 1891-1904.

Kelley, J. C., 1968, Least squares analysis of tectonite fabric data: Geol. Soc. America Bull., v. 79, p. 223-40.

Lamplugh, G. W., 1911, On the Shelly moraine of the Sefström Glacier and other Spitsbergen phenomena illustrative of British glacial conditions: Proc. Yorkshire Geol. Soc., v. 17, p. 216-41.

Lundqvist, G., 1949, The orientation of block material in certain species of flow earth: Geog. Annaler, H. 1-2, p. 335-47.

Nye, J. F., 1959, A method of determining the strain rate tensor at the surface of a glacier: Jour. Glaciology, v. 3, no. 25, p. 409-19.

Palosuo, E., and Schytt, V., 1969, Till Nordostlandet med den Svenska Glaciologiska Expeditionem in Terrasta: no. 1, p. 1-19.

Price, R. J., 1969, Moraines, sandur, kames and eskers near Breidamerkurjökull, Iceland: Trans. Inst. Brit. Geogr., no. 46, p. 17-43.

Rees, A. I., 1965, The use of anisotropy of magnetic susceptibility in the estimation of sedimentary fabric: Sedimentology, v. 4, p. 257-83.

Richter, K., 1936, Gefugestudien im Engebrae, Fondalsbrae und ihren Vorlandsedimenten: Zeitschr. für Gletschrk., v. 24, p. 22-30.

Sharp, R. P., 1949, Studies of superglacial debris in valley glaciers: Amer. Jour. Sci., v. 247, no. 5, p. 289-315.

Steinmetz, R., 1962, Analysis of vectorial data: Jour. Sed. Petrology, v. 32, no. 4, p. 801-12.

West, R. G., and Banham, P. H., 1968, Short field meeting on the north Norfolk coast: Proc. Geol. Assoc., v. 79, pt. 4, p. 493-512.

West, R. G., and Donner, J. J., 1956, The glaciations of East Anglia and the east Midlands. A differentiation based on stone orientation measurements of the tills: Quart. Jour. Geol. Soc. London, v. 112, p. 69-91.

Young, J. A. T., 1969, Variations in till microfabric over very short distances: Geol. Soc. America Bull., v. 80, no. 11, p. 2343-52.

Evidence for Ablation and Basal Till in East-Central New Hampshire

Lon Drake

ABSTRACT

Two superimposed tills are readily identifiable in east-central New Hampshire. The upper till is stony, friable, and contains many small yellow and yellow-brown lenses. The lower till is compact, homogeneous, and gray in appearance, and develops a platy structure when exposed. Three hypotheses for the origin of the upper till are considered: (1) a postglacial oxidized zone below the soil developed from the lower till material, (2) a separate till produced by a glacial readvance, and (3) an ablation till developed on stagnant ice from the basal load of the ice. A total of 78 properties for each of 42 drift samples were measured and evaluated in terms of the three hypotheses. Only those properties which lend support to, or are incompatible with, one or two of the hypotheses are included. Evidences that the lower till is basal till are: strongly aligned pebbles, consistently parallel to striations on the bedrock; mechanical analyses and internal structures showing a lack of washing; and pebbles that are flattened but smoothed between the basal shear planes, with minor breakage and weathering. Evidences that the upper till is ablation till are: weakly developed, inconsistent fabrics which do not parallel bedrock striations or topographic trends; mechanical analyses and internal structures showing evidence of washing; and pebbles that are noticeably angular, equidimensional, and broken and decomposed by frost action and weathering during the ablation process. Shallow lower tills (extending two to five feet below surface) and deep upper tills (extending 15 feet below surface) show that postglacial frost action (presently to five feet) has had little effect except to modify the fabrics of the shallow lower tills. The differences between the two tills are not due to lithologic variations. The soil hypothesis and the multiple-advance hypothesis received some support, but received considerably more opposition, and are therefore considered untenable. The basal-ablation till hypothesis has much supporting evidence as the explanation for the two uppermost till units found over much of interior New England.

TWO TILLS IN NEW ENGLAND

Many studies in interior New England have described two tills; an upper, loose, sandy till, and a lower, compact, silty-clayey till (Chute, 1940; Denny, 1940, 1941a, 1941b; Currier, 1941; Moss, 1943; White, 1947; Hansen, 1956; Flint, 1961; Schafer and Hartshorn, 1965; R. P. Goldthwait, 1968). In all of these studies, the lower till is described as gray to grayish-brown, more silty or clayey than the upper till, very compact, and commonly "fissile." The upper till is described as yellow to yellowish-brown to gray, sandy, friable,

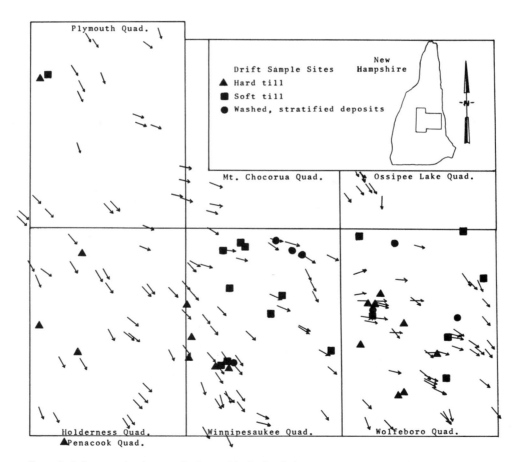

Figure 1. Index map showing sample sites and bedrock striations.

and never "fissile." Genetic interpretations of these two tills differ. Suggested mechanisms include: multiple glaciation, degree of soil development, flow till over basal till, englacial till over basal till, ablation till over basal till, groundwater staining, and various combinations of these.

Multiple till sheets are common in coastal New England (for example, Kaye, 1961, 1964a, 1964b; Schafer, 1961) as far north as southern Maine (Caldwell, 1959). Multiple till sheets are also reported for the St. Lawrence Lowland (MacClintock, 1958) and they extend into the northern green Mountains of Vermont and the White Mountains of New Hampshire (Stewart and MacClintock, 1964). These tills representing multiple ice advances are not included in the "two-till problem" considered in this paper.

The region surrounding Lake Winnipesaukee (Fig. 2), in east-central New Hampshire, was selected for study of the two-till problem because: (1) the bedrock types are distinctive (Fig. 2) and already mapped (Bill-

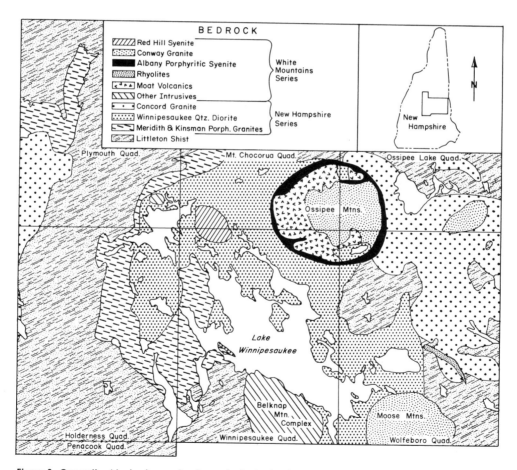

Figure 2. Generalized bedrock map showing major bedrock units.

ings, 1956; Kingsley, 1931; Moke, 1946; Quinn, 1937, 1941, 1953; Smith and others, 1939), (2) some of the small plutons are unique indicator rocks for till pebbles, and (3) the glacial deposits of the eastern part of the area were recently mapped (Goldthwait, 1968).

PROCEDURE

Till data were all collected from sites where sampling could be more than five feet deep (with the exception of three noted in the text), to avoid the effects of present-day frost action. In order to avoid, as much as possible, an interpretational bias, non-genetic terms were selected for field descriptions of the two tills. Tills which are readily excavated with a shovel were designated "soft tills," whereas those requiring a pick to remove small fragments were termed "hard tills." The soft tills are obviously sandy; they range from yellow to yellowish-brown to brownish-gray and are usually composed of many lenses of variable composition and color (Fig. 3). By contrast, the hard tills appear to be less sandy and more compact than the soft tills, are olive gray and uniform in color and structure, and where weathered, show the development of a platy structure or fissility (Fig. 4). All but one till sample readily fell into one of the two categories (this single sample was later included with the hard tills, based on laboratory data). Samples were also taken from washed and stratified deposits, including kames, eskers, and alluvial fans. Of the 42 samples studied in detail, 19 were hard till, 16 were soft till, and seven were washed deposits.

Seventy-eight properties or characteristics were compiled for each till sample, only the most meaningful of which are presented here (see Drake, 1968). Each property or group of properties was evaluated in terms of the potential origin of the till. Three of the most plausible hypotheses for the origins of the two tills were considered:

A. The soft till is a soil (oxidized zone C) developed from the hard till after deglaciation ("soil hypothesis").

B. The soft till is a separate till from the hard till, produced by a glacial readvance, and is superposed upon the hard till ("multiple-advance hypothesis").

C. The soft till is an ablation till developed on stagnant ice, during deglaciation, by modification of the basal load ("Ablation hypothesis").

Many of the properties measured could be interpreted as supporting all three of the above hypotheses, and therefore contribute nothing to the

Figure 3. Typical soft till. Note many lenses of varying composition. Scale calibrated in inches and centimeters.

Figure 4. Typical hard till. Surface at base of chalkboard has weathered for 5 years. Platy structure grading downward into massive till. Scale calibrated in inches and centimeters.

solution of this problem. Therefore, for the sake of brevity, these properties are not considered further in this report. Only those properties which lend support to, or are incompatible with, one or two of the above three hypotheses have been included (except striation index). A complete listing of all 78 properties, their characteristics for each sample, and their interpretation is available (Drake, 1968). An evaluation of the most significant properties follows.

FABRICS

Regional evidence for the flow direction of the last ice advance is present and is of two kinds.

1. When the Red Hill Syenite boulder train is contoured, the long axis parallels the bedrock striations shown on Figure 1. The boulders

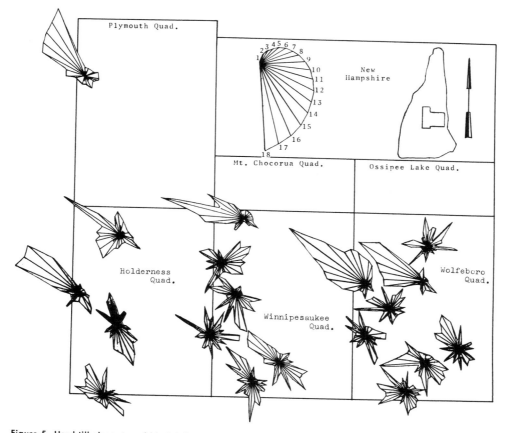

Figure 5. Hard-till elongate-pebble fabrics.

counted in the train were mostly derived from the upper, soft till, yet the soft-till fabrics as a group (Fig. 6) show no consistent parallelism to the boulder-train axis. This can be explained readily if the upper till has been derived from the basal load by ablation.

2. The long axes of "rock drumlins" are dominantly northwest-southeast, parallel to the bedrock striations and hard-till fabrics. Many of the lakes are also elongated northwest-southeast.

Rose diagrams showing orientations of elongated pebbles ($A:B \geq 1.7:1$) in each hard- and soft-till sample are shown in Figures 5 and 6. The strongest trend in the hard-till fabrics is clearly northwest-southeast, whereas no pattern is readily discernible among the soft tills. The three most irregular hard-till fabrics are from the three sites mentioned earlier where samples were collected from above present frostline and which have probably been modified by frost action. The orientation of the hard-till fabrics parallels

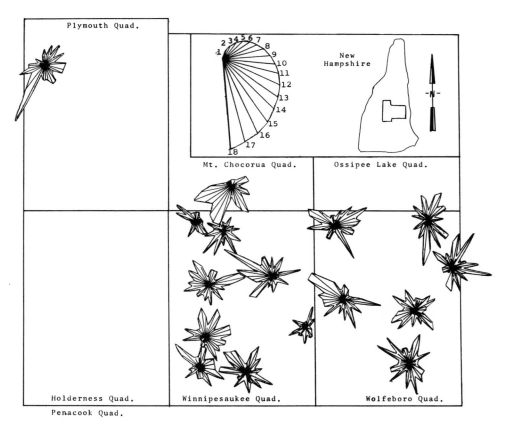

Figure 6. Soft-till elongate-pebble fabrics.

that of the striations on the local bedrock, but there is no good parallelism between the soft-till fabrics and the bedrock striations. Contoured stereo nets were prepared for 100 elongate pebbles in each sample by Corbató's (1965) computer method. Using the same contour interval for all nets, the resulting plots for the hard-till samples show a more marked fabric (average 11.7 contour units, maximum 19 units) than do those of the soft tills (average 8.7 contour units, maximum 12 units). These fabrics are compatible with the interpretation of the hard till being basal till, with a strongly pre-ferred orientation paralleling the bedrock striations, and soft till being ablation till, with the fabrics reduced in intensity and realigned from their original englacial or subglacial orientation during washing and reworking (hypothesis C). MacClintock and Dreimanis (1964) have shown that a readvance of ice in the St. Lawrence Valley left the upper till of that area with stronger fabrics and disturbed the fabrics of lower till, the reverse of the New Hampshire condition.

The uphill and downhill azimuths of the topography at each sample site were compared to the elongate-pebble fabric maxima of both hard and soft tills. No parallel or perpendicular relationships to these azimuths could be found for either till group. It is suggested that, if the upper till had been reworked during postglacial soil development (hypothesis A), the fabric should be realigned perpendicular or parallel to the local slope. However, if the upper till had been reworked during deglaciation, on the surface of ablating ice, then fabrics would be realigned with the local ice surface, which has subsequently disappeared and left the soft till with a fabric inconsistent with present topographic trends.

MECHANICAL ANALYSES

Mechanical analyses were performed on all 42 samples. Initial sample sizes ranged from 12 to 316 pounds, but much smaller splits were used for the sand-to-clay sizes. Cumulative curves for all the samples are shown in Figure 7, and are reasonably distinctive for the two tills, despite some over-lap. The mean grain size (average of the graphic mean sizes; Folk, 1958) for the hard tills is in the fine-sand range ($+2.7\phi$), in the very-coarse-sand range (-0.3ϕ) for the soft tills, and in the granule range (-1.9ϕ) for the washed deposits. This suggests that the soft till was a partially washed ablation till, with the abundant kames, eskers, and fans representing the more extreme washing, thus supporting the ablation hypothesis (C). Had a second ice advance been responsible for the upper, soft till (hypothesis B), there should have been a supply of previously ground material (lower till) and perhaps lake sediments, which should have produced a finer grained upper till. On the other hand, in the normal process of soil (oxidation) development (hypothesis A), material mechanically washed from upper layers (A horizon) would accumulate in the B horizon. However, only the normal B horizon of a podzol was found, a few feet below the surface, in all the tills.

When the relative percentages of the major size groups in the average soft till are compared to those of the average hard till, a regular trend

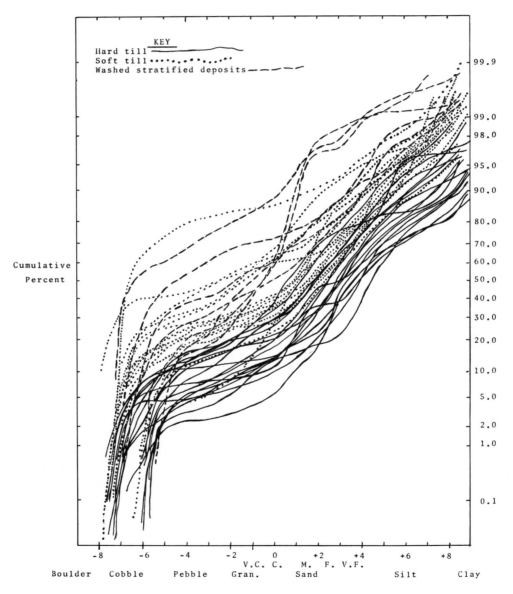

Figure 7. Cumulative curves for till grain-size distributions.

appears of increasing proportions with increasing grain sizes. The soft tills have about 0.18 as much clay, 0.78 as much silt, 0.85 as much sand, 1.32 as much granules, 1.84 as much pebbles, and 6.52 as much cobbles as do the hard tills. This regular sequence reflects the possibility of the small grains being washed out of the soft tills by meltwater reworking the basal load. Hjulstrom (1939) has shown that clays are less susceptible to erosion than are silts. However, his study was apparently done with clay-sized clay minerals. In New Hampshire, the clay-sized fraction is predominantly rock flour — 80-90 percent quartz and feldspar — as shown by X-ray studies of both hard and soft tills. These minerals may lack the bonding strengths of clay minerals and may erode more readily than do the minerals composing the silt-sized particles. Further evidence supporting this suggestion was obtained experimentally, by filling a portable cement-mixer with hard till and briefly churning it with enough water to overflow from the drum. Mechanical analyses of a series of samples taken from this progressively washed hard till revealed the greatest loss in the clay-sized material and the greatest (relative) gain in the cobbles, with a continuous transition between.

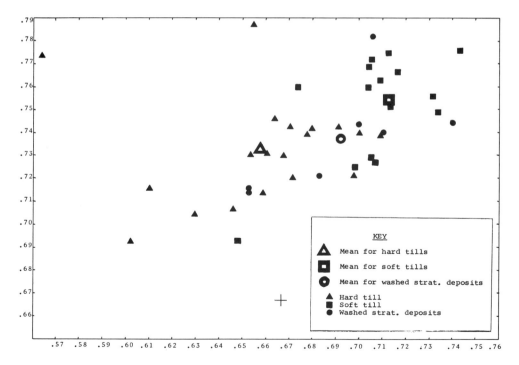

Figure 8. Partial Zingg diagram of till pebble shapes. Each point represents the mean value of 100 pebbles from each site.

PEBBLE SHAPE

The shapes of 100 pebbles collected from each site from which a till sample had been obtained were measured by Krumbein's (1941) triaxial method. The average shape of each 100 pebbles was plotted as a point on a Zingg (1935) diagram on Figure 8. The Zingg shapes are known to be almost independent of roundness; for example, a cube and most other equidimensional polyhedrons are spheres on the Zingg diagram. Figure 8 shows that the pebbles from the hard tills tend to be slightly more blade-shaped and less spherical than the pebbles from the soft tills. This is compatible with the ablation-till hypothesis (C), which proposes that the pebbles in the basal till are flattened by abrasion between shear planes in basal ice and/or till. In the ablation till, these pebbles may have been broken by frost action into more equidimensional shapes.

PEBBLE ROUNDNESS

The same pebbles (100 per sample) used for the shape measurements were also compared to Krumbein's (1941, plate I) roundness scale (0.1 = very angular, 1.0 = very well rounded). The mean values for the individual hard and soft till samples grouped well, with the gross mean roundness values for hard till being 0.48, for soft till being 0.37, and for washed deposits being 0.52. The higher rounding of the hard-till pebbles (with their lower sphericity) can also be attributed to abrasion in the basal ice or till, whereas the soft till pebbles were probably broken by frost action, thus becoming more angular (ablation hypothesis, C). The pebbles of the washed deposits were most highly rounded, due to fluvial abrasion.

In a study of superglacial debris, Sharp (1949, p. 289-315) notes the presence of material apparently broken by frost action. He also shows local accumulations of well-rounded pebbles which have probably been carried in meltwater streams. In the mild climate of the Finger Lakes Region, New York, Taber (1950) found that, on wave-worked beaches, rocks of certain lithologies (especially limestone) become more angular, due to frost action, whereas other lithologies (like sandstone) continue to be rounded. These cases are cited to show that frost action is capable of producing angular pebbles. The roundness data initially might be considered as supporting soil hypothesis (A). However, three of the hard tills were sampled at very shallow depths, well within the range of present-day frost action (about five feet). The pebbles in these samples are also well rounded, while pebbles from some soft-till samples, nine to 12 feet deep, are angular. It would appear that present-day frost action is not sufficiently frequent and/or intense enough to produce measurable pebble breakage at shallow depths.

Therefore, the roundness data are incompatible with the soil hypothesis (A), because pebble roundness does not appear to be related to depth.

Opposition to the multiple-advance hypothesis (B) is also evident. Drake (1968) has shown that pebbles are rapidly rounded during ice transportation and that over one-half of them attain a roundness of 0.5 or greater within a few miles of their source. If a readvance had occurred, the pebbles of the upper till should be at least as round as those of the lower till.

PEBBLE BREAKAGE, WEATHERING, AND STRIATIONS

Other pebble characteristics that were measured were breakage, weathering, and striations. A measure of pebble breakage was carried out on the same first 100 pebbles from each sample. If a pebble showed more than one surface of breakage and abrasion, and the most recent break was less than 0.3 on Krumbein's (1941) scale of roundness, the pebble was considered freshly broken. The average values for pebbles of the hard tills are 26 percent freshly broken; soft tills, 37 percent; and washed deposits, 17 percent.

Pebble weathering was also recorded. A fresh pebble was defined as one showing no signs of weathering. Pebbles were classified as rotten if they could be crumbled by hand. The hard tills averaged 69 percent fresh and five percent rotten pebbles, and the soft tills averaged 63 percent fresh and 11 percent rotten pebbles.

For the striation data, a simple count was made of the number of pebbles with at least one striation on them. Of the pebbles in the hard tills, an average of 2.9 percent showed striations, but only 0.1 percent of the soft-till pebbles were striated.

These data are all compatible with the ablation hypothesis (C), according to which pebbles in the ablation till should be frost broken and weathered, and the striated surface eroded. An ice readvance (hypothesis B) should have produced equal numbers of striated pebbles, because the lithologies involved are similar in both tills. Diligent search of soft-till sites usually revealed only a few pebbles that possessed eroded remnants of striations on the surface.

STONE COUNTS

The comparison of the lithologic contents of the two tills is most important, because lithologic variations could indicate both a different source area (favoring hypothesis B) and also be responsible for many of the differences recorded for pebble striations, weathering, shape, etc. Twenty-nine distinctive rock types were recognizable in the underlying bedrock and the till (Fig. 2). Stone counts from both hard and soft tills, at nearby sample sites, showed similar lithologies. Most of the stone counts at the hard- and soft-till sites were dominated by the lithology of the underlying bedrock or by other lithologies located no more than a mile or two to the northwest.

To further evaluate the effects of lithologic differences between the two tills, four indices of four specific properties were established. These indices — of weatherability, tenacity, erosion, and striation — were measured for pebbles in both tills, in order to allow direct numerical comparison.

Weatherability Index

This index allowed comparison of the weatherability of the total lithologic content of the two tills. The percentages of soft and rotten pebbles of each lithology were totaled for each of the till samples. The 29 lithologies involved were then placed in a weatherability order ranging from 1 to 25, with the lithology having the lowest percentage of weathered pebbles in position 1 and that with the greatest percentage in position 25 (five lithologies were represented by no weathered pebbles and were all assigned scale position 1, hence the total of 29 lithologies resulted in only 25 scale divisions). The pebble content of the hard tills, on this scale, had an average scale position of 15.3, while the pebbles in the soft tills had an average position of 14.9. These numbers suggest that the pebbles in each of the tills were about equally susceptible to weathering, and that the gross lithology had little effect on the weathering differences between the two tills.

Tenacity Index

This index is a measure of how easily a lithology can be crushed. Boulders of the 20 most common lithologies were crushed with a sledge hammer and ranked on a subjective scale from 1 (easily crushed) to 20 (most difficult). On this scale, the hard-till pebbles had an average index of 7.5, while the soft-till pebbles had an average rank of 7.9. This suggests that the pebbles in both tills are about equal in their susceptibility to crushing.

Erosional Index

This index was designed to rate the lithologic content of the till pebbles in terms of their over-all resistance to erosion. The over-all durability of a bedrock unit is revealed in its topographic position, with the most durable rocks forming the mountains and least durable rocks forming the lowlands. Four categories were used to distinguish the topographic position of different kinds of bedrock and their content in the form of pebbles in the hard and soft tills, and the extent of their present outcrop on the map area (see index map).

The following chart presents the results of this survey, showing the topographic expression of bedrock making up the given percentage of the pebbles in each of the tills and the given percentage of bedrock outcrop in the area studied:

	Lowland	Low Hills	High Hills	Mountains
Hard till	48%	29%	13%	10%
Soft till	56%	16%	18%	11%
Map area	51%	24%	15%	10%

The above figures indicate that the lithologies of the pebble contents of of both the hard and soft tills are reasonably representative of the proportions of lithologies making up the bedrock exposed in the map area. The lithologies of the pebbles composing the two tills are also about equal in their resistance to erosion.

Striation Index

Most of the striated pebbles in both tills were of only a few fine-grained, durable lithologies (especially basalts and felsites). The hard tills contained an average of 8.7 percent of susceptible lithologies and the soft tills averaged 8.5 percent. Therefore, lithologic differences are not responsible for the much higher percentage of striated pebbles in the hard till. All approaches used to evaluate the stone counts showed no significant differences in overall lithologic content between the two tills.

TILL STRUCTURE

The hard till is massive when freshly exposed. After exposure, platy structure appears to develop. Figure 4 shows a vertical cut in hard till, with well-developed platy structure grading downward into massive till. The horizontal surface at the bottom of the chalkboard was cut about 5 years before the photo was taken.

The soft tills were composed of many small lenses in all stages of sorting, ranging from material resembling the hard till to well-washed layers of clean sand, with most layers intermediate between these two extremes. Note the similarity between Figure 3 and Boulton's (1968, p. 409) photo of ablation till on the Stubendorfbreen Glacier. At several sites, where exposed in the same cut, soft till graded gradually into kame gravel with no discernible break.

The hard and soft tills are generally separated by a thin layer (two to eight inches thick) of material intermediate in appearance between the hard and soft tills. It is suggested that this is englacial till.

CONCLUSIONS

The two tills of east-central New Hampshire are readily differentiated. Evidences that the lower "hard" till is basal till are: strongly aligned pebbles, consistently parallel to the bedrock striations; mechanical analyses and internal stuctures that show a lack of washing; pebbles that are flattened but smoothed, with minor breakage and weathering. Evidences that the upper "soft" till is ablation till are: weakly developed, inconsistent fabrics which do not parallel either bedrock striations or topographic trends; mechanical analyses and internal structure showing evidence of washing; pebbles that are noticeably angular, and equidimensional. Observations of hard tills observed near the surface (two to five foot deep) and of soft tills at greater depths (15 feet) show that postglacial frost action (presently to five feet) has had little effect except to modify the fabrics of the shallow hard tills. It has been shown that lithologic variations are not related to the differences between the two tills.

Data obtained in this effort to explain the origins of these two tills provide only limited support of the soil hypothesis (A) and the multiple-advance hypothesis (B); most of the data oppose these hypotheses, so that they are considered untenable. Of the total of 78 properties of the tills that were originally measured (Drake, 1968), many could be interpreted as supporting all three of the till-origin hypotheses. Only those which lent support to, or were incompatible with, one or two of the above three hypotheses have been included here. None of the 78 properties opposed the ablation-till hypothesis.

DISCUSSION

The literature abounds with descriptions of two tills, over much of interior New England, that would seem to be equivalent to the two tills described in this paper. From my brief observations in traverses across New Hampshire, Vermont, Massachusetts, and Connecticut, it appears that the uppermost two tills of these areas are equivalent to those of east-central New Hampshire. The multiple-till sequences of coastal New England and in the St. Lawrence Valley are excluded in this comparison. The ablation-till hypothesis is suggested as the most reasonable explanation for the uppermost two tills found over much of interior New England.

At the time of the Till Symposium, in May, 1969, the author's interpretation of the two-till problem appeared to be in disagreement with the views of Pessl and of Schafer, both of whom favored a multiple-advance hypothesis for the two tills observed in western Connecticut. As a result of a joint field trip in November, 1969, to exposures in western Connecticut (F. Pessl and P. Schafer), southwestern New Hampshire (C. Koteff), and east-central New Hampshire (L. Drake), the conclusion tentatively arrived at by the group was that the upper till unit described by Pessl, Schafer, and Koteff encompasses both till units described by Drake. The lower till unit of Pessl, Schafer, and Koteff was not observed by the author in his field area. Joint research by the above authors is continuing in an effort to clarify the Pleistocene stratigraphy across New England.

ACKNOWLEDGMENTS

This study is part of the author's Ph.D. dissertation at the Ohio State University, written under the guidance of Dr. Richard P. Goldthwait. I am particularly indebted to Dr. Goldthwait for his suggestions in the field and for his aid in obtaining N.S.F. Grant Number GA-945, which supported part of this study. Mr. Stephen Forster's careful assistance in the field during the summer of 1967 is appreciated. I also wish to thank Dr. Charles E. Corbató for aid in debugging my computer programs and in providing the subroutine for the contoured stereo nets. Carl Koteff, Fred Pessl, and Phil Schafer provided lively discussion and valid criticisms of the manuscript, which were appreciated by the author.

REFERENCES

Billings, M. P., 1956, The Geology of New Hampshire, Part II, Bedrock Geology: State of New Hampshire, Dept. Res. and Econ. Dev. 203 p. and maps.

Boulton, G. S., 1968, Flow tills and related deposits on some Vestspitsbergen glaciers: Jour. Glaciology, v. 7, no. 51, p. 391-412.

Caldwell, D. W., 1959, Glacial lake and glacial marine clays of the Farmington area, Maine: Maine Geol. Survey Spec. Geol. Study 3, 48 p.

Chute, N. E., 1940, Preliminary report on the geology of the Blue Hills Quad., Mass.: Mass. Dept. Public Works, U.S. Geol. Surv. Coop. Project Bull. 1, p. 1-52.

Corbató, C. E., 1965, Fabric diagrams by computer: manuscript, Dept. of Geology, Univ. of Calif., Los Angeles, Calif.

Currier, L. W., 1941, Tills of eastern Massachusetts (Abst.): Geol. Soc. America Bull., v. 52, p. 1895.

Denny, C. S., 1940, Glacial deposits of the Canaan area, New Hampshire (Abst.): Geol. Soc. America Bull., v. 51, p. 1924-25.

———, 1941a, Glacial drift near Canaan, New Hampshire (Abst.): Geol. Soc. America Bull., v. 52, p. 1898.

———, 1941b, Glacial features near Canaan, New Hampshire (Abst.): Geol. Soc. America Bull., v. 52, p. 2013.

Drake, L. D., 1968, Till studies in New Hampshire: Ph.D. dissertation: Ohio State Univ., 106 p.

Flint, R. F., 1961, Two tills in southern Connecticut: Geol. Soc. America Bull., v. 72, p. 1678-92.

Folk, R. L., 1958, Petrology of Sedimentary Rocks: Austin, Texas, Hemphills, 154 p.

Goldthwait, J. W., 1925, The geology of New Hampshire: New Hampshire Acad. of Sci. Handbook no. 1.

———, 1938, The uncovering of New Hampshire by the last ice sheet: Amer. Jour. Sci., v. 36, ser. 5, p. 345-72.

Goldthwait, R. P., 1968, Surficial geology of the Wolfeboro-Winnipesaukee area, New Hampshire: N. H. State Dept. of Resources and Economic Development, 60 p.

Hansen, W. R., 1956, Geology and mineral resources of the Hudson and Maynard quadrangle, Massachusetts: U. S. Geol. Survey Bull. 1038, 104 p.

Hjulstrom, F., 1939, Transportation of detritus by moving water, *in* Recent Marine Sediments: Amer. Assoc. Petroleum Geologists, p. 3-31.

Kaye, C. A., 1961, Pleistocene stratigraphy of Boston, Massachusetts: U. S. Geol. Survey Prof. Paper 424B, p. 73-76.

———, 1964a, Outline of Pleistocene geology of Martha's Vineyard, Massachusetts: U. S. Geol. Survey Prof. Paper 501-G, p. 134-39.

———, 1964b, Illinoian and Early Wisconsin moraines of Martha's Vineyard, Massachusetts: U. S. Geol. Survey Prof. Paper 501-C, p. 140-43.

Kingsley, L., 1931, Cauldron-subsidence of the Ossippee Mountains: Amer. Jour. Sci., 5th ser., v. 22, p. 139-68.

Krumbein, W. C., 1941, Measurement and geological significance of shape and roundness of sedimentary particles: Jour. Sed. Petrology, p. 64-72.

MacClintock, P., 1958, Glacial geology of the St. Lawrence Seaway and power project: New York State Mus. and Sci. Service.

MacClintock, P., and Dreimanis, A., 1964, Reorientation of till fabric by overriding glacier in the St. Lawrence Valley: Amer. Jour. Science, v. 262, p. 133-42.

Moke, C. B., 1946, Geology of the Plymouth Quadrangle: State of New Hamp., Div. of Econ. Dev., 21 p.

Moss, J. H., 1943, Two tills in the Concord Quadrangle, Mass. (Abst.): Geol. Soc. America Bull., v. 54, p. 1826.

Quinn, A., 1937, Petrology of the alkaline rocks at Red Hill, New Hampshire: Geol. Soc. America Bull., v. 48, p. 373-402.

———, 1941, Geology of Winnepesaukee Quadrangle: State of New Hamp., Div. of Econ. Dev., 22 p.

———, 1953, Geology of the Wolfeboro Quadrangle: State of New Hamp., Div. of Econ. Dev., 24 p.

Sayles, R. W., and Mather, K. F., 1938, Multiple Pleistocene stages in southern Maine (Abst.): Geol. Soc. America Proc. 1937, p. 109-110.

Schafer, J. P., Correlations of end moraines in southern Rhode Island: U. S. Geol. Survey Prof. Paper 424-D, p. 68-70.

Schafer, J. R., and Hartshorn, J. H., 1965, The Quaternary of New England, p. 117-128, *in* The Quaternary of the United States, Wright, H. E., Jr., and Frey, D. G. (eds.), Princeton University Press, 422 p.

Sharp, R. P., 1949, Studies of superglacial debris on valley glaciers: Amer. Jour. Sci., v. 264, p. 289-315.

Smith, A. P., Kingsley, L., and Quinn, A., 1939, Geology of Mt. Chocorua Quadrangle: State of New Hamp., Div. of Econ. Dev., 14 p.

Stewart, D. P., and MacClintock, P., 1964, The Wisconsin stratigraphy of Northern Vermont: Amer. Jour. Sci., v. 262, p. 1089-97.

Taber, S., 1950, Intensive frost action along lake shores: Amer. Jour. Sci., v. 248, p. 784-93.

White, S. E., 1947, Two tills and the development of glacial drainage in the vicinity of Stafford Springs, Connecticut: Amer. Jour. Sci., v. 245, p. 734-78.

Zingg, Th., 1935, Beitrag zur Schotteranalyses: Schweiz. Min. u. Pet. Mitt., v. 15, p. 38-140.

Till Fabrics and Till Stratigraphy in Western Connecticut*

Fred Pessl, Jr.

ABSTRACT

Two texturally and structurally distinct tills are widespread in the crystalline-rock uplands of southern New England, but their ages and modes of deposition have been controversial. Because of their consistent stratigraphic position, they are referred to as the upper and the lower tills.

Strong fabric lineation and the presence of the lower till as the major constituent in most drumlins suggest that this till is a subglacial deposit. The orientation of drumlin axes and of till fabrics indicate that the lower till was deposited by ice flowing from north-northwest.

Preferred orientation of single-fabric samples from the upper till at widely separated localities indicates that locally it was deposited by ice flowing from northeast. However, other directional data for this till, such as striations and indicator fans generally indicate ice-flow from north-northwest. Till fabrics from one deep exposure of upper till suggest that a gradual easterly shift in ice-flow direction from north-northwest to northeast may have occurred during deposition of the upper till. This shift could explain the apparent contradiction between the common northeast-preferred orientation of single-fabric samples from the upper till and the relative scarcity of other northeast directional data.

The upper till probably includes subglacial and superglacial facies. Strongly developed fabric lineation and nearly massive structure at some exposures indicate subglacial, lodgement deposition. Deformed water-laid interbeds, coarser texture, and weak fabric lineation in other parts (commonly upper parts) of the upper till suggest superglacial deposition.

*Prepared in cooperation with the State of Connecticut Geological and Natural History Survey. Publication approved by the Director, U.S. Geological Survey.

INTRODUCTION

Two texturally and structurally distinct tills have long been recognized in southern New England (Upham, 1878; Flint, 1930, p. 12-16, 24, 42, 71-72; Judson, 1949; Hansen, 1956, p. 60, 61). Wherever exposed in superposition, the two tills show a consistent stratigraphic relationship. The upper till is olive gray to light olive gray, sandy, and moderately compact to friable, and overlies the very compact, jointed, less-sandy lower till, in which the matrix color grades from olive brown to olive gray in an upper, oxidized zone to dark gray in the nonoxidized zone beneath. Because of this consistent stratigraphic position and because the terms are well established in the literature, the two tills are referred to here as the upper till and the lower till.

Although the occurrence of these two tills is widely recognized, their relative ages and modes of deposition remain controversial. One view considers the two tills to be contemporaneous deposits laid down by a single ice sheet, the lower one as lodgment till, the upper as ablation till (Goldthwait, 1916; Denny, 1958, p. 76-82; Chute, 1966, p. B48-B50; Drake, 1971, this volume). According to this hypothesis, physical differences between the tills are considered to reflect differences in mineral composition, mode of deposition, and oxidation by texturally controlled circulation of groundwater. The other view interprets the two tills as deposits of separate ice sheets, differing in age and possibly also in mode of deposition (Currier, 1941; Moss, 1943; White, 1947; Flint, 1961; Pessl, 1966). The contrast in color and staining of the tills is explained, in this view, as the result of subaerial weathering of the lower till before deposition of the upper till. In both hypotheses, the lower till is considered to be predominantly lodgement till. The upper till, on the other hand, is considered to be an ablation deposit by the single-ice-sheet proponents, but is interpreted to be a more complex deposit, perhaps containing both lodgement and ablation facies, by the two-ice-sheet proponents. In recent literature, this two-till problem is discussed by Schafer and Hartshorn (1954, p. 117-18), and by Pessl and Schafer (1968). The latter reference includes a summary table in which the physical characteristics of the two tills are compared. This paper presents till fabrics and other directional data from western Connecticut, where the controversial tills are well exposed, and considers the possible modes of deposition in light of these data.

Western Connecticut is an area of moderate topographical relief, bounded on the east by the Connecticut Valley Lowland, and underlain for the most part by igneous and metamorphic rocks. Till-fabric samples were taken from exposures in the east- and south-central parts of the crystalline-rock uplands of western Connecticut (Fig. 1). Fabric data were collected over a period of several years during the course of quadrangle mapping by U. S. Geologi-

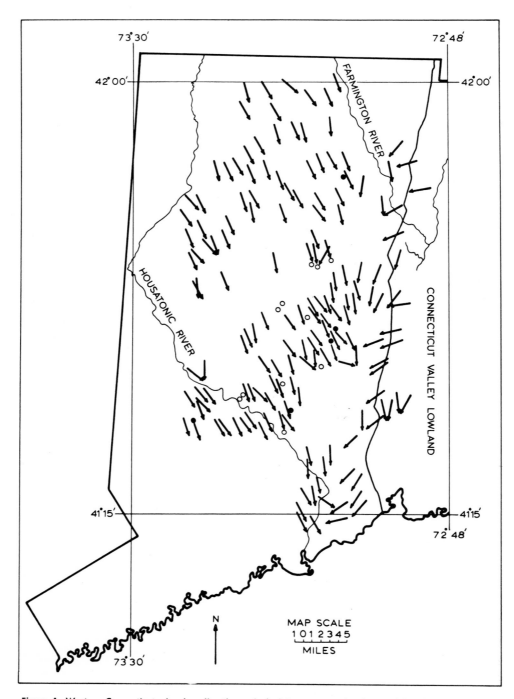

Figure 1. Western Connecticut, showing directions of glacial grooves and striae, and locations of till-fabric localities. Arrows show directions of grooves or striae (after Colton, R.B., 1969, unpublished data); position of observation at tip of arrow. Open circle indicates single till-fabric measurement. Solid circle indicates multiple till-fabric measurements. Line trending irregularly north-south marks the west margin of area of Triassic-age rocks in the Connecticut Valley Lowland.

cal Survey personnel in cooperation with the State of Connecticut Geological and Natural History Survey. The data reflect a considerable variety in sampling technique and site location, and thus do not accommodate rigorous statistical analysis of within-site and between-site variations, as has been convincingly advocated by Andrews and others (1966, 1968). However, several working hypotheses may be formulated from a general discussion of the data at hand, and these in turn may serve to stimulate more detailed study of till fabrics in southern New England.

DIRECTIONAL DATA

Glacial striations in western Connecticut (Fig. 1) show a prominent northwest-southeast trend. The orientation of the striations is commonly the same, regardless of which till directly overlies the striated bedrock. Along the east border of the area, northeast-southwest to nearly east-west striae are also present. This directionally complex zone, parallel to the west border of the Connecticut Valley Lowland and mirrored on the east side of the Lowland, probably reflects increased lobation of the wasting ice sheet in response to local topographic control by the lowland, and does not specifically relate to the tills in the uplands.

Drumlin axes (Fig. 2) in western Connecticut show directional trends similar to those indicated by the striae. In the uplands the dominant trend is northwest-southeast, and along the border with the Connecticut Valley Lowland both northwest-southeast and northeast-southwest orientations occur. No exclusively upper-till drumlins are recognized from western Connecticut; the lower till comprises the bulk of the drumlins; however, upper till does overlie lower till in some drumlins, either as a collar of till around the drumlin flanks or as a continuous, thin mantle covering the entire drumlin.

An indicator fan and a boulder train composed of erratic boulders exposed at the surface in areas underlain by till (Fig. 2) emanate from restricted outcrop areas of distinctive rock types within the crystalline terrain. The smaller outcrop area (Fig. 2) is underlain by a very coarse-textured kyanite schist, in which single kyanite crystals as much as 24 inches long are prominent. The larger outcrop area (Fig. 2) outlines a small fault valley underlain by igneous and sedimentary rocks of Triassic age. Distribution of erratics from each of these source areas, both in the lower till and in the upper till, indicates glacial transport to the southeast.

Thus the dominant ice-flow direction in western Connecticut, as indicated by the orientation of striations and drumlin axes, and the distribution of erratics, is from northwest to southeast.

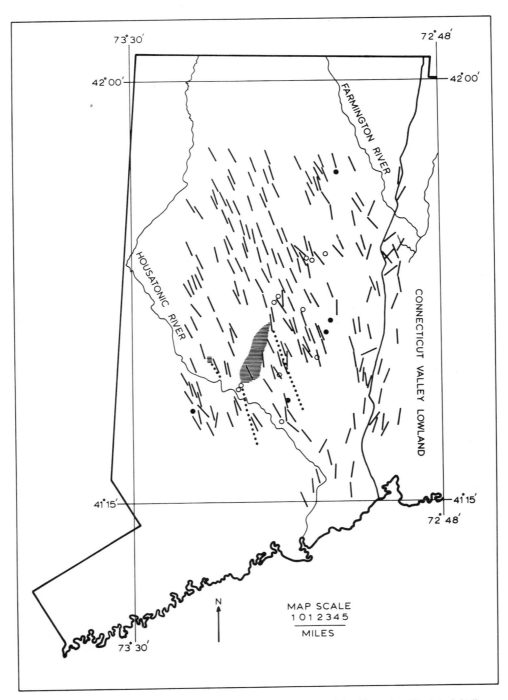

Figure 2. Western Connecticut, showing drumlin axes, indicator fans, and boulder trains. Short straight lines indicate drumlin axes (after Colton, 1969, unpub. data). Dots delineate limits of indicator fans and boulder trains; ruled pattern indicates outcrop area of rocks composing indicator fans or boulder trains. Other symbols are as used in Fig. 1.

TILL-FABRIC DATA

Fabric data from 16 till localities are shown as two-dimensional diagrams in Figure 3. Each diagram represents a 50-stone sample from which the bearings of long axes of stones are grouped in 10-degree classes. For each sample, stones were marked with reference lines prior to removal from the exposure and later reoriented and measured by means of a goniometer in

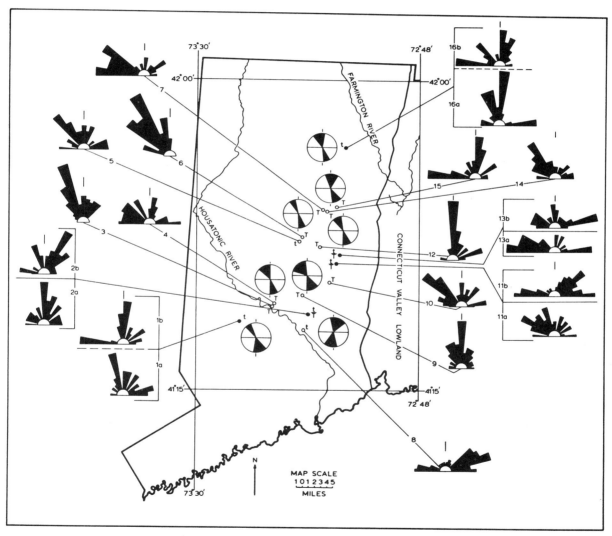

Figure 3. Two-dimensional till-fabric diagrams from western Connecticut. Each diagram represents a 50-stone sample with azimuths of a-axes divided into 10 degree classes. Pie diagrams indicate range in orientation of striations (upper half) and drumlin axes (lower half) in vicinity of fabric locality. T = lower till; t = upper till. Where two fabrics were measured at a simple exposure, with either one sample taken from each of two superposed tills, or both samples taken from a single till unit, "a" indicates lower fabric sample, "b" indicates upper fabric sample. Other symbols are as used in Fig. 1.

the manner described by Karlstrom (1952). In the case of multiple-sample localities (solid circles in Fig. 3), superposition of diagrams indicates superposed fabric samples taken in vertical sequence from the face of the exposure. Pie diagrams (Fig. 3) indicate the range of striation bearings (upper half) and drumlin orientations (lower half) in the vicinity of the till exposures.

Fabric data from the lower till (Fig. 3) generally indicate a north-northwest to south-southeast ice-flow direction, and there is reasonably good agreement between the principal modes of the fabric diagrams and the other local directional data. This agreement tends to support the hypothesis of a lodgement origin for the lower till. Upper-till fabrics, on the other hand, appear more diverse in directional trend, although a northeast-south-west-oriented principal mode is common. Sample 5 is a notable exception and is discussed later. Agreement between ice-flow direction as indicated by upper-till fabrics and the other directional data is less good than that indicated by the lower-till fabrics.

Diagrams from two-till localities (samples 2a, 2b, 11a, 11b, 13a, 13b, Fig. 3) show the principal modes in the upper-till diagrams to be east of the principal mode in the lower-till fabrics. Similarly, diagrams from upper-till localities where two vertically-spaced fabric samples were taken from the same till unit (samples 1a, 1b, 16a, and 16b, Fig. 3) show the principal mode in the upper fabric east of the principal mode in the lower fabric.

Three-dimensional fabric diagrams for some of the samples shown in Figure 3 are presented in Figure 4. Sample-numbers are the same for both figures. The general directional trends indicated in Figure 3 are also apparent in most of the three-dimensional diagrams, but the details of the fabric patterns are more clear in Figure 4, and in some cases, such as sample 5, the directional significance of the fabric (or, as in this case, the lack thereof) is more clearly seen in the three-dimensional diagrams. A multimodal distribution, i.e., a distribution in which both A and B lineations[1] both containing elements with opposing dips, is present in several of the contoured diagrams, especially those from the lower till (samples 13a and 15, Fig. 4). Upper-till samples, on the other hand, show a greater range in fabric pattern. Some, such as samples 1a and 1b (Fig. 4), have well-defined multimodal distributions; others, such as samples 5 and 13b (Fig. 4), show diffuse or girdled distributions.

The two-dimensional diagram for sample 5 (Fig. 3) shows a preferred northwest-southeast orientation, in contrast to the more common northeast-southwest orientation of other upper-till fabrics. Sample 5 was collected from a very friable, sandy, stony upper-till facies of rather restricted lateral

---

1. A lineation refers to the preferred orientation (principal mode) of the fabric; B lineation is a secondary mode, commonly normal to the A lineation, sometimes referred to as a "cross fabric."

extent, which has irregular, knobby topography with moderate local relief. An ablation origin for this till is suggested by these field observations and is supported by the diffuse pattern of the three-dimensional fabric diagram (sample 5, Fig. 4).

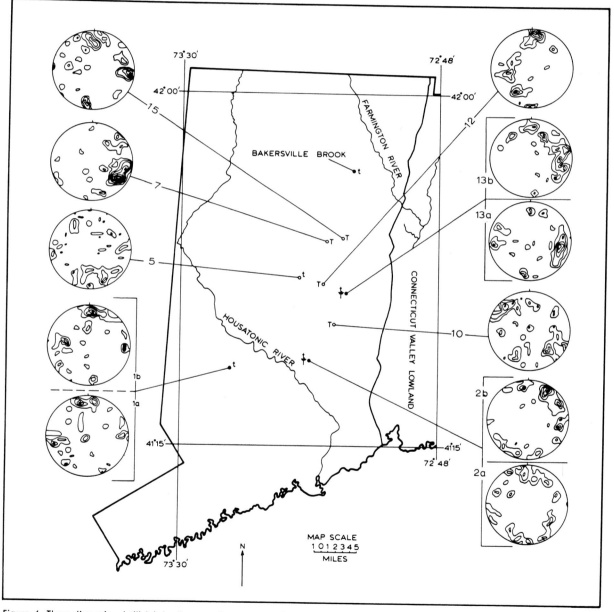

Figure 4. Three-dimensional till-fabric diagrams from western Connecticut. Each diagram shows distribution of a-axis orientations from a 50-stone sample, projected from lower hemisphere. Contour interval 2% per 1% area. Sample numbers are same as in Fig. 3.

Fabric samples 1a and 1b (Figs. 3 and 4), in contrast, are from less friable, more massive upper till exposed at the south end of a small, isolated hill. The well-developed fabric lineation and the multimodal distribution in the three-dimensional diagrams for these samples seem to indicate a mode of deposition different from that of sample 5. Strong lineation and multimodal distribution are common in lower-till fabrics and perhaps are characteristic of lodgement till.

BAKERSVILLE BROOK LOCALITY

In order to examine more carefully the differences in fabric pattern and fabric orientation among samples from the upper till, 10 fabric samples were collected from an unusually thick section of this till exposed along Bakersville Brook near Torrington, Connecticut (Fig. 4). The exposure is 30 to 35 feet deep and about 200 feet long. Three units can be distinguished, each of which is described below.

Basal unit: about 15 feet thick, composed of massive, stony, olive-gray (5Y 6.5/2 to 5.5/3) till containing minor interbeds and lenses of well-sorted, crossbedded, medium- to coarse-grained, water-laid sand. Discontinuous sandy partings produce a texturally controlled mottling with lighter colored sandy zones and darker colored siltier zones.

Middle unit: about 10 feet thick, composed of layered, stony, light-olive-gray (5Y 6/2) till with prominent interbeds of well-sorted, crossbedded, coarse- to medium-grained sand and granule gravel. Small-scale folds and thrusts, and moderately contorted sand layers interbedded with lenses of pebble gravel and stony, till-like masses are present in some parts of the unit. In other parts, undeformed layers of well-sorted, crossbedded, medium- to coarse-grained sand extend continuously across the exposed width of the section.

Upper unit: about 5 feet thick, composed of very stony, coarse-textured, nonlayered, very poorly sorted, light-olive-gray to light-gray (5Y 6.5/2) till with strongly contorted, disrupted, and discontinuous remnants of fluvial interbeds. Overlying this upper unit is a thin mantle, 1 to 3 feet thick, of eolian sand and silt mixed with material from the underlying till.

Two suites of fabric samples were taken from this exposure. Each suite consists of five 50-stone samples collected in a vertical sequence about 150 feet apart. The generalized stratigraphic column, Figure 5, shows the three units described from the exposure and the location of the fabric samples.

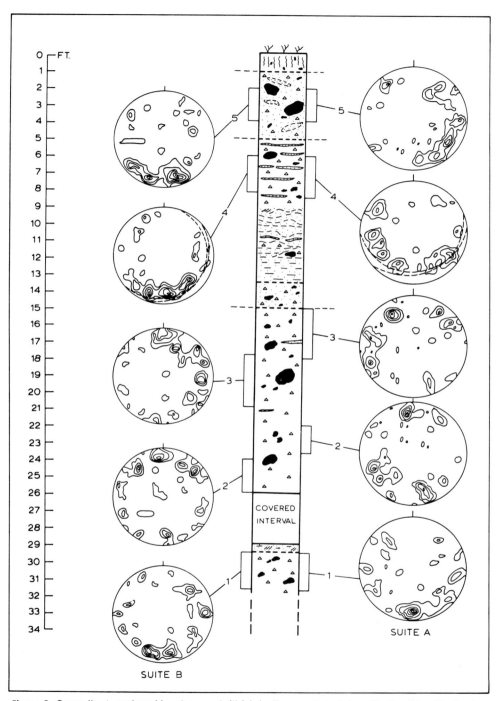

Figure 5. Generalized stratigraphic column and till-fabric diagrams from Bakersville Brook locality, Torrington, Connecticut. Each diagram shows distribution of a-axis orientations of 50 pebbles, projected from lower hemisphere. Contour interval 2% per 1% area. Till matrix is shown by small triangles, sand is shown by dots, conspicuous stones are in black, and traces of bedding planes are shown by small dashed lines.

In Suite A, the fabric pattern of the lower three samples, from the basal unit of the section, contrasts sharply with that of samples A4 and A5, from the middle and upper units, respectively. Samples A1 to A3 show variously developed multimodal distributions, with prominent A lineations oriented north-south to northwest-southeast. The B lineation, oriented east-west, is rather weakly developed in sample A1, but appears to grow in strength and to shift to a northeast-southwest orientation in samples A2 and A3. In sample A3, it is difficult, in fact, to distinguish between A and B lineations. A systematic shift in the direction of the preferred dip of A-lineation components from south-dipping to north-dipping occurs upward through samples A1-A3.

Sample A4, from the middle unit and, to a lesser degree, sample A5, from the upper unit, show a girdled distribution. The dashed lines in sample A4 (Fig. 5) indicate the range of bedding-plane attitudes within the layered, middle unit of the till. Attitudes were measured from the water-laid interbeds; the fabric sample was taken from stones in the till matrix between the fluvial layers. The rough coincidence of contour maxima with the lines of bedding-plane attitudes indicates that many of the till stones lie approximately in the plane of the fluvial bedding, with only slightly steeper dips. This suggests that fabric sample A4 is secondary, reflecting reworking by fluvial processes, rather than a primary fabric indicating ice-flow direction.

In suite B, the contrast in fabric pattern between samples B1 to B3 from the lower unit and samples B4 and B5 from the middle and upper units is less noticeable than in suite A (Fig. 5), but sample B1 is strikingly similar in pattern to sample A1, and the girdled pattern of sample B4 is similar to that of A4. The systematic change in dip direction from south to north, as noted in samples A1 to A3, is similarly present in samples B1 to B3. The relationship between contour maxima and bedding-plane attitudes in sample 4 is also similar to both suites. On the other hand, the fabric of B5 differs noticeably from that of A5. Sample A5 has a rough girdled distribution, while sample B5 has a strong south-dip component and is rather diffuse in the upper half of the diagram.

There is a progressive increase in the concentration of points in the northeast quadrant from sample B1 to sample B3. Sample B2 shows a diffuse distribution and appears transitional between samples B1 and B3. This shift in fabric orientation perhaps reflects the same change in the stress field that controlled the increased expression of the northeast-southwest lineation in samples A1 to A3.

Fabric diagrams (Fig. 5) and the field relations at the Bakersville Brook locality suggest the following interpretations:

1. The lower unit in the section is a lodgement deposit. The massive structure, minor amounts of water-laid material, and finer texture

relative to the texture of the overlying units are all compatible with this interpretation. The multimodal distribution of fabrics from particularly the lowermost portions of this basal unit is similar to that of the fabric patterns from many exposures of lower till, which is generally considered to be of lodgement origin.

2. The upper unit is a superglacial ablation deposit. The coarse texture, very poor sorting, disrupted remnants of fluvial interbeds, and laterally disordered fabric orientations all support this hypothesis.

3. The fluvial interbeds and layered structure of the middle unit are characteristics often associated with ablation drift. However, the lateral continuity of the fluvial layers seems to contradict an interpretation of superglacial origin, in which distorted and disrupted bedding due to collapse and flowage are common. The coincidence of long-pebble axes lying nearly in the plane of the fluvial bedding suggests that the fabric from the middle unit is a secondary feature, perhaps the result of reworking by subglacial streams. I suggest that this middle unit is a subglacial deposit which reflects a change in the thermal regimen of the ice, resulting in greater basal melting and an increase in the activity of free-flowing, basal meltwater. Perhaps this unit can be considered a subglacial ablation deposit of the sort described by Elson (1960, p. 17).

CONCLUSIONS

If the preceding interpretations are valid, they suggest the following regional considerations:

1. The inferred occurrence of both lodgement and ablation facies, or subglacial and superglacial facies, in the upper till supports the two-ice-sheet explanation of the two-till problem, for it is considered unlikely that a single ice sheet would produce a till in which the subglacial facies would have such different characteristics as are exhibited in the upper and lower tills.

2. A gradual shift in ice-flow direction from west of north, or north, to east of north may have occurred late in the depositional history of the lodgement facies of the upper till. This means, then, that most of this lodgement till was deposited by ice flowing from the north or northwest, and would explain the relative scarcity of northeast-southwest-oriented directional data. Fabrics with northeast-southwest-preferred orientation, locally observed in the upper

till, probably were taken from the upper portion of the lodgement facies, or perhaps reflect secondary orientations within the ablation facies of the upper till.

REFERENCES

Andrews, J. T., and Shimizu, K., 1966, Three-dimensional vector technique for analyzing till fabrics: discussion and FORTRAN program: Geog. Bull., v. 8, no. 2, p. 151-65.

Andrews, J. T., and King, C. A. M., 1968, Comparative till fabrics and till fabric variability in a till sheet and a drumlin: a small-scale study: Proc. Yorkshire Geol. Soc., v. 36, pt. 4, no. 25, p. 435-59.

Chute, N. E., 1966, Geology of the Norwood quadrangle, Norfolk and Suffolk Counties, Massachusetts: U. S. Geol. Survey Bull. 1163-B, p. B1-B78.

Currier, L. W., 1941, Tills of eastern Massachusetts (Abstr.): Geol. Soc. America Bull., v. 52, no. 12, pt. 2, p. 1895-96.

Denny, C. S., 1958, Surficial geology of the Canaan area, New Hampshire: U. S. Geol. Survey Bull. 1061-C, p. 73-100.

Elson, J. A., 1960, Geology of glacial tills: 14th Canadian Soil Mech. Conf., Niagara Falls, Ontario.

Flint, R. F., 1930, The glacial geology of Connecticut: Connecticut Geol. and Nat. History Survey Bull. 47, 294 p.

———, 1961, Two tills in southern Connecticut: Geol. Soc. America Bull., v. 72, no. 11, p. 1687-92.

Goldthwait, J. W., 1916, Geology of the Hanover district: Hanover, New Hamp. (private printing).

Hansen, W. R., 1956, Geology and mineral resources of the Hudson and Maynard quadrangles, Massachusetts: U. S. Geol. Survey Bull. 1038, 104 p.

Judson, Sheldon, 1949, The Plestocene stratigraphy of Boston, Massachusetts, and its relation to the Boylston Street fishweir, in The Boylston Street fishweir II, a study of the geology, paleobotany, and biology of a site on Stuart Street in the Back Bay District of Boston, Massachusetts: Papers of the Robert S. Peabody Found. for Archaeology, v. 4, no. 1, p. 7-48.

Karlstrom, T. N. V., 1952, Improved equipment and techniques for orientation studies of large particles in sediments: Jour. Geology, v. 60, no. 5, p. 489-93.

Moss, J. H., 1943, Two tills in the Concord quadrangle, Massachusetts (Abstr.): Geol. Soc. America Bull., v. 54, p. 1826.

Pessl, Fred, Jr., 1966, A two-till locality in Northeastern Connecticut: U. S. Geol. Survey Prof. Paper 550-D, D89-D93.

Pessl, Fred, Jr., and Schafer, J. P., 1968, Two-till problem in Naugatuck-Torrington area, western Connecticut: New England Intercollegiate Geological Conference, Trip B-1, State Geol. and Nat. History Survey of Connecticut Guidebook No. 2, p. 1-25.

Schafer, J. P., and Hartshorn, J. H., 1965, The Quaternary of New England, *in* Wright, H. E., Jr., and Frey, D. G., (eds.), The Quaternary of the United States: Princeton Univ. Press, p. 113-28.

Upham, Warren, 1878, Modified drift in New Hampshire, *in* Hitchcock, C. H., The geology of New Hampshire, pt. 3: Concord, N. H., Edward A. Jenks, State Printer, p. 3-176.

White, S. E., 1947, Two tills and the development of glacial drainage in the vicinity of Stafford Springs, Connecticut: Amer. Jour. Sci., v. 245, p. 754-78.

Till / a Symposium

Ablation Till in Northeastern Vermont

David P. Stewart and Paul MacClintock

ABSTRACT

The surface drift in northeastern Vermont is the Shelburne Till, a loose, sandy, bouldery material with a pebble orientation indicating deposition by ice invading from the northeast. Frontal (end) moraines in the St. Johnsbury region are composed of this material. Despite a superficial resemblance, this till cannot be superglacial till because of its strong northeast fabric, and because it is not found overlying a basal till with the same fabric. Material beneath this till is commonly either bedrock or a till with a northwest-southeast fabric (the Bennington).

Genesis of the Shelburne Till, with its loose, sandy texture and strong northeast-southwest fabric, is believed to be as ablation till, formed by the gentle let-down of debris as stagnant ice of a peripheral zone slowly melted. The meltwater flow was enough to transport the fines, but not sufficient to remove sand or to destroy the orientation of the fabric.

The occurrence of frontal (end) moraines composed of this ablation till prohibits a hypothesis of complete stagnation. It is suggested that stagnation occurred around the periphery, while active ice persisted farther from the ice edge. Advances of this active ice against the stagnated peripheral ice caused compressive flow and shearing of the active ice, resulting in stacking and in increased thickness in the resulting ablation till forming the frontal moraines.

INTRODUCTION

The region which this paper concerns encompasses approximately 1,000 square miles in northeastern Vermont (Fig. 1). It extends about 50 miles

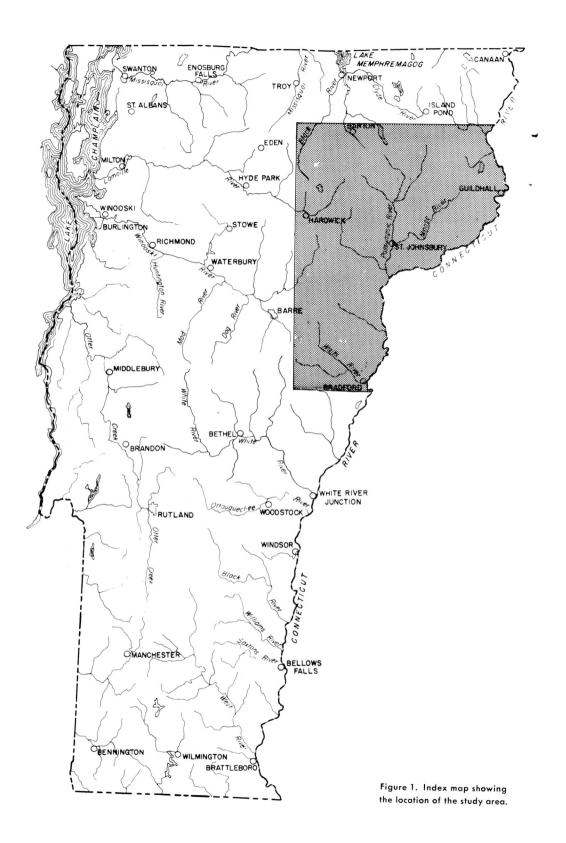

Figure 1. Index map showing the location of the study area.

north and south between Bradford and Barton and westward from the Connecticut River to the longitude of Greensboro (Fig. 2). Two different ice sheets, both believed to be Wisconsin in age, have been identified in this area (Stewart and MacClintock, 1964). The surface till, designated the Shelburne, was deposited by ice advancing from the northeast, while the underlying, older drift, the Bennington, was deposited by a glacier moving from the northwest (Stewart and MacClintock, 1964, 1969). Striae on the bedrock in this area (Fig. 2), beneath the lower till, also trend both northeast and northwest, recording ice advances from these same two directions.

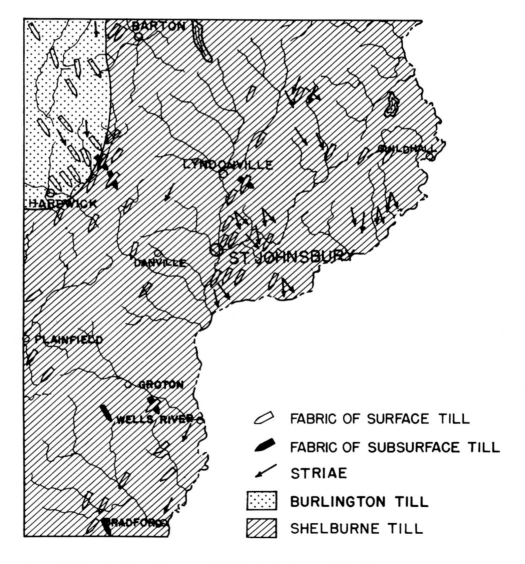

Figure 2. Fabric orientations of the surface tills in the St. Johnsbury region (map).

A young till, the Burlington, deposited by ice moving into the area from the northwest, covers the extreme northwest corner of the area shown in Figure 2 (Stewart and MacClintock, 1969). Throughout the area south of the Burlington Till, in the area of the Shelburne drift, both outwash and till are present at the surface, forming ground moraine, frontal (end) moraines, kames, kame terraces, and a well-developed 24-mile-long esker.

The Shelburne Till has a characteristic loose, sandy texture, which suggests that it might be superglacial till, as described by Upham (1891) and Chamberlin (1894). On the other hand, the pebbles in this till have a strong northeast fabric orientation, a feature normally absent in superglacial tills. It is the purpose of this paper to examine the characteristics of this surface till and, on the basis of these characteristics, to interpret the genesis, both of the till in general and of some frontal moraines in the vicinity of St. Johnsbury that are composed of this till.

NATURE OF THE SURFACE TILL

The till at the surface throughout most of this area is the Shelburne, a loose, sandy, bouldery till that resembles outwash, particularly on rain-washed exposures, although, when excavated, it displays the structure and compactness normally associated with till. Conspicuous characteristics of this till are: (1) a high content of sand, with very little clay; (2) irregularly distributed masses or crude layers of sand and/or fine gravel; (3) a tannish, reddish, or brownish color, due to oxidation, throughout its total thickness; (4) angular cobbles and boulders; (5) a high content of local bedrock in most places; and (6) a strong northeast fabric orientation of the pebbles (Fig. 2). According to our interpretation of the reports of other workers (Currier, 1941; White, 1947; Judson, 1949; Denny, 1958; Flint, 1961; Eschman, 1966), till with these characteristics is common at the surface over much of New England.

ORIGIN OF THE SURFACE TILL

The Shelburne Till, because of its loose, sandy nature, fits well the physical description of superglacial till as given by Upham (1891) and by Chamberlin (1894). In their discussion, however, superglacial till meant drift composed of debris which had accumulated *on top of* glacial ice, debris that was subsequently let down onto a lower deposit of more compact till from an englacial and/or subglacial source. In the process of being let down, any fabric orientation should have become disordered or completely lost by freezing and thawing, wetting and drying, and slumping. The Shelburne

Till of Vermont, however, has a well-oriented regional till fabric (Fig. 2) and, in addition, has not been found lying on a related basal till of subglacial or englacial origin. In contrast, it most commonly lies either directly on the bedrock or on an older till with a distinctly different fabric; lacustrine or fluvial sediments frequently occur between the surficial Shelburne Till and the underlying till.

An example of such a section occurs at the hamlet of West Norwich, four and one-half miles due south of South Stafford and approximately 25 miles southwest of the area shown in Figure 2. Here a road cut exposes 10 to 15 feet of calcareous till (Shelburne) with a northeast fabric at the surface. Beneath this till are nine to eleven feet of leached and oxidized basal till (Bennington) with a northwest fabric. This leached till grades downward into fresh, unleached till, more than 15 feet thick, which also has a north-west pebble orientation. This stratigraphic sequence near the St. Johnsbury area is believed to record a fairly lengthy interval of non-glaciation preceding the deposition of the Shelburne Till (Stewart and MacClintock, 1969, p. 70). Thus, no true superglacial drift was recognized in this area unless it was present in the very thin zone right at the surface, a zone avoided in this survey because of the potential danger of disturbance of surface activities.

Our explanation for the loose, sandy surface till is as ablation till, meaning (*ablatio* — to carry or wash away) material left as the ice melted, or washed away both the meltwater itself and the fine materials (silt and clay) from the till. Thus, ablation till results when a mass of stagnant ice melts so slowly that the contained debris is let down to the ground sufficiently gently that the preferred orientation of the pebbles, produced by the flow of the earlier active ice, before stagnation, is retained (Harrison, 1957). The fines, the clay and silt, are washed out as the meltwater slowly oozes away. The theory of a stagnant zone proposed by Flint (1929, 1933) would yield the conditions necessary to produce such ablation till, and would give rise both to the loose, sandy texture of the Shelburne Till on the Vermont uplands and the vast quantities of lacustrine silt and clay in the valleys.

ABLATION-TILL MORAINES

In the vicinity of St. Johnsbury, thick accumulations of Shelburne abla-tion till form frontal (end) moraines that collectively form a system com-posed of several units (Fig. 3). The surface expression of the moraines is that of a characteristic, random distribution of small, rounded hills and undrained depressions, commonly called simply morainic topography. Ledges of bedrock, so representative of the Vermont uplands, are infre-quent, or completely absent, indicating a considerable thickness of drift.

The largest of these accumulations, the Danville Moraine, extends north-ward from the vicinity of Bradford to near Glover. It is 50 miles long and varies from one-half to four miles in width. Several pits in this moraine expose 20 to 40 feet of ablation till, in contrast to a till thickness on the adjacent uplands that generally averages about ten feet. The moraine has conspicuous relief, though postglacial erosion has rendered the relief diffi-cult to measure in places. Three other morainic belts with similar character-istics are also present, except that they have more subdued relief, a trend that is more or less east-west, and a curvature of their trend that suggests

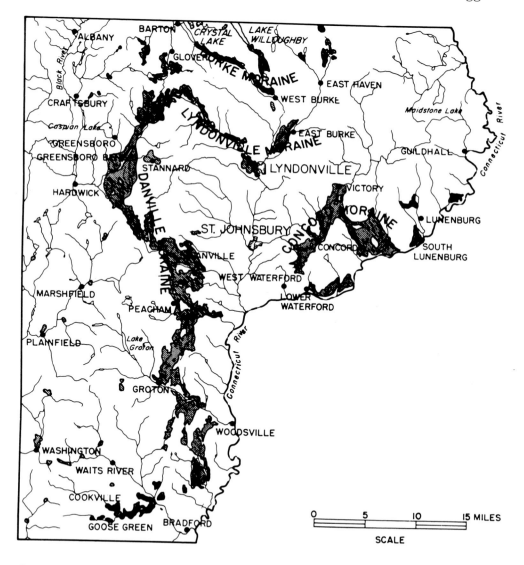

Figure 3. Moraines of Shelburne age in the St. Johnsbury region (map).

deposition along the margin of an ice lobe. All of these moraines contain a fabric that is believed to be that developed during deposition of the till.

The presence of the till in morainic belts, together with morainic topography and an appearance of having been deposited along an ice margin, all imply depositional conditions other than those of a stagnant ice mass, as was required above to explain the origin of the surficial Shelburne Till. We believe that moraines composed of ablation till require a situation in which there is a peripheral zone of stagnant ice and an inner zone of active ice. Active ice impinging against stagnant ice would lead to compressive shearing of the active ice at the contact, resulting in compressive flow (as described by Goldthwait, 1951, and by Nye, 1952), and an upward imbricate shearing, resulting in the stacking of layers of sheared-up basal ice, one upon another (Fig. 4). Anderson (1931) used this general idea to explain what he called "Winter Moraines" in Denmark, but he said nothing about the till, or the till fabric, of the moraines. Each layer that was moved upward by shearing would carry its load of debris intact both in lithology, texture, and structure (fabric). Similar shear plates in glacial ice in Alaska were photographed and described by Chamberlin (1928).

If these conditions remained the same for some time, such shear stacking could cause considerable localized thickening of the drift, after it had been let down by the melting of the ice. Ablation till with a regional fabric would form if the melting had been active enough to remove the fine sediment, but gentle enough to permit the fabric to be retained.

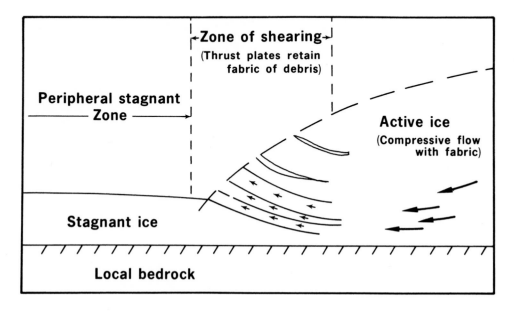

Figure 4. Active ice shearing over stagnant ice in peripheral zone of a glacier.

A second theory proposed by Nye (1952) contends that glaciers moving against rising terrain also generate compressive flow with resulting upward shear. Clayton and Freers (1967) utilized this mechanism to explain the rise of debris up to a superglacial position over the Missouri Coteau. In Vermont, ice moving from northeast to southwest, across the region of the ablation till, must have crossed the Connecticut River valley, at an elevation of approximately 1,000 feet and then moved up the sloping terrain west of the valley to the upland level at an elevation of 3,000 to 3,500 feet. Compressive flow due to the rise in elevation may account for the moraines along the summit of the upland in the St. Johnsbury region.

The combination of compressive flow and shearing of the ice might well also explain the abundance of angular fragments of local bedrock in the ablation till. The active ice could have acquired the fragments from beneath the glacier at the origin of the shear planes; since transportation was short, there was little chance for wear and rounding of the fragments. Continued shearing and stacking of the ice, while transporting debris from a near source, could account for the concentration of the material.

REFERENCES

Anderson, S. A., 1931, Eskers and terraces on the basin of River Susaa: Danish Geol. Surv. II Raekke, no. 54 (English summary, p. 171).

Chamberlin, R. T., 1928, Experimental work on the nature of glacier motion: Jour. Geol., v. 36, p. 1-30.

Chamberlin, T. C., 1894, Proposed genetic classification of Pleistocene glacial formations: Jour. Geol., v. 2, p. 517-38.

Clayton, Lee, and Freers, T. F., 1967, Glacial geology of the Missouri Coteau and adjacent areas: Guidebook, Midwest Friends of the Pleistocene, 1967, North Dakota Geol. Survey Misc. ser. 30, 170 p.

Currier, L. W., 1941, Tills of eastern Massachusetts (abs.): Geol. Soc. Amer. Bull., 52, p. 1895-96.

Denny, C. S., 1958, Surficial geology of the Canaan area, New Hampshire: U.S. Geol. Survey Bull. 1061-C, p. 71-101.

Eschman, D. F., 1966, Surficial geology of the Athol Quadrangle, Worcester and Franklin counties, Massachusetts: U.S. Geol. Survey Bull. 1163-C, 20 p.

Flint, R. F., 1929, Stagnation and dissipation of the last ice sheet: Geog. Rev., v. 19, p. 256-89.

————, 1933, Late Pleistocene sequences in the Connecticut valley: Geol. Soc. Amer. Bull., v. 44, p. 965-88.

————, 1961, Two tills in southern Connecticut: Geol. Soc. Amer. Bull., v. 72, p. 1687-92.

Goldthwait, R. P., 1951, Development of end moraines in east-central Baffin Island: Jour. Geol., v. 59, p. 567-77.

Harrison, P. W., 1957, Fabric of debris in glacial ice: Jour. Geol., v. 65, p. 98-105.

Judson, S. S., Jr., 1949, The Pleistocene stratigraphy of Boston, Massachusetts, and its relation to the Boylston Street Fishweir: Peabody Archaeology Papers, v. 4, 133 p.

Nye, J. F., 1952, The mechanics of glacial flow: Jour. Glaciology, v. 2, p. 82-91.

Stewart, D. P., and MacClintock, Paul, 1964, The Pleistocene stratigraphy of northern Vermont: Amer. Jour. Sci., v. 262, p. 1089-97.

————, 1969, The surficial geology and Pleistocene history of Vermont: Vermont Geol. Survey, Bull. 31, 251 p.

Upham, Warren, 1891, Criteria of englacial and subglacial drift: Amer. Geologist, v. 8, p. 376-85.

White, S. E., 1947, Two tills and the development of glacial drainage in the vicinity of Stafford Springs, Connecticut: Amer. Jour. Sci., v. 245, p. 754-78.

3

Thickness and Structure

*Influence of Irregularities
of the Bed of an Ice Sheet on
Deposition Rate of Till*

L. H. Nobles and J. Weertman

ABSTRACT

Till will be deposited at the bottom surface of an ice sheet if: (1) the lower layers of ice contain englacial debris, and (2) the temperature gradient at the bottom of the ice sheet is insufficient to drain off by thermal conduction both the geothermal heat and the heat produced by the sliding of the ice sheet over its bed. If this temperature gradient is too small, some ice at the bottom of the ice sheet must melt and deposition takes place.

"Dirty" ice layers have been observed directly at the bottom of both the Greenland and the Antarctic ice sheets. It is not unreasonable to assume that the bottom layers of Pleistocene ice sheets also contained englacial debris.

Consider a stage in the history of a Pleistocene ice sheet in which ice is melted over appreciable areas of its bottom surface despite the fact that the temperature within the ice mass is below the freezing point. The larger the bottom temperature-gradient, the smaller is the rate of melting and thus the rate of till deposition. The bottom temperature gradient will fluctuate with position in horizontal directions if the ice sheet slides over its bed and if the bed contains irregularities in wave length of the order of or smaller than the ice thickness. For an active ice sheet, the thermal gradient is larger over a "hill" in the bed than over a "hollow." Therefore the rate of till deposition is larger over depressions in the bed than over elevations. This effect on the rate of till deposition under an active ice sheet is to smooth out irregularities of the bed.

INTRODUCTION

This paper examines the problem of the rate of deposition of till at the bottom surface of an ice sheet and the variation of this rate produced by irregularities of the bed surface. Till will be deposited at the bottom surface if: (1) the lower layers of ice contain englacial debris, and (2) the temperature gradient at the bottom of an ice sheet is insufficient to drain off by thermal conduction both the geothermal heat and the heat produced by the sliding of the ice sheet over its bed. "Dirty" ice layers have been observed directly at the bottom of both the Greenland and the Antarctic ice sheets. The lower 17 meters of ice of the Greenland ice sheet beneath Camp Century contain englacial debris (Hansen and Langway, 1966); the lower 5 meters of ice beneath Byrd Station in Antarctica are dirty (Gow, Ueda, and Garfield, 1968). It is not unreasonable to assume that the bottom layers of Pleistocene ice sheets contained englacial debris.

It is easy to explain (Weertman, 1961; 1966) how debris might be incorporated into the bottom of an ice sheet at some stage in its history. Figure 1 shows a hypothetical cross-section of a Pleistocene ice sheet. The accumulation and ablation zones are labeled at the top of the figure. At the bottom of the figure are indicated zones in which ice is melted from, or water is frozen to, the bottom surface.

Ice is melted from the bottom surface wherever the vertical temperature gradient $\nabla T = \delta T/\delta y$ (where T is the temperature and y is distance in the vertical direction) is so small that the heat conducted toward the surface is smaller than the sum of the geothermal heat Q and the heat produced by

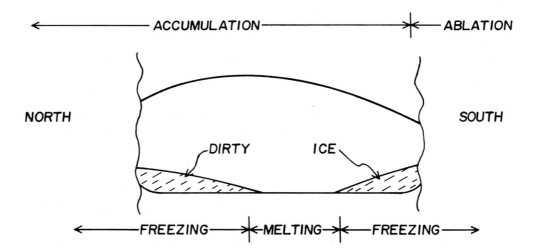

Figure 1. Cross section of a Pleistocene Ice Sheet at a stage in its history in which the freezing of ice to its bottom surface produces ice layers which contain morainal debris.

sliding motion of the ice sheet (namely $JV\sigma$, where J is the mechanical equivalent of heat, V is the sliding velocity, and σ is the shear stress acting at the bottom surface). Thus for melting to take place,

$$Q + JV\sigma > |K\nabla T|\,, \tag{1}$$

where K is the thermal conductivity of ice.

Obviously water in contact with the bottom surface of an ice sheet can become frozen to it if

$$Q + JV\sigma < |K\nabla T|\,. \tag{2}$$

In Figure 1, the water which is melted from the bottom surface of the ice sheet will flow, because of the pressure differential, toward the edge of the ice sheet. If, near the edge of the ice sheet, expression (2) is satisfied, then the outward-flowing water will become frozen to the bottom surface of the glacier. During the freezing process, loose debris lying on the underlying surface is very likely also to be incorporated into this newly formed ice, making it dirty.

The thickness of the newly formed, dirty ice layer under accumulation zones for an ice sheet in a steady state condition is easily shown (Weertman, 1966) to approach the value given by the equation

$$h_d = \frac{ha_b}{(a_t + a_b)}\,, \tag{3}$$

where h_d is the thickness of the dirty ice layer, h is the total ice thickness, a_b is the thickness of the ice layer being frozen in a unit time to the bottom surface, and a_t is the thickness of the ice layer added in a unit time at the top surface. Under ablation zones near the edge of the ice sheet, the thickness of the dirty ice layer becomes equal to the total ice thickness (not shown in Figure 1).

AVERAGE DEPOSITION RATE OF TILL

Suppose, in another stage in the history of the ice sheet of Figure 1, the temperature gradients under a portion of the ice sheet which contain englacial debris satisfy expression (1). Melting of debris-carrying ice must occur. Figure 2 shows such a situation.

The average rate of till deposition is equal to

$$\dot{D} = [c/(1-c)]\,[(Q + JV\sigma - |K\nabla T|)/L], \tag{4}$$

where $\dot{D}$ is the thickness of till deposited in unit time, c is the volume fraction of debris within the ice, and L is the latent heat of fusion of ice.

The temperature gradient at the bottom of an ice sheet in a steady-state condition depends not only on the temperature at the bottom and top surfaces of the glaciers, but also on the deformation rate of the ice. Figure 3a shows the steady-state temperature profile through a stagnant ice sheet. No deformation of ice occurs within it. The absolute value of the temperature gradient $|\nabla T|$ is equal to $(T_0 - T_1)/h$ where T_0 is the bottom surface temperature, T_1 is the upper surface temperature, and h is the total ice thickness. Under accumulation areas, the temperature profile is modified to the general shape shown in Figure 3b (Robin, 1955; Weertman, 1968; Dansgaard and Johnsen, 1969).

The temperature gradient at the bottom surface is approximately equal to

$$(T_0 - T_1)\,/h',\tag{5}$$

$$\text{where } h' \approx (2\kappa h/V)^{1/2},$$

where κ is the coefficient of thermal diffusivity of ice, and V is the accumulation rate of snow (in equivalent of high-density ice) at the upper surface. Under ablation areas, the general shape of the temperature profile should be that shown in Figure 3c. The quantity $(h-h^*)$ in this figure is approximately equal to

$$h-h^* \approx (2\kappa h/V^*)^{1/2},\tag{6}$$

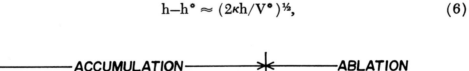

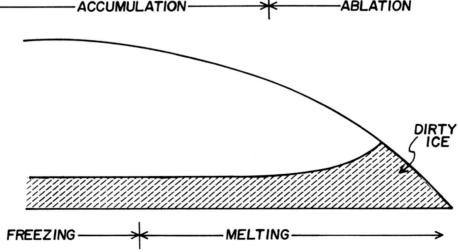

Figure 2. A stage in the history of the ice sheet shown in Figure 1 in which melting of dirty ice occurs at the bottom of the ice surface.

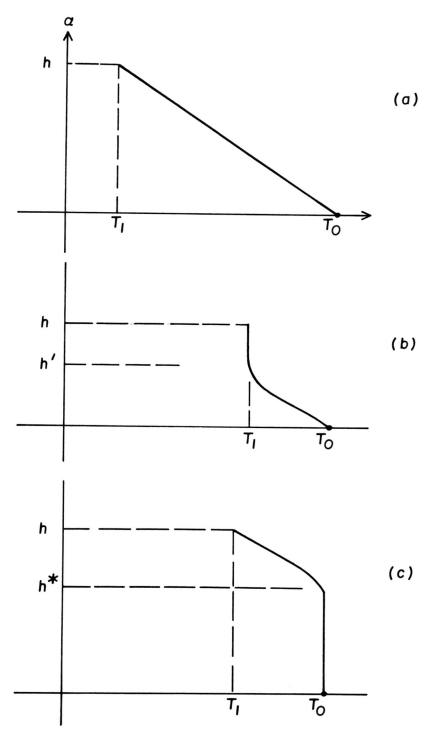

Figure 3. Steady-state temperature profiles through ice sheets, under the conditions of: (a) no extending or compressive flow; (b) extending flow (accumulation area); and (c) compressive flow (ablation area).

where V^* is the ablation rate of ice at the upper surface. The temperature gradient at the bottom surface in Figure 3c is equal to zero.

If, in Figure 3, the actual temperature gradient at the bottom of the ice surface were known, then through Equation (4), the till-deposition rate can be calculated. Since the paleo-temperature gradients are not known, only limits can be set on $\dot{D}$. A lower limit is, obviously, $\dot{D} = 0$. An upper limit is of the order of $\gamma c / (1-c)$, where $\gamma \approx 0.5$ cm/year to 3 cm/year. The upper limit is established by noting that, if all the geothermal heat Q is absorbed by the melting process, then approximately 0.5 cm³ of ice is melted per year from an area of 1 cm² at the bottom surface. The heat generated by the sliding motion of the ice over the bed can increase this volume of melted ice to 2 to 3 cm³ per year. (However, if an ice sheet were to surge, the melting rate could temporarily be increased to 300 cm³ of ice per year per 1 cm² area [Weertman, 1969].)

VARIATION OF THE DEPOSITION RATE OF TILL PRODUCED BY IRREGULARITIES OF THE BED SURFACE

Little more can be concluded about the average rate of deposition of till

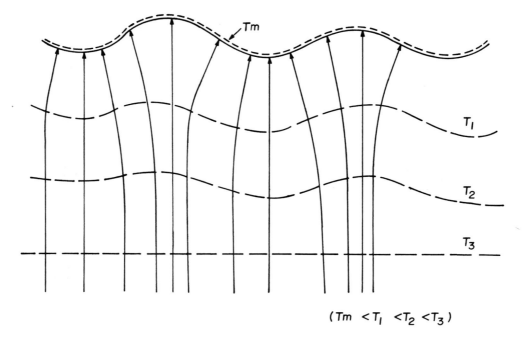

$(Tm \; < T_1 \; < T_2 \; < T_3)$

Figure 4. Cross section showing isotherms and paths of heat flow beneath the bed of an ice sheet which is irregular in form. Tm = melting temperature.

from the bottom surface of Pleistocene ice sheets other than the limits estimated in the last section. However, some important conclusions can be
made about the influence of irregularities of the bed surface on the local,
rather than the average, deposition rate of till.

At least two effects exist which will cause till to be deposited at a faster
rate over "valleys" in the bed than over "hills." This effect on the rate of till
deposition under an active ice sheet is to smooth out irregularities of the
bed.

One mechanism by which the till-deposition rate is changed is shown in
Figure 4. The bed surface is shown according to convention in cross-section
as a wave. (It may also take on a wave form in the perpendicular cross-
section.) Since the bottom surface is at the melting temperature (otherwise,
till would not be deposited), the melting temperature T_m isotherm must
follow the bed topography as shown in Figure 4. The isotherms shown
below the bed have wave amplitudes which diminish the further they are
below the bed.

The heat flux follows paths which are normal to the isothermal surfaces
(assuming that the thermal conductivity is uniform and isotropic). Figure 4
also shows the paths of heat flow. The heat-flow paths converge into the
hollows of the bed. Thus more ice is melted and more till deposited there
per unit time than at the hills. It is immaterial to this mechanism whether
the temperature profile in the ice is given by Figure 3a, 3b, or 3c.

A second mechanism which can cause a difference in the till-deposition
rate operates if a temperature gradient ∇T exists at the bottom ice surface
of an active ice sheet. The mechanism is illustrated in Figure 5, which shows

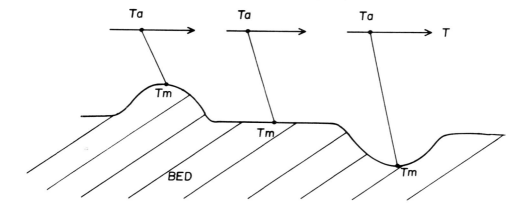

Figure 5. Temperature gradients within an ice sheet immediately above a bed of irregular shape. $Tm =$
melting temperature; $Ta =$ temperature of the upper surface of the ice.

a bed containing a hill and a hollow. As the ice moves across this bed, the temperature gradient in the ice immediately above the bed has a different value above a hill than over a hollow. The origin of the difference in ∇T is easily explained, through use of Figure 6. If the block of material shown in Figure 6a has the lower surface maintained at temperature T_m and its upper surface temperature maintained at T_a, a temperature gradient ∇T will be set up in it. If, as shown in Figure 6b, the block is suddenly compressed (and the temperature rise produced by this mechanical working is ignored), the temperature gradient in it will be increased. If the block is extended, as shown in Figure 6c, the gradient is decreased in value.

The ice shown flowing over a hollow is extended in the same direction. Hence the temperature gradient over a hill is larger than over a hollow.

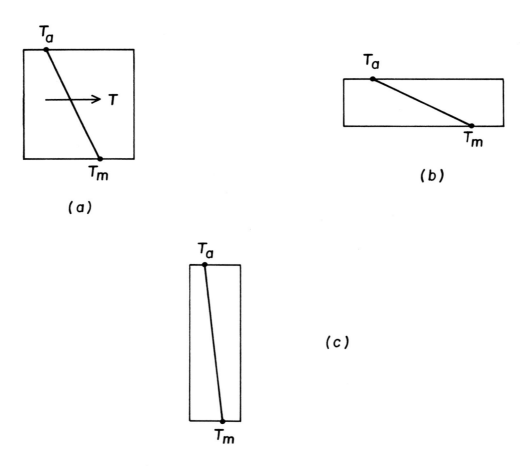

Figure 6. Change of temperature gradient in block (a) produced by fast compression in (b) and fast extension in (c).

The amount of ice melted at the bottom surface is smaller, the larger the temperature gradient. Thus the rate of till deposition is smaller over a hill than over a hollow.

DISCUSSION

Two mechanisms have been presented to demonstrate that the till-deposition rate at the bottom of an ice sheet is greater over depressions in the bed than over elevations. These mechanisms operate at any wave-length scale of bed irregularities up to wave lengths approaching, say, $h/10$, where h is the thickness of an ice sheet. No formal calculation of the actual difference in till-deposition rate over a hill, as contrasted to till-deposition rate over a hollow, was attempted. Is is obvious, however, that differences of the order of 20 percent in deposition rate could easily exist for undulations in the bed whose amplitudes are of the same order of magnitude as their wave lengths. Thus irregularities in the bed tend to be smoothed out by this difference in till-deposition rate. (We have not examined the process of erosion of till by sliding ice motion. Erosion conceivably could lead to either roughening of the bed or to further smoothing of the bed surface.) The theory presented in this paper offers no light on understanding the process of drumlin formation. On the contrary, our theory emphasizes the enigma that is the drumlin.

The smoothing out of the bed surface through differences in the till-deposition rate depends on the ice sheet being in an active state. Should the ice sheet become stagnant, the rate of till deposition at various localities of the bed is no longer an interesting problem. The total amount of basal till that will be deposited directly from the melting ice, from the moment the ice sheet becomes stagnant until its disappears, is simply equal to the total amount of debris contained in the ice column which is directly above the particular point at the bed. (Flowing water at the base of stagnant ice will greatly complicate the depositional process.) However, when the ice sheet is active, the dirty ice-layer above the bed remains in horizontal motion over the bed. The supply of debris for the till deposited at nearby localities remains constant, regardless of differences in deposition rates, until the dirty-ice layer ceases to exist.

REFERENCES

Dansgaard, W., and Johnsen, S. J., 1969, Comment on paper by J. Weertman, "Comparison between measured and theoretical temperature profiles of the Camp Century, Greenland, borehole": Jour. of Geophys. Research, v. 74, p. 1101-10.

Gow, A. J., Ueda, H. T., and Garfield, D. E., 1968, Antarctic Ice Sheet: Preliminary results of first core hole to bedrock: Science, v. 161, p. 1011-13.

Hansen, B. L., and Langway, C. C., Jr., 1966, Deep core drilling in ice and core analysis at Camp Century, Greenland, 1961-1966: Antarctic Jour. of the U. S., v. 1, p. 207-208.

Robin, G. de Q., 1955, Ice movement and temperature distribution in glaciers and ice sheets: Jour. Glaciology, v. 2, p. 523-32.

Weertman, J., 1961, Mechanism for the formation of inner moraines found near the edge of cold ice caps and ice sheets: Jour. Glaciology, v. 3, p. 965-77.

————, 1966, Effect of a basal water layer on the dimensions of ice sheets: Jour. Glaciology, v. 6, p. 191-207.

————, 1968, Comparison between measured and theoretical temperature profiles of the Camp Century, Greenland, borehole: Jour. of Geophys. Research, v. 73, p. 2691-2700.

————, 1969, Water lubrication mechanism of glacier surges, Canadian Jour. of Earth Sciences, v. 6, p. 929-42.

Glaciotectonic Structures in Drift

Stephen R. Moran

ABSTRACT

Glaciotectonic structures of three types are present throughout the glaciated portion of North America.

"Simple *in situ* deformation" involves small-scale folds and faults produced by ice push and bed shear in both bedrock and drift. These structures have long been used to reconstruct direction of ice advances.

"Large-scale block inclusion" is the incorporation of large intact blocks of bedrock and drift into younger drift. The incorporated blocks were sheared from bed material having low shear strength as a result of groundwater discharge in response to high heads produced by the glacier. Thick permafrost may have helped to reduce shear strength locally. Shearing occurred most readily where the compressive-flow regime was favored, as near ice margins or where the ice flowed up scarps. Failure to recognize repetition of glacial sequences and the occurrence of erratic blocks of older drift or bedrock in younger drift will lead to major errors in interpreting glacial sequences. Ridges produced by large-scale block inclusion may be erroneously interpreted as end moraines.

"Transportational stacking within single till sheets" by sporadic differential movement along shear planes in the debris-charged basal zone of the ice may produce disturbed sequences. Continual deposition and re-erosion along different planes of failure cause the original sequence within the till to be shuffled, so that slices from the base of the drift-rich zone may be deposited anywhere throughout the till sheet. This process may produce, by faulting and stacking in one till sheet during deposition, complex patterns resembling normal sequences of several till sheets.

INTRODUCTION

Observations in three areas that are widely separated, both geographically and in type of glacial processes and features, indicate that structural deformation of both bedrock and glacial drift is common. Deformed drift and bedrock have been observed in many places in exposures along strip-mine highwalls in the glaciated Allegheny Plateau of eastern Ohio and northwestern Pennsylvania. Numerous examples of such deformation have been seen in exposures in Illinois and Indiana. Repeated examination of active strip mines near Danville, Illinois, as stripping proceeded from 1965 to 1969, has revealed several examples of thrust faulting in both bedrock and drift. Studies of surface exposures in the Hudson Bay Area (63 D and C) in east-central Saskatchewan have revealed other examples of deformed bedrock and drift (Moran, 1969). Test drilling by the Saskatchewan Research Council in the Hudson Bay Area encountered two major glacio-tectonic structures.

For purposes of discussion, glaciotectonic structures have been subdivided into three classes on the basis of the type and magnitude of deformation: (1) "simple *in situ* deformation," (2) "large-scale block inclusion," and (3) "transportational stacking within single till sheets."

SIMPLE IN SITU DEFORMATION

Description

The first class of deformation involves bedrock or glacial drift which has been contorted by ice push or bed shear into folds and faults with only minor displacements. Structures in this class, which are generally small, consist of simple folds in till, stratified drift, and bedrock, and faults of visibly small displacement. The attitude of beds prior to deformation can be readily reconstructed because of the simplicity of the structures or proximity to the site of *décollement*. Deformation of this type is generally of small enough scale that the entire structure is visible in a single outcrop. This type of deformation has been recognized for many years, because of its generally small-scale proportions and because it is common near the bedrock-drift contact where contrasting lithologies are present to delineate structures. Most of the common reported occurrences of glacial deformation are included in this class.

Around the turn of the century, F. W. Sardeson (1905, 1906) described several folds and faults in dolomite and shale in and around Minneapolis which he attributed to glacial disturbance. He demonstrated that these structures were not the result of frost heave, as had been claimed by some, but were the result of horizontal compression associated with Pleistocene glaciation.

Bluemle (1966) described blocks of shale sheared up into till along thrust faults in Cavalier County, North Dakota. T. C. Brown (1933) reported minor structures associated with glacial deformation of an over-ridden esker in Massachusetts. Deformed Pennsylvanian bedrock in Iowa and Kansas had been considered to be the result of ice push by Lammerson and Dellwig (1957) and by Dellwig and Baldwin (1965).

In the quarry of the Carbon Limestone Company about five miles southeast of Youngstown, Ohio, near Hillsville, Pennsylvania, minor folding of shale just below the bedrock surface was observed by the author. These folds, which die out a few feet below the base of the drift, are the result of simple *in situ* deformation by subglacial shear. In a strip mine along the Pennsylvania Turnpike one mile northeast of New Galilee, Beaver County, Pennsylvania, a coal bed up to one foot thick and approximately one hundred feet long occurs at the base of the Titusville Till overlying strongly oxidized glacial gravel. Other occurrences of coal beds sheared up into till were observed elsewhere in the region. Along the east wall of the quarry of the New Castle Limestone Company, two miles northeast of Lowellville, Mahoning County, Ohio, a four-foot-thick bed of Vanport Limestone (Pennsylvanian) overlies six inches of Titusville Till for a horizontal distance of several hundred feet. The limestone was incorporated into the basal drift of the advancing glacier and transported intact along a horizontal shear plane. Two well-developed folds exposed in the upper part of the Titusville Till on the same wall were formed during transportation of the till. On the west wall of the north pit of the same quarry, the upper part of the Titusville Till is repeated along two low-angle thrust faults. The two- to four-foot-thick thrust slices are clearly marked by the threefold repetition of brown oxidized till overlying gray unoxidized till.

Additional evidence of this type of simple *in situ* deformation was seen

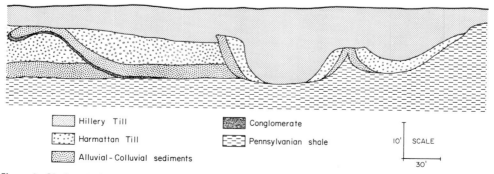

Figure 1. Glaciotectonic structures exposed in Harmatton strip mine near Danville, Illinois, June, 1969.

in strip mines west of Danville, Illinois. Evidence of movement along bedding planes in Pennsylvanian shales below the lowermost till was commonly observed. As stripping proceeded from 1965 to 1969, continually exposing new highwall, thrust faults carrying gray shale up into reddish-pink till or carrying the reddish-pink till up into the gray till overlying it have been observed. Along a fault exposed in 1966, red till could be traced for a horizontal distance of 100 feet and for a vertical distance of up to 15 feet above the base of the overlying gray till. Figure 1 shows a structure exposed in the mine during the spring of 1969.

Mechanism of Formation

Many writers have attributed glacial structures of the simple *in situ* deformation type to ice push or shearing along the bed of the ice. The author agrees that these small-scale structures were in fact probably caused by the normal stresses exerted on the upstream faces of protuberances from the bed (ice push), shear exerted by moving ice along its bed, or a combination of both.

Significance

Small-scale *in situ* deformation of both bedrock and drift has long been used as evidence of post-depositional overriding by glaciers. The direction of overturning of folds and of movement of thrust blocks has been used as an indicator of the direction of glacial flow. Because structures in this class are small and usually marked by beds of distinctive lithology, they are readily recognized and therefore generally do not pose significant problems in interpretation. A small mass of bedrock whose base is not exposed in shallow exposures might be interpreted as being *in situ*, while in fact it overlies drift, thereby causing an erroneous concept of drift thickness and makeup. Other than local cases such as this, these structures are not significant for stratigraphic interpretation.

LARGE-SCALE BLOCK INCLUSION

Description

The second class of glacially derived structures involves incorporation of large masses of bedrock or pre-existing glacial drift into the ice and transportation of the intact block away from the site of *décollement*. It is evident that any boundary between simple *in situ* deformation and large-scale block inclusion is placed arbitrarily, because both the scale of deformation and distance of transport are gradational from one class to the other. In general, if the original location and attitude of the block involved cannot be reconstructed, the deformation is considered to belong in the second class. Because of their very different stratigraphic implications, structures in this class of deformation are divided on the basis of lithology into two subgroups: bedrock structures and glacial drift structures.

Like simple *in situ* deformation, bedrock structures in this group have been recognized for many years. Sardeson interpreted several large anomalous outliers of Cretaceous rock in Minnesota as being blocks transported intact by glacier ice (Sardeson, 1898). Wolford (1932, p. 362-67) reported a large erratic block of limestone, several acres in area, buried in the drift

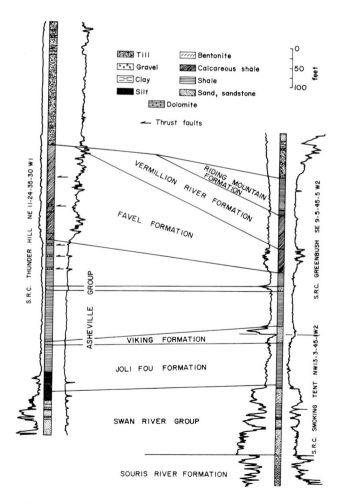

Figure 2. Repetition of sequence in Upper Cretaceous sediments resulting from large-scale block inclusion, Thunder Hill, eastern Saskatchewan.

in central Ohio. Wickenden (1945, p. 75-76) interpreted anomalous structures down to a depth of several hundred feet in a well on Thunder Hill in east-central Saskatchewan as the result of glacial deformation. Rutten (1960) and Brinkmann (1953) discuss the origin of glacially derived bedrock structures in northern Europe. Kupsch (1962) and Byers (1960) described contorted bedrock along the Missouri Coteau of Saskatchewan and attributed the deformation to glacial action. Hopkins (1923) concluded that the faults, folds, and steeply dipping rocks exposed in the Mud Buttes and Tit Hills of Alberta were the result of glacial deformation. Slater recorded numerous ice-thrust bedrock structures in Britain, Europe, and North America (Slater, 1926, 1927a, 1927b, 1927c, 1927d, 1927e, and 1929).

In the Hudson Bay Area of east-central Saskatchewan, a testhole drilled by the Saskatchewan Research Council on Thunder Hill (NE 11-24-35-30 W1)[1] encountered a complexly faulted sequence consisting of at least six overthrust slices of Cretaceous shales with Pleistocene gravels and till (?) included along at least one of the faults (Figs. 2 and 3). Another SRC testhole near Steen, Saskatchewan, encountered 18 feet of glacial drift over

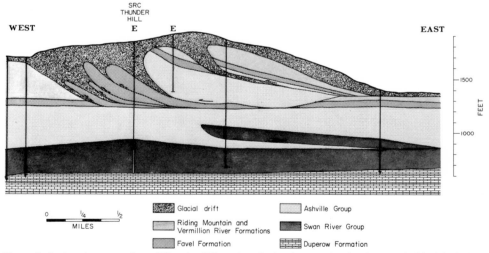

Figure 3. East-west cross section of Thunder Hill, eastern Saskatchewan, showing large-scale block inclusion of Cretaceous sediments. "E" denotes interpretation based on electric logs.

113 feet of clayey silt of the Riding Mountain Formation (Cretaceous), which in turn overlay five feet of till resting on Riding Mountain Formation *in situ* (Fig. 4). This large block of shale incorporated into the drift is expressed at the surface as a series of isolated hills rising above a flat to gently rolling lake plain.

1. For an explanation of the National Topographic System of land location used in Canada, see Christiansen, 1971; this volume, p. 169, Fig. 2.

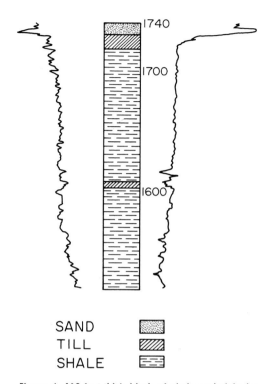

SAND

TILL

SHALE

Figure 4. 110-foot-thick block of shale underlain by till, encountered in a Saskatchewan Research Council testhole near Steen, Saskatchewan (NW 15-31-42-11 W2).

Large-scale block inclusion of glacial drift has not been generally recognized, but as more detailed stratigraphic information becomes available, more and more instances of this type of structural deformation are being reported. Repetition of the same peat bed and of till sequences under end moraines has been observed in borings in central Illinois (J. P. Kempton, personal communication, 1968).

In Saskatchewan, testholes drilled in areas of hummocky moraine often encounter four or five oxidized zones, whereas testholes in adjacent areas encounter only one, or at most two, such weathered zones (E. A. Christiansen, personal communication, 1969). Large-scale deformation involving large blocks of drift has been encountered in several locations throughout Saskatchewan, particularly along the Missouri Coteau (E. A. Christiansen, W. A. Meneley, S. H. Whitaker, personal communications, 1969). These blocks of drift commonly rest on younger drift, causing repetition of sequence, and, in a few cases, are completely inverted so that progressively younger drift is encountered with depth.

Mechanism of Formation

The origin of large-scale block inclusion structures has been the subject of considerable debate. Most of those who have discussed the problem have considered it necessary that the material be frozen in order to be incorporated (e.g., Mathews and MacKay, 1960; Kupsch, 1962; Rutten, 1960). Undoubtedly, in some instances, the included blocks have been frozen at the time of their incorporation, but this is not a necessary prerequisite for incorporation if the porewater pressure in the subjacent beds was sufficiently high.

The shear strength of a soil material is directly related to the effective normal stress on the soil. The effective normal stress is defined as the total normal stress minus the porewater pressure (Terzaghi and Peck, 1967).

$$\overline{\sigma} = \sigma - U, \qquad \text{where:} \quad \begin{aligned} \sigma &= \text{total normal stress} \\ \overline{\sigma} &= \text{effective normal stress} \\ U &= \text{porewater pressure} \end{aligned}$$

It follows that if the porewater pressure is increased an amount greater than any accompanying increase in the total normal stress (the total load), then the effective normal stress will be decreased. This decrease in effective normal stress produces a corresponding decrease in the shear strength of the material. Under certain conditions, it is possible for the porewater pressure to be sufficiently increased to permit shear failures in materials overridden by glaciers. Two general sets of conditions must be met to permit the incorporation of large intact blocks into an advancing glacier. First, the soil or rocks which are overridden must contain beds of permeable material confined by less permeable material. In most instances, the permeability discontinuities also coincide with changes in strength of the materials. Secondly, the ground-water flow must be modified so that water moves into these confined beds, generating elevated porewater pressures.

In addition to the geohydraulic conditions, the incorporation of large blocks of soil or rock into glaciers requires conditions of compressive flow within the glacier itself. Under the compressive-flow regime, the ice is moving, either by plastic deformation or by shear, upward from the base. This flow regime is favored in regions where ablation is in excess of accumulation and where the bed of the glacier is concave upward, as where the ice flows up over an obstruction on its bed. Under such conditions, large blocks of material sheared into the ice as a result of the decreased effective stress are carried up into the glacier and may be transported some distance essentially intact.

One or more geologic situations may have existed to generate the necessary confined conditions to produce large-scale block inclusion in any given instance. The common interbedding of sands or sandstones with clays, silts, or shales in bedrock and glacial sequences produced the necessary stratigraphic situation for this type of failure. Where the permeable beds in such a sequence were confined laterally, the migration of ground water was prevented, and porewater pressures built up. Facies changes from sand to clay in either bedrock or glacial sequences would cause such confinement. Truncation of a permeable bed by an erosional unconformity over which an impermeable material was deposited would also generate such a condition. Where the permeable beds were cut into less permeable material, as in a gravel valley-fill cut into older shales or clays, the valley sides would act as the obstruction to lateral ground-water migration. The presence of ground ice or permafrost in the outcrop area of a permeable bed might prove sufficiently confining to permit the development of high porewater pressures.

Several different mechanisms probably operated to produce the ground-water flow necessary to create the elevated porewater pressures. As was the case with the geologic controls, a given case of deformation may have resulted from the operation of one or more of these mechanisms.

The occurrence of continental glaciers in low areas caused the discharge areas of large, regional-scale, ground-water flow systems to be blocked. The directions of ground-water flow were modified or reversed, producing new flow systems which discharged beyond the limits of the glacier. Where such a modified flow system interacted with one of the geologic situations described above, abnormally high porewater pressures may have resulted. The operation of this mechanism has not yet been fully investigated and therefore little is known of the magnitude of porewater pressure which can be developed.

Generation of elevated porewater pressures by the rapid formation of a permafrost layer at the time of glaciation is discussed by Mathews and MacKay (1960). It is not clear whether such a permafrost layer can form sufficiently rapidly to prevent porewater pressure conditions from coming to equilibrium with each successive increase in permafrost thickness. If such rapid formation of permafrost is in fact possible, then this mechanism may have played a significant role in the formation of large-scale block-inclusion structures.

Where a glacier overrode unfrozen cohesive sediments, the load of the glacier would tend to produce consolidation of the sediments. If drainage from the clay or silts into sands or gravels was possible, the cohesive sediments would consolidate and the water forced into the permeable beds would flow toward areas of lower head, near the margin of the ice sheet. Where the permeable beds were confined as described above, such consolidation of cohesive sediments back under the ice could produce sufficiently elevated porewater pressure nearer the margin of the glacier to permit the incorporation of large blocks of material (Fig. 5).

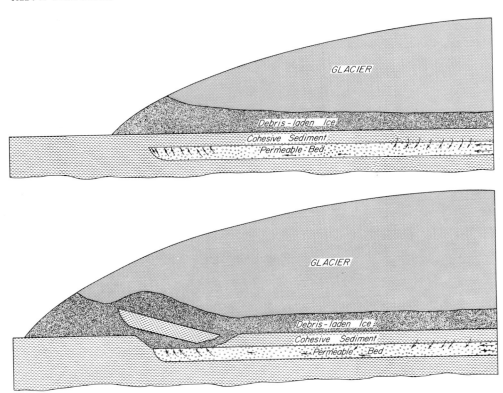

Figure 5. Schematic drawing of large-scale block inclusion resulting from elevated pore-water pressures.

High porewater pressures would also be generated where a glacier advanced over debris containing buried ice blocks remaining from a former glacier. The ice blocks would serve as source layers of water as they melted, maintaining anomalously high porewater pressures in the manner described by Hanshaw and Bredehoeft (1968). The high fluid pressures thus generated would cause a zone of weakness at the interface between the buried ice and the overlying debris. The shear stress resulting from the overriding glacier could cause failure along this surface, thereby incorporating the entire debris layer into the younger glacier.

Significance

Large-scale block inclusions of both bedrock and drift material can have profound effects on the interpretation of glacial stratigraphy. Failure to recognize the presence of large blocks of bedrock, some of which may be over one hundred feet thick, as in SRC Steen (NW 15-31-42-11 W2), within a drift sequence can lead to very great errors in determination of drift thickness, as well as of the configuration of the bedrock surface. For this reason, anomolously high bedrock elevations should be suspect.

Large-scale block inclusion of drift poses especially serious problems, because it is usually very difficult to detect. In the absence of detailed knowledge of the regional stratigraphy it may be impossible to recognize the presence of such blocks. Because blocks of drift carried upward in this manner can be redeposited above younger drift, it is immediately obvious that considerable confusion may occur as the result of the presence of older drift and of weathering zones occurring above younger drift. In the absence of knowledge of regional lithostratigraphy, studies of small areas must be undertaken with great care, and correlations based solely on the occurrence of a paleosol without knowledge of the stratigraphic framework involved are questionable.

Large-scale block inclusion has probably played a major role in the formation of many if not most of the classical end moraines of midwestern United States. Because many of the natural and man-made exposures throughout the midwest are located in morainic belts, the recognition of the presence of large-scale block inclusion of drift becomes very important in the correct interpretation of these outcrops. The land forms mapped throughout the midwest as end moraines can be separated into three types: true end moraines, moraines resulting from bed configuration, and moraines resulting from hydrodynamics in the bed.

True End Moraines

The true end moraines include both terminal and recessional moraines which formed at or very near the margin of an ice sheet. Large-scale block inclusion probably played a major role in the formation of many true end moraines. Rapid changes in porewater pressure resulting from the rapid expansion of a glacier over ice-free marginal areas would favor the incorporation of large blocks by one of the mechanisms described above. In areas where the stress environment in the ice favored compressive flow (Nye, 1952, p. 89), large blocks incorporated into the ice by any of these three processes would be carried upward in the ice and stacked. Optimum conditions favoring compressive flow occur near the ice margin in the zone of ablation where the bed is concave upward.

The combination of these two processes at or near glacial margins — the common occurrence of large-scale block inclusion and the compressive flow regime to lift and stack the blocks — probably accounts for a large proportion of the volume of most true end moraines. Because this block stacking may have involved glacial drift and/or bedrock material of varying ages, great care must be taken in interpreting glacial sequence within true end moraines.

Moraines Resulting from Bed Configuration

In areas where the bed of the glacier was concave upward, such as where the ice flowed up a cuesta scarp, the compressive-flow regime was favored, and large-scale block inclusion would be expected. In such areas, a thicker deposit consisting of large blocks as well as smaller debris would be formed by the resultant stacking. Because the state of stress over this slope or obstruction continued to favor compressive flow as long as the ice continued to advance, this greater accumulation of material persisted after the margin had passed the area. When ablation occurred, the thicker debris would be let down to form a ridge or moraine which was the result of bed configuration (Fig. 6). The coincidence of several moraines in Illinois with bedrock

A. Compressive Flow

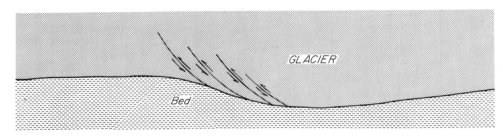

B. Imbrication of Slabs of Bed Material

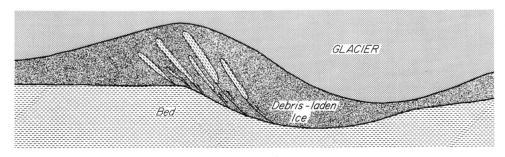

C. Ridge after Ablation of the Glacier

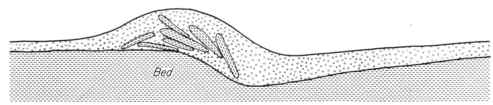

Figure 6. Formation of a "moraine" ridge by thrust faulting in a zone of compressive flow.

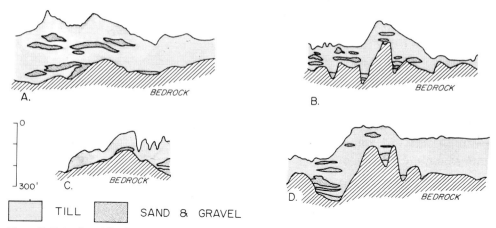

Figure 7. Examples of "end moraines" in Illinois, associated with large-scale bedrock obstructions. (Modified from Piskin and Bergstrom, 1967, Plate 2).

uplands or scarps suggests that they may have formed in this manner (Fig. 7).

Moraines Resulting from Hydrodynamics of Bed

If, for one of the reasons discussed above, the hydraulic uplift in an area under a glacier should become greater than the superincumbent load of ice and rock debris, the material would be lifted up into the ice. Such a condition would continue to produce large-scale blocks until the uplift pressure

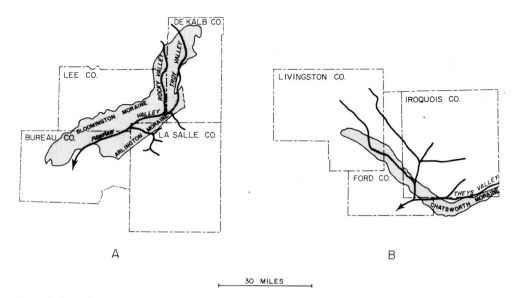

Figure 8. Examples of moraines in Illinois associated with buried bedrock valleys.

became sufficiently reduced. When ablation occurred, a ridge would result in the same manner as in the case above. The coincidence of a number of moraines in Illinois with large bedrock valleys suggests the operation of such a mechanism (Fig. 8).

TRANSPORTATIONAL STACKING WITHIN A SINGLE TILL SHEET

Description

The third class of structural deformation involves differential movement and overthrusting of elements within a single till sheet while it is being transported and deposited. Although the existence of this type of deformation has not been previously demonstrated, two authors have suggested such a mechanism to explain observed characteristics of tills.

From his studies of the fabric of clay till, Harrison (1957, p. 300-301) concluded the debris-laden zone at the base of a glacier advanced by often sporadic but generally continuous differential movement along more or less horizontal shear planes. He stressed that when friction along a particular plane became too great because of the abundance of debris, movement ceased and temporary deposition occurred. Renewed movement could begin on any of the shear planes within the stationary debris-laden ice mass when friction along that plane was overcome by greater shear stress exerted on it.

Virkkala (1952) reported the presence of thin partings within till sequences in Finland which he referred to as bed limits. The bed limits were finer textured than the surrounding till and were generally contorted. In all sections containing more than one bed limit, "the deformation of the bed limits increases downward" (Virkkala, 1952, p. 105). Virkkala (1952, p. 107-9) concluded that the bed limits reflected the shear planes within englacial drift along which movement had occurred. He suggested a model of the base of a glacier consisting of beds of drift-free ice interbedded with drift-rich ice. Continued concentration of debris in the dirty ice decreased the ability of the ice to flow until movement ceased. Re-erosion of deposited debris would occur, Virkkala suggested, by movement along the already-present shear planes, causing further distortion of the planes (bed limits) each time they were utilized.

In both of these models, the base of the ice was considered to consist of a series of discrete slabs of debris-laden ice moving sporadically along horizontal shear planes. Through the continual process of deposition and re-erosion of material along different shear planes, the sequence of these slices could become thoroughly disordered. Material originally located at the base of the debris-laden zone could be brought fairly readily to the top of the zone by a very few episodes of deposition and re-erosion. The operation of this mechanism should, therefore, result in till deposits in which the original sequence of slices is nearly everywhere out of order.

The author's work in the Allegheny Plateau of Ohio lends support to this hypothesis by demonstrating the presence of stacking of elements within a single till sheet. Gross and Moran (1971; this volume, p. 251) reported a fairly regular decrease in feldspar content with increased depth below the top of the Titusville Till throughout the Plateau. However, only four of the seventeen sections near Youngstown, Ohio, studied by the author (Moran, 1967) display this trend. Most of the remaining sections display the expected gradation, except for major breaks in the sequence at which feldspar content decreased abruptly only to continue to increase above the break (Fig. 9). Although these abrupt changes in feldspar content were originally interpreted as representing contacts between separate till sheets (Moran, 1968), it is now suggested that these breaks indicate the presence of thrust faults along which the sequence has been duplicated. This interpretation is strongly supported by the thickness of the Titusville Till in the two groups of sections. In the four sections which display the normal trend of increasing feldspar content upward, the mean thickness of the Titusville Till is 17.1 feet (13.4 feet, excluding associated stratified drift). In the thirteen sections which are believed to contain one or more thrust faults, the mean thickness of the Titusville Till is 26.25 feet (20.6 feet, excluding associated stratified drift).

The textural composition and/or potassium feldspar content of the till in two sections in the Allegheny Plateau corroborates the interpretation of the breaks in feldspar content trend as thrust faults. In section 80, the lowermost part of the sequence appears to be repeated once (Table 1). Marked similarities occur between samples 83 and 85, and between 84 and 86. Except for the sand content in the second pair, all differences appear to be the result of random variation. Table 2 gives the mean values of the reconstructed sequence, with the influence of the shear plane between samples 84 and 85 removed. Section 71 does not contain any apparent trend in

TABLE 1

COMPOSITION OF TITUSVILLE TILL IN SECTION 80[*]

Sample	% Sand	% Clay	% Feldspar	% K Feldspar
80	33	22	16	42
81	37	24	12	42
82	40	20	15	42
83	39	24	14	56
84	62	13	10	46
85	38	23	14	56
86	47	15	10	44
87	28	28	8	28

[*]Located in an old Bessemer Limestone Company quarry (½ mile south of Ohio Rt. 630, 150 yds. east of State Line Rd., in northern Beaver Twp., Lawrence Co., Pennsylvania).

TABLE 2

RECONSTRUCTION OF SECTION 80*

Sample	% Sand	% Clay	% Feldspar	% K Feldspar
80-82	37	22	14	42
83,85	38	24	14	56
84,86	55	14	10	45
87	28	28	8	28

*For location see note to Table 1.

TABLE 3

COMPOSITION OF TITUSVILLE TILL IN SECTION 71*

Sample	% Sand	% Clay	% Feldspar
71	39	15	16
72	38	15	15
73	39	19	10
74	37	21	13
75	31	22	15
76	39	19	13
77	36	21	12
78	32	22	14

*Located in American Fire Clay Company quarry (NW¼ SE½, Sec. 6, Beaver Twp., Mahoning Co., Ohio).

feldspar composition, but investigation of textural data (Table 3) indicates the possibility of a repetition of the base of the section on a thrust fault between samples 75 and 76. Removal of the effect of this duplication of section produces the reconstructed sequence shown in Table 4. The apparent reversal of trend in feldspar is believed to be the result of error variation and is not significant though it may reflect a reversal of sequence by thrust faulting.

The position of postulated thrust planes in a number of the remaining sections is shown in Figure 9. None of these sections contain additional proof of repetition of particular beds, but the existence of the trends in feldspar content is considered a sufficiently valid basis to interpret these sections in this manner. If the model proposed by Virkkala (1952) and Harrison (1957) is considered, it seems obvious that clear-cut sequences containing only a single shear plane should be the exception rather than the rule. The data themselves are by no means conclusive and could be subjected to alternate interpretations, but in view of the clear evidence of

TABLE 4

RECONSTRUCTION OF SECTION 71*

Sample	% Sand	% Clay	% Feldspar
71-72	38.5	15	15.0
73,76	39.0	19	11.5
74,77	36.5	21	12.5
75,78	31.5	22	14.5

*For location see note to Table 3.

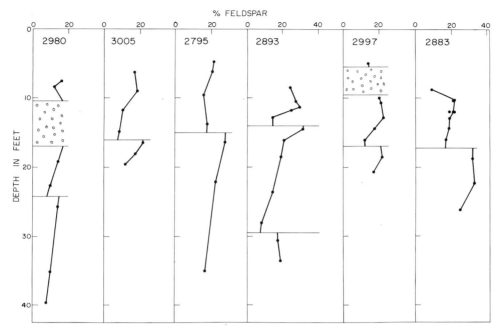

Figure 9. Plot of feldspar content of the Titusville Till (Youngstown, Ohio, area) against depth. Discontinuities in the trend of upward-increasing feldspar content resulting from the presence of thrust faults.

stacking shown in some of the data, the author feels justfiied in interpreting them in this manner.

Evidence for a number of thrust planes within a single till was observed in a section in the strip-mining area west of Danville, Illinois, in SW¼ SW¼ Section 33, T. 20 N., R. 12 W. Table 5 contains textural data from the

TABLE 5

COMPOSITION OF BLOOMINGTON-TYPE TILL, EMERALD POND SECTION*

Sample	% Sand	% Clay	Element
11	40	26	A
12	40	25	A
13	40	21	B
14	26	20	C
15	39	21	B†
16	29	20	C
17	39	20	B
18	39	21	B
19	41	20	B
20	40	21	B
21	27	20	C
22	40	17	B
23	38	19	B
24	26	19	C
25	37	19	B

*Located west of Danville, Ill. (in SW¼ SW¼, Sec. 33, T.20N, R.12W).
†Lines indicate position of thrust faults.

Bloomington-type till (Johnson and others, 1971) in this section, called the Emerald Pond Section. Examination of these data reveals a series of three distinct textural groupings within this till unit. These three groups have sand, silt, and clay percentages of approximately 40-35-25, 40-40-20, and 27-53-20, respectively. The position of the thrust faults necessary to produce the reconstructed sequence shown on Table 6 is indicated on Table 5. Two of these faults were clearly marked by lenticular masses of sand located along the fault plane. As was the case in the sections through the Titusville Till, this section is considerably thicker (18 feet) than are nearby sections of the same unit which exhibit no evidence of stacking (mean of 8 feet for three sections).

TABLE 6

RECONSTRUCTION SEQUENCE OF BLOOMINGTON-TYPE TILL,
EMERALD POND SECTION*

Element	% Sand	% Clay
A(2)	40.0	25.5
B(9)	39.2	19.9
C(4)	27.0	19.8

*For location see note to Table 5.

Mechanism of Formation

Transportational thrust faulting is believed simply to be the product of differential movement along discrete shear planes within the debris-laden basal zone of ice, such as was postulated by Harrison (1957) and by Virkkala (1952). It is generally agreed that the majority of the differential movement occurring within an ice sheet occurs near the base, and it is reasonable to assume that this movement occurs along discrete shear planes in this zone, because the large concentration of debris should tend to decrease other forms of movement within the ice.

Significance

Where vertical compositional changes occur in till sheets, transportational thrust faulting becomes quite significant. As described above, the studies of Gross (1967) and of Gross and Moran (1971) indicate that feldspar content progressively increases toward the top of tills in the Allegheny Plateau of Ohio and Pennsylvania. Gross (personal communication, 1969) found vertical gradations in clay content within single till sheets in northeast Illinois. Elson (1961) suggested that a single ice sheet should deposit four types of till, each of which differs from the others in texture and/or mineralogical composition, as well as in origin: superglacial ablation till, subglacial ablation till; deformation till, and comminution till. Although he indicated that it is not likely for all four types of till to occur in any single section, several of them may occur together. The four types of till should occur in the same order as listed above if they all occurred together. It is evident, in the cases of both of the above types of compositional change within a single till sheet, that the operation of the Harrison-Virkkala transportation-deposition model may produce complex patterns resembling normal sequences in several till sheets by faulting and stacking during one ice advance. In this case, as in the case of large-scale block inclusion, confusion and errors in interpretation can be prevented by a knowledge of the regional stratigraphic pattern and of the potential of the deformational mechanism.

SUMMARY

Glacial tectonic structures of three types occur generally throughout Pleistocene glacial sequences.

1. Simple *in situ* deformation involves small-scale folds and faults produced by ice push on upstream faces of protuberances and by shear exerted on the bed by an advancing glacier.
2. Large-scale block inclusion of bedrock and drift includes shearing of large masses of material up into the ice and transporting them more or less intact away from the area of *décollement*. This process involves a combination of the proper hydrogeologic and hydrologic conditions in the bed below the glacier to produce low shear strength and compressive flow in the basal zone of the glacier.
3. Transportational thrust faulting involves the deposition and re-erosion of slices of debris-laden ice in the basal zone of the glacier. The use of different shear planes during each re-erosion causes the original pattern of slices to be shuffled. As a result, slices which originally occurred at the base of the drift zone may occur anywhere throughout the till sheet and may overlie slices from the top of the original sequence.

Failure to recognize large-scale block inclusion and transportational thrust faulting can result in the creation of nonexistent local units which are actually the result of repetition of other beds in the sequence. It can also lead to very erroneous models of the thickness and make-up of glacial-drift sequences where older beds are encountered near the surface.

Structures of all three types are believed to be ubiquitous throughout the glaciated part of North America, but can be expected to be most prevalent where the ice flowed up over scarps, over confined aquifers which were continuous back under the ice, and at ice-marginal positions. End moraines, in particular, were the sites of extreme glaciotectonic activity. Many if not most of the moraine-like ridges throughout the midwest may have been formed by glaciotectonic activity and thus never represented, during a glaciation, ice-marginal positions.

ACKNOWLEDGMENTS

The ideas presented here have been formulated over the past four years from observations by the author in Ohio, Illinois, and Saskatchewan. S. H. Whitaker and W. A. Meneley of the Saskatchewan Research Council, Frank Patton of the University of Illinois, and J. A. Cherry of the University of Manitoba, who have been working on similar problems, have been instrumental in clarifying and modifying the ideas presented. Discussion with G. W. White, W. H. Johnson, J. P. Kempton, P. B. DuMontelle, Lee Clayton, E. A. Christiansen, D. L. Gross, and others have also been very helpful to the author.

REFERENCES

Bluemle, J. P., 1966, Ice thrust bedrock in northeast Cavalier County, North Dakota: Proc. North Dakota Acad. Sci., v. 20, p. 112-18 (North Dakota Geol. Survey Misc., Series, No. 33).

Brinkmann, R., 1953, Uber die diluvialen Storumgen aus Rugen: Geol. Rudsdrau, v. 41, Sanderband, p. 231-41.

Brown, T. C., 1933, The waning of the last ice sheet in central Massachusetts: Jour. Geology, v. 41, no. 2, p. 144-58.

Byers, A. R., 1960, Deformation of the Whitemud and Eastend Formations near Claybank, Saskatchewan: Royal Soc. Canada, Trans., v. 53, ser. 3, sec. 4, p. 1-16.

Christiansen, E. A., 1971, Tills in southern Saskatchewan, Canada, *in* this volume.

Dellwig, L. F., and Baldwin, A. D., 1965, Ice-push deformation in northeastern Kansas: Kansas Geol. Survey Bull. 175, part 2, 16 p.

Elson, J. A., 1961, The geology of tills: *in* Proc. 14th Canadian Soil Mechanics Conf. 13 and 14 October 1960, Nat. Res. Council, Canada, Assoc. Comm. on Soil and Snow Mechanics, Tech. Memorandum No. 69, p. 7-13.

Gross, D. L., 1967, Mineralogical gradations within Titusville Till and associated tills in northwestern Pennsylvania: M.Sc. thesis, University of Illinois, 77 p.

Gross, D. L., and Moran, S. R., 1971, Textural and Mineralogical gradations within tills of the Allegheny Plateau: *in* this volume.

Hanshaw, B. B., and Bredehoeft, J. D., 1968, On the maintenance of anomalous fluid pressures: II source layer at depth: Geol. Soc. America Bull., v. 79, no. 9, p. 1107-22.

Harrison, P. W., 1967, A clay-till fabric: its character and origin: Jour. Geology, v. 65, no. 3, p. 275-308.

Hopkins, O. B., 1923, Some structural features of the plains area of Alberta caused by Pleistocene glaciation: Geol. Soc. America Bull., v. 34, p. 419-30.

Johnson, W. H., Gross, D. L., and Moran, S. R., 1971, Till stratigraphy of the Danville region, east-central Illinois: in this volume.

Kupsch, W. O., 1962, Ice-thrust ridges in western Canada: Jour. Geology, v. 70, p. 582-94.

Lammerson, P. R., and Dellwig, L. F., 1957, Deformation by ice push of lithified sediments in south-central Iowa: Jour. Geology, v. 65, no. 5, p. 546-50.

Mathews, W. H., and MacKay, J. R., 1960, Deformation of soils by glacier ice and the influence of pore pressure and permafrost: Royal Soc. Canada Trans., v. 54, ser. 3, sec. 4, p. 27-36.

Moran, S. R., 1967, Stratigraphy of Titusville Till in the Youngstown region, eastern Ohio: M.Sc. thesis, University of Illinois, 73 p.

————, 1968, Stratigraphic divisions of the Titusville Till near Youngstown, Ohio (abs.): Geol. Soc. America, Abstracts for 1967, p. 393-94.

————, 1969, Geology of the Hudson Bay area, Saskatchewan: Ph.D. dissertation, University of Illinois, 194 p.

Nye, J. F., 1952, The mechanics of glacier flow: Jour. Glaciology, v. 2, p. 82-93.

Piskin, K., and Bergstrom, R. E., 1967, Glacial drift in Illinois: thickness and character: Illinois Geol. Surv. Circ. 416, 33 p.

Rutten, M. G., 1960, Ice-pushed ridges, permafrost and drainage: Amer. Jour. Sci., v. 258, no. 4, p. 293-97.

Sardeson, F. W., 1898, The so-called Cretaceous deposits in southeastern Minnesota: Jour. Geology, v. 6, p. 679-91.

————, 1905, A peculiar case of glacial erosion: Jour. Geology, no. 105, no. 4, p. 351-57.

————, 1906, The folding of subjacent strata by glacial action: Jour. Geology, v. 14, p. 226-32.

Slater, George, 1926, Glacial tectonics, as reflected in disturbed drift deposits: Geologists' Assoc. Proc., v. 37, 392-400 p.

————, 1927a, The structure of the disturbed deposits of the Hadleigh Road area, Ipswich: Geologists' Assoc. Proc., v. 38, p. 183-261.

————, 1927b, The structure of the disturbed deposits of Moens Klint, Denmark: Proc. Roy. Soc. Edinburgh, v. 60, part 2, no. 12, p. 289-302.

————, 1927c, The disturbed glacial deposits in the neighborhood of Lonstrip, near Horving, North Denmark: Proc. Roy. Soc. Edinburgh, v. 60, part 2, no. 13, p. 303-315.

————, 1927d, The structure of the disturbed deposits in the lower part of the Dipping Valley near Ipswich: Proc. Geol. Assoc., v. 28, p. 157-182.

————, 1927e, Structure of the Mud Buttes and Tit Hills in Alberta: Geol. Soc. America Bull. v. 38, p. 721-30.

————, 1929, The structure of the drumlins exposed on the south shore of Lake Ontario: New York State Mus. Bull. 281, p. 3-19.

Terzaghi, K., and Peck, R. B., 1967, Soil mechanics in engineering practice: New York, John Wiley and Sons, Inc., 729 p.

Virkkala, K., 1952, On the bed structure of till in eastern Finland: Komm. Geol. de Finlande, Bull. no. 57, p. 97-109.

Wickenden, R. T. D., 1945, Mesozoic stratigraphy of the eastern plains, Manitoba and Saskatchewan: Geol. Survey Canada Mem. 239, 87 p.

Wolford, J. J., 1932, A record size glacial erratic: Amer. Jour. Sci., v. 224, p. 362-367.

Thickness of Wisconsinan Tills in Grand River and Killbuck Lobes, Northeastern Ohio and Northwestern Pennsylvania

George W. White

ABSTRACT

The Woodfordian tills of the Grand River Lobe in northeastern Ohio and northwestern Pennsylvania (Hiram, Lavery, and Kent Tills) and the correlative tills of the adjacent Killbuck Lobe in Ohio (Hiram, Hayesville, and Navarre Tills), have median thicknesses of from four to six feet. These data are based on 568 measurements at localities where both top and bottom of a till sheet could be clearly identified. Zero thickness of the Navarre Till occurs at 30 of 90 measured sections where the Hayesville Till lies directly on Millbrook Till.

The Altonian, Titusville (Grand River Lobe), and Millbrook (Killbuck Lobe) Tills are much thicker. The Titusville Till, composed of from three to five units, ranges from 0 to 110 feet; its median thickness is 16 feet. The upper unit of the Titusville Till and its underlying sand have median thicknesses of nine and one feet, respectively.

It is of economic importance for pedologists, environmental and engineering geologists, and engineers, who advise on engineering structures built on or in tills, to recognize that individual till sheets, with different properties, are generally only a very few feet thick. The contact planes — interfaces — between these separate till sheets, with or without associated sand or silt, provide pathways for fluid flow.

INTRODUCTION

Laterally and vertically extensive exposures of glacial materials, mainly till, have been made over the past 20 years in superhighway construction and in greatly increased strip mining for coal and limestone in the Allegheny

Plateau. Many hundreds of exposures of the complete sequence of glacial deposits on the Pennsylvanian and Mississippian bedrock have been examined in detail. The deposits consist of a series of till sheets which can be identified and traced for scores of miles as rock-stratigraphic units.

The location and extent of the different tills at the surface are shown in Figure 1. An Illinoian ice sheet and the earlier Wisconsinan ice sheets generally advanced farther than did the later ones, so that the tills at the surface exist in a series of belts from one to 10 miles or more wide. Each till can be traced for tens of miles northward beneath the next younger till.

The stratigraphic classification of the tills is shown in Figure 2. The tills can be identified by characteristic color, hardness, texture, and mineralogy, by differences in weathering, and by the presence of paleosols between some of the tills. The tills have been named and the evidence for their continuity and separation described in detail (White, 1960, 1961, 1963, 1967, 1969; White, Totten, and Gross, 1969).

As more and more sections were measured, it was seen that not only were the surface tills usually only a few feet in thickness, but also that other till sheets in the subsurface were also thin and even missing in some places. Diagrams of such sections have now been published (White, 1963, 1967; White, Totten, and Gross, 1969). The lower Wisconsinan tills in both lobes — the Titusville Till of the Grand River Lobe and the Millbrook Till of the

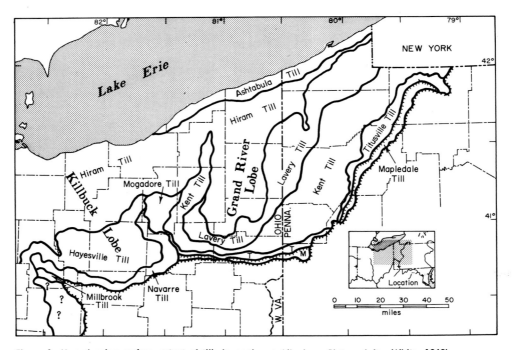

Figure 1. Map showing surface extent of tills in northwest Allegheny Plateau (after White, 1969).

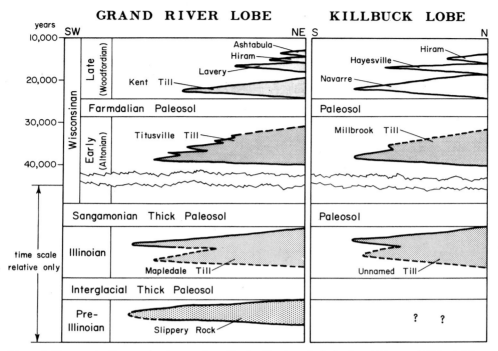

Figure 2. Time-space diagram showing stratigraphic classification of tills in Grand River and Killbuck lobes (after White, 1969). Thickness of each "wedge" indicates relative time duration, not till thickness.

Killbuck Lobe — were seen to be generally thicker than the overlying tills and to make up the bulk of the drift.

In order to make quantitative statements about the thicknesses of the tills, all measured sections that showed the top and bottom of one or more tills were reviewed and the measurements tabulated in the histograms, Figures 3, 4, 5, and 6, and summarized in Table 1.

TABLE 1
THICKNESS OF WISCONSINAN TILLS

	No. of Measurements	Range Ft.	Mean Ft.	Q₁ Ft.	Q₂ (Median) Ft.	Q₃ Ft.
Grand River Lobe						
Hiram	84	0- 25+	5.8	3	4	7
Lavery	37	0- 11+	5.0	3	4	6
Kent	131*	0- 25+	6.8	4	5	8
Titusville						
(Total)	73	0-110+	20.2	11	16	27
Titusville I	25	3- 16	9.2	6	9	12
Titusville I						
Sand below	25	0- 25	3.1	0	1	2½
Killbuck Lobe						
Hiram	41	0- 22	8.3	4	6	11
Hayesville	113	0- 20	5.2	4	5	6
Navarre	90†	0- 18	6.9	4	6	9
Millbrook	20	0- 54	12.6	5	8	15

*Only 116 finite thicknesses used for quartiles and mean.
†Only 60 finite thicknesses used for quartiles and mean.

Figures for thicknesses of pre-Wisconsinan tills are not presented. These tills, the Illinoian Mapledale Till and the pre-Illinoian Slippery Rock Till of the Grand River Lobe, are usually preserved beneath later tills only in depressions in the bedrock surface and are absent at most exposures. The pre-Wisconsinan tills of the Killbuck Lobe are less well-known because of fewer exposures to bedrock, as extensive strip mines are not common in this lobe.

GRAND RIVER LOBE

Hiram Till

The clay-rich Hiram Till (White, 1960, p. A-8) has a mean thickness of 5.8 feet and a median thickness of 4 feet (Fig. 3). At many places a five-foot soil auger will go all the way through it and penetrate the underlying unit. In some places the till is so thin that its entire thickness is incorporated in the soil and it cannot be identified with certainty. Within the area of surface outcrop of the Hiram Till are some tracts where that till is absent and the surface soils are not Mahoning or Ellsworth soils, derived from the Hiram Tills, but are Rittman-Wadsworth soils from the older Lavery Till, or even from Wooster-Canfield soils from the still older Kent Till, locally exposed (White, 1952). In this manner, soils maps can be used to help locate "windows" through the Hiram Till. The Defiance Moraine has classically been associated with the Hiram Till. However, at most places, the Hiram Till forms only a thin veneer on the surface of the Defiance Moraine, the bulk of which is made of older drift.

Locally the Hiram Till is thicker. At two places the Hiram Till is 19 and 25 feet thick, respectively, and it is more than 20 feet thick at several sites in the Ashtabula Moraine system. These greater thicknesses of the till sheet are interpreted as the result of thrust stacking, as discussed by Moran (1971; this volume, p. 127).

Lavery Till

The silty Lavery Till (Shepps and others, 1959) is now known to have a wider surface extent than formerly mapped (White, Totten, and Gross, 1969). In the outer five to ten miles of its surface extent, it is very thin and discontinuous. Where it is present at the surface or beneath the Hiram Till, it has a mean thickness of 5.0 feet and a median thickness of 4 feet (Fig. 3). The greatest thickness recorded where the base was exposed was 12 feet, but at one locality in Stark County, Ohio, 18 feet of Lavery Till was exposed. In northern Lake and Ashtabula Counties, Ohio, the Lavery Till is more than 20 feet in thickness. In these localities, where it is overlain by the

Hiram and Ashtabula Tills, it may form a considerable part of the bulk of the drift of the Ashtabula Moraine system (Lake Escarpment moraine system of Leverett, 1902, pl. 15, p. 651).

Kent Till

The sandy, yellow-brown-weathering Kent Till (White, 1960, p. A-5) has a wide surface extent in the Grand River Lobe and can be identified below later tills as far north as the shore of Lake Erie (White, 1968). The Kent Till has a mean thickness of 6.8 feet and a median thickness of 5 feet in those 116 sections where the entire thickness of the till is exposed (Fig. 3). At 15 exposures, Lavery Till lies directly on Titusville Till and Kent Till is absent (for examples, see White, Totten, and Gross, 1969, Fig. 8). At 20 exposures, this till has thicknesses of from nine to 13 feet, and single exposures showing thicknesses of 17, 18, 20, and 25 feet are recorded. These greater thicknesses are interpreted as the result of thrust stacking. It is

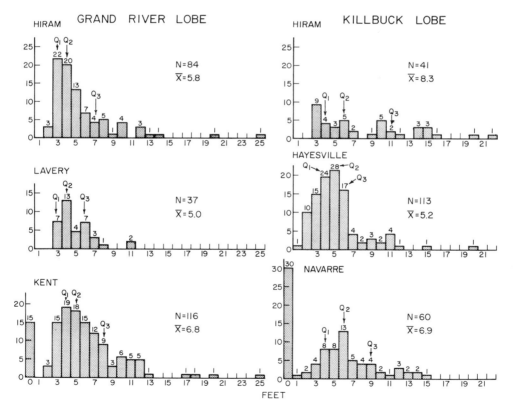

Figure 3. Histograms at one-foot increments showing thicknesses of Woodfordian ("late Wisconsinan") tills. Note that zero thickness (till absent) is shown only for Kent and Navarre Tills. Quartiles and means for these two tills are calculated with zero thicknesses excluded.

important to note that the Kent Till in the Kent Moraine (Shepps and others, 1959, map; White, 1963, map; also in Flint and others, 1959) is at most places no thicker than elsewhere, and thus it only forms a veneer that extends over the moraine and one to three miles beyond it!

Titusville Till

The sandy, olive-brown-weathering Titusville Till (White and Totten, 1965; White, Totten, and Gross, 1969) is the surface till in a narrow belt around the Grand River Lobe. North and west of the outcrop belt, the Titusville Till in the subsurface forms the bulk of the drift and can be traced almost to Lake Erie.

At many exposures the Titusville Till consists of three or more sheets separated by sand layers. Moran (1967) has shown that these separate till sheets have rather subtle variations in composition. The total thickness of the Titusville Till is markedly greater than that of the later tills. The histogram, Figure 4, has been constructed for five-foot increments, as the 72 measurements range from two to 50 feet (a single measurement of 110 feet, base not seen, is not included). The mean thickness of the Titusville Till is 20.2 feet and the median is 16 feet. In Figure 4, measurements of similar five-foot increments in the Kent Till are presented for comparison; these are also shown converted to percentages to make the comparison easier between the thin Kent Till, as a representative of the Woodfordian tills, and the thicker Titusville Till.

The uppermost Titusville till sheet, here called Titusville I, and the persistent sand underlying it, can be traced over a wide area. Its thickness was recorded at 25 localities; the average thickness is 9.2 feet and the median is 9 feet (Fig. 5).

The sand below Titusville I till ranges in thickness from two inches to 25 feet, but at only five localities was the thickness greater than two and one-half feet. At those localities where the sand is 10 feet, 17 feet, and 25 feet in thickness, the expanded sand masses are interpreted as buried kames, as illustrated in diagrams by White, Totten, and Gross (1969, Figs. 27, 35, and 36). The sand layer transmits water considerable distances. The presence of this persistent sand layer must be taken into account in strip mining, because it transmits water and induces slumping of pit walls. Similar slumping may also occur at this level in highway excavations.

KILLBUCK LOBE

Hiram Till

The Hiram Till of the Killbuck Lobe is continuous with that of the Grand

River Lobe, and has, in general, the same character, but with some lateral differences in texture (Shepps, 1953). In the Killbuck Lobe, the Hiram Till has a mean thickness of 8.3 feet and a median thickness of 6 feet. Thick-

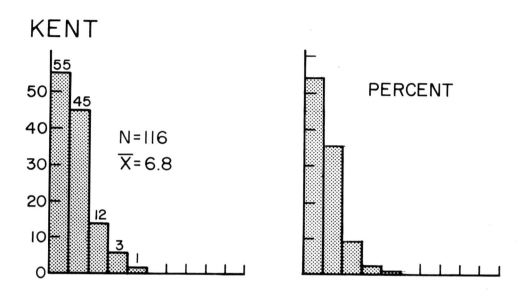

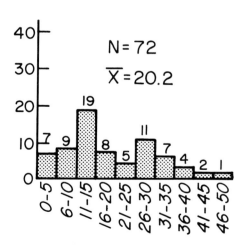
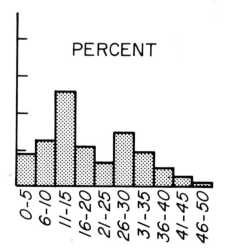

Figure 4. Histogram of Titusville Till thicknesses at five-foot increments; Kent Till at same increments shown for comparison.

nesses of the Hiram Till of more than 9 feet probably represent thrust stacking. The margin of the Hiram Till in Wayne County and in southern Medina County, Ohio, is at the Wabash Moraine. However, exposures in the moraine show that the Hiram Till is only a veneer, generally from two to five feet thick, and that the moraine is composed of older tills (White, 1967; Totten, 1969, Fig. 4). The two measurements of 21 and 25 feet were obtained in southeast-central Medina County in the Fort Wayne Moraine.

The Hiram Till is thicker than the median at proportionately more locations in the Killbuck Lobe than in those in the Grand River Lobe (Fig. 3). This is explained as follows: in the Grand River Lobe, the Hiram ice travelled about 50 miles from the margin of the main Erie Lobe, as southward flow was favored in the Grand River lowland. In the Killbuck Lobe, the extension from the main Erie Lobe was 20 miles or less, generally over a slowly rising surface, so that thrust stacking probably took place much more frequently.

Hayesville Till

The Hayesville Till has a large surface extent. Despite this fact, it is very thin; many exposures in shallow roadcuts and in auger borings reveal its whole thickness. The contact with the till below is especially evident, because of marked color and textural differences.

The mean thickness of the Hayesville Till is 5.2 feet and the median is 5 feet. At only two exposures was the till more than 12 feet thick; at one it was 15 and at another 20 feet thick. Even in the Wabash and Fort Wayne Moraines in northwestern Wayne and southwestern Medina Counties, the till is usually less than 10 feet thick. The till appears to be of greater thickness in the Wabash and Fort Wayne Moraines in central and eastern Medina County, but sections permitting clear measurements were not available.

The Hayesville Till, as is also true of the Lavery Till, its correlative in the Grand River Lobe, has the lowest mean thickness of any of the tills and also the lowest number of "abnormally" thick till exposures. Thus almost no thrust stacking of this till must have taken place, except possibly in the moraines in eastern Medina County.

Navarre Till

The Navarre Till (White, 1961) has only a narrow belt of surface outcrop, but it is present under cover over a large area. However, it is not always present in exposures where it should occur, the underlying Millbrook Till then forming the surface material.

In the 60 exposures where the Navarre Till is present, it has a mean thick-

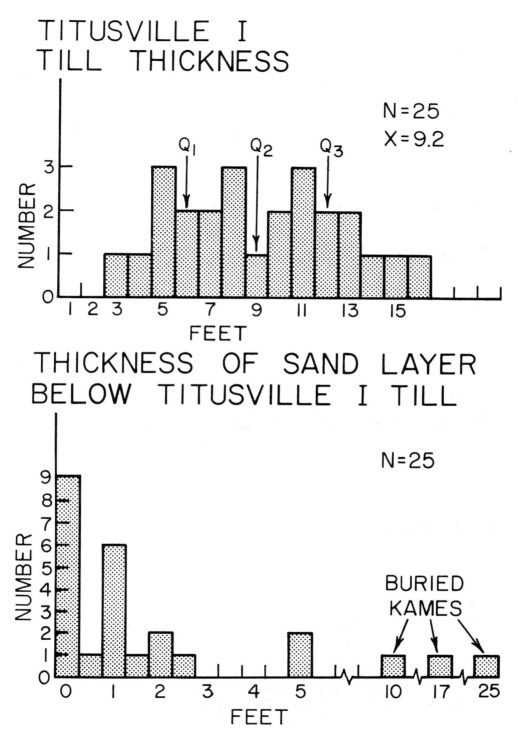

Figure 5. Histogram of thicknesses of Titusville I Till at one-foot increments and of thicknesses of sand layer below the till at one-half-foot increments.

ness of 6.9 feet and a median thickness of 6 feet (Fig. 3). The 30 exposures in which the Navarre Till is not present are all ones in which the Hayesville Till lies directly on the Millbrook Till, so that the absence of the Navarré Till can be clearly seen. The nine localities where the Navarre Till is from 11- to 15-feet thick probably represent thrust stacking.

Millbrook Till

The sandy, olive-brown-weathering Millbrook Till (White, 1961) in the Killbuck Lobe is the correlative of the Titusville Till of the Grand River

MILLBROOK — "greater than"

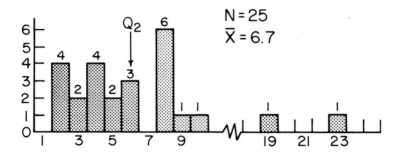

MILLBROOK

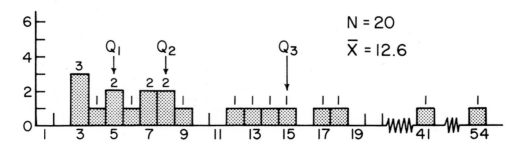

Figure 6. Lower: Histogram of thicknesses of Millbrook Till at one-foot increments. Upper: Thicknesses of Millbrook Till where base not exposed, so that only "greater than" thicknesses are shown; total thicknesses not known.

Lobe (White and Totten, 1965). It is exposed at the surface only in a very small area in southern Richland County (Totten, 1962). However, wherever there are sections where the Navarre Till should be exposed at the surface but is missing, the Millbrook Till is present. The Millbrook Till forms the bulk of the drift south of the northern boundaries of Wayne and Ashland Counties and probably also of the drift north of that line. It is present in more than one sheet (White, 1967, Fig. 5), but the uppermost sheet is not as obvious as is the uppermost Titusville (Titusville I) sheet.

Because of the absence of strip mines in most of this lobe and the location of superhighway construction only well north in the lobe at the time the field studies were made, not nearly as many exposures showed the base of the Millbrook Till as showed the base of the Titusville Till in the Grand River lobe. Of the 20 exposures revealing the complete thickness of the Millbrook Till, the mean thickness is 12.6 feet and the median 8 feet. Thrust stacking is believed to account for the thicknesses of 41 and 54 feet recorded for the Millbrook Till in Figure 6.

At 25 localities where the Millbrook till was exposed, the base was not seen. These thicknesses, which are entered on Figure 6 as finite data, are actually "greater than" values.

Mogadore Till

The Mogadore Till (White, 1960) is correlative with the Millbrook Till of the Killbuck Lobe and the Titusville Till of the Grand River Lobe (White, Totten, and Gross, 1969). However, it is slightly less weathered than are the Titusville and Millbrook Tills (White, 1967, Fig. 3) in their areas of surface outcrop, and it probably belongs to one of the later (upper) sheets of the Altonian sequence. No data on its thickness are presented here; till at this horizon to the east is included under Titusville and to the west under Millbrook.

ENVIRONMENTAL IMPLICATIONS

Conceptions of anatomy of glacial deposits in the Allegheny Plateau, especially of till deposits, have ranged from the naive assumption that "till is till" and that the till of the whole deposit is like that at the surface, through the assumption that the exposures where two or more tills may be seen are unusual and that thin tills are the exception, to the realization that till thickness of any of the upper till sheets (Woodfordian) of more than 5 or 6 feet is abnormal. The data presented here show that:

1. In almost any excavation in the glaciated northwestern Allegheny Plateau, a till different from the surface till will be encountered, and in an excavation of 15 feet or more, several till sheets of different ages are to be expected.

2. Upon reaching the Altonian till — Titusville or Millbrook — which will generally (but by no means always) be the lowest till lying on the bedrock, it cannot be assumed that this till is in a single sheet. It is almost certain to be made up of two or more sheets.

3. The tills vary in texture, composition, compactness, permeability, and in joint spacing. The detailed investigation of these properties for engineering purposes is a promising and indeed necessary field for immediate investigation. It is certain that tills vary in ease of excavation, load-bearing capacity, slope stability, and other properties (White, 1969).

4. The interfaces separating the till sheets may be quite regular or may be somewhat irregular. The till sheets may be separated by a sand layer or a silt layer of varying thickness. As shown in many of the diagrams by White, Totten, and Gross (1969), unweathered till may lie upon weathered till, or a paleosol, or another unweathered till (the most difficult contact to identify).

5. The interfaces between units are important for lateral movement of fluids. The movement through sand layers is evident and recognized, but fluid movement also takes place along interfaces between tills where no sand is present, as shown by "weeping" along vertical walls of strip mines and other excavations. These extensive discontinuities must be taken into account in permeability studies for water supply. They are especially important in waste-disposal studies, as fluid wastes may travel long distances along interfaces.

6. Some tills are much more jointed than others. Frequently a lower till may be the most jointed one in the sequence. Fluids can therefore travel laterally along an interface and then through joints in a till sheet to another interface, either above or below the first one.

7. While these studies and conclusions are based on gross and fine till anatomy in the Allegheny Plateau, it is probable that they apply to other regions as well. The designer of engineering structures in, on, or through a till mass several tens or scores of feet thick, who assumes this material to be a single homogeneous thick deposit, proceeds at his peril.

SUMMARY AND CONCLUSIONS

The Woodfordian tills of the Grand River Lobe — Hiram, Lavery, and Kent — and of the Killbuck Lobe — Hiram, Hayesville, and Navarre — have a consistent pattern of thickness. Their mean thickness ranges from five to eight feet and their medians from four to six feet.

At much fewer than one-fourth of the exposures (except for the Hiram Till of the Killbuck Lobe) is any Woodfordian till two or more times thicker than the median. These "abnormally" thick tills are believed to be thicker because of thrust stacking (Moran, 1971; this volume, p. 127).

The median thickness of the Hiram Till is about five feet in the two lobes. The greater number of thicker Hiram Till locations in the Killbuck Lobe is explained by more frequent stacking in that lobe.

Confirmation of complete absence of a till sheet, where that till is the surface material in the general area, is difficult, because a till sheet only one or two feet thick may be incorporated so completely in the present soil that its identification is uncertain (for examples see Winslow and White, 1966, Figs. 8 and 30). At 30 of 90 exposures, where the Hayesville Till was at the surface, the Navarre Till was absent and the Hayesville Till was in direct contact with the Millbrook Till.

The Altonian Tills — Titusville of the Grand River Lobe and Millbrook of the Killbuck Lobe — present a thickness pattern quite different from that of the Woodfordian Tills (Fig. 4). Not only is the median thickness of the total Titusville Till more than three times that of any Grand-River-Lobe Woodfordian Till, but the median thickness of just the uppermost Titusville Till sheet, Titusville I, is about twice that of any Woodfordian Till.

The median thickness of the Millbrook Till of the Killbuck Lobe is distinctly greater than that of any of the Woodfordian Tills, although the difference is not so marked as that between the Titusville Till and the younger tills in the Grand River Lobe. In both lobes, the bulk of the till is Altonian in age.

To a certain extent, the greater thickness of the Titusville Till can be explained as due to the presence of several sheets of similar tills. However, the uppermost sheet, "Titusville I," which is demonstrably a single sheet, is generally almost twice as thick as any Woodfordian Till. This invites the speculation that the glaciological conditions in the Allegheny Plateau in Altonian time were different from those in Woodfordian time. What these differences may have been are as yet unknown.

The realization that many tills are only a few feet in thickness is important for engineering and environmental studies and design. Deep excavations will probably encounter several till sheets having different properties and will intersect the possible fluid-transmitting interfaces between them.

ACKNOWLEDGMENTS

I am grateful to my associates, Dr. David L. Gross and Dr. Stephen R. Moran, for commenting on this study at various stages and for their encouraging me to present these data from our records. The penultimate copy of the manuscript has been much improved by the comments of Professor W. Hilton Johnson and Dr. Gross. A part of the field work for this study was supported by NSF Grant 2675.

REFERENCES

Flint, R. F., and others, 1959, Glacial map of the United States east of the Rocky Mountains: Geol. Soc. America.

Leverett, Frank, 1902, Glacial formations and drainage features of the Erie and Ohio basins: U. S. Geol. Survey, Mon. 41, 802 p.

Moran, S. R., 1967, Stratigraphy of Titusville Till in the Youngstown region, eastern Ohio: M. S. thesis, Univ. of Illinois, 73 p.

————, 1971, Glacio-tectonic structures in till, *in* this volume.

Shepps, V. C., 1953, Correlation of tills of northeastern Ohio by size analysis: Jour. Sed. Petrology v. 23, p. 34-48.

Shepps, V. C., White, G. W., Droste, J. B., and Sitler, R. F., 1959, Glacial geology of northwestern Pennsylvania: Pa. Geol. Survey Bull. G 32, 54 p.

Totten, S. M., 1962, Glacial geology of Richland County, Ohio: Ph.D. thesis, Univ. of Illinois, 136 p.

————, 1969, Overridden recessional moraines of north-central Ohio: Geol. Soc. America Bull. 80, p. 1931-46.

White, G. W., 1952, Discontinuities in till sheets: Geol. Soc. America Bull., v. 63, p. 1312.

————, 1960, Classification of Wisconsin glacial deposits in northeastern Ohio: U. S. Geol. Survey Bull. 1121-A, p. A1-A12.

————, 1961, Classification of glacial deposits in the Killbuck lobe, northeast-central Ohio: U. S. Geol. Survey Prof. Paper 424 C, p. 96-8.

————, 1963, Glacial geology of Stark County, Ohio: Ohio Geol. Survey Bull. 61, p. 118-56.

————, 1967, Glacial geology of Wayne County, Ohio: Ohio Div. Geol. Survey. Rep. Invest. 62, 39 p.

————, 1968, Age and correlation of Pleistocene deposits at Garfield Heights (Cleveland), Ohio: Geol. Soc. America Bull., v. 79, p. 749-52.

————, 1969, Pleistocene deposits of the northwestern Allegheny Plateau, U.S.A.: Quart. Jour. Geol. Soc. London, v. 124, pt. 2, p. 131-51.

————, 1969, Anatomy of till deposits—engineering implications: Geol. Soc. America Abstracts with Programs for 1969, Part 7, p. 235-236.

White, G. W., and Totten, S. M., 1965, Wisconsinan age of the Titusville Till (formerly called "Inner Illinoian"), northwest Pennsylvania: Science, v. 148, p. 234-35.

White, G. W., Totten, S. M., and Gross, D. L., 1969, Pleistocene stratigraphy of northwestern Pennsylvania: Pennsylvania Geol. Survey, 4th ser., Bull., G 55, 88 p.

Winslow, J. D., and White, G. W., 1966, Geology and ground-water resources of Portage County, Ohio: U. S. Geol. Survey Prof. Paper 511, 80 p.

4

Stratigraphic Correlations

Tills in Southern Saskatchewan, Canada

E. A. Christiansen

ABSTRACT

Four tills ranging in thickness from one to 250 feet have been recognized and traced in cored and electric-logged testholes for more than 200 miles in southern Saskatchewan. The two older tills (presently unnamed) have been assigned to the Sutherland Group, whereas the younger tills (Floral and Battleford Formations) have been assigned to the Saskatoon Group. The Groups are separated and correlated primarily on the basis of carbonate content of their till units, the tills of the Saskatoon Group having higher carbonate contents than do the tills of the Sutherland Group. Indicators of weathering include oxidation and translocation of gypsum, iron and manganese oxides, and carbonates. The Battleford Formation (youngest till) is Woodfordian and was deposited 12,000 to 20,000 years ago. The increase in carbonate and the decrease in clay in the younger tills is attributed to progressive stripping of Upper Cretaceous clay beds, which progressively decreased the source area for clayey sediments and progressively increased the area of Paleozoic carbonates available for glacial erosion.

INTRODUCTION

Several tills have been recognized in numerous sections, and in about 1,000 testholes which have been drilled in southern Saskatchewan during the last eight years. The youngest till can be traced readily in exposures over much of southern Saskatchewan, whereas the older tills can be traced, for the most part, only by test drilling. Tills have been differentiated and traced by this method for hundreds of miles in southern Saskatchewan, which comprises an area of about 100,000 miles (Fig. 1). The southern boundary, which is the International Boundary between Canada and the

United States, is located about fifty miles north of the terminus of glaciation. The location-numbering system is shown in Figure 2.

The purpose of this paper is to describe the test drilling and coring methods by which the subsurface data were obtained; to show the mineralogy, texture, age, and source of the tills in southern Saskatchewan; and to describe some of the factors which complicate the correlation of these tills.

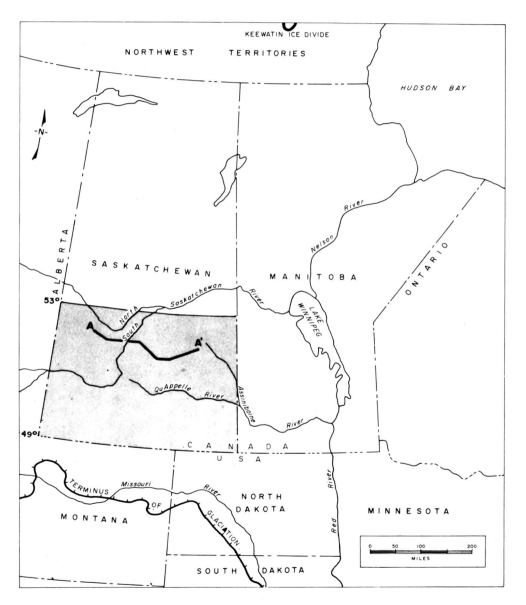

Figure 1. Location of study area and index for cross section.

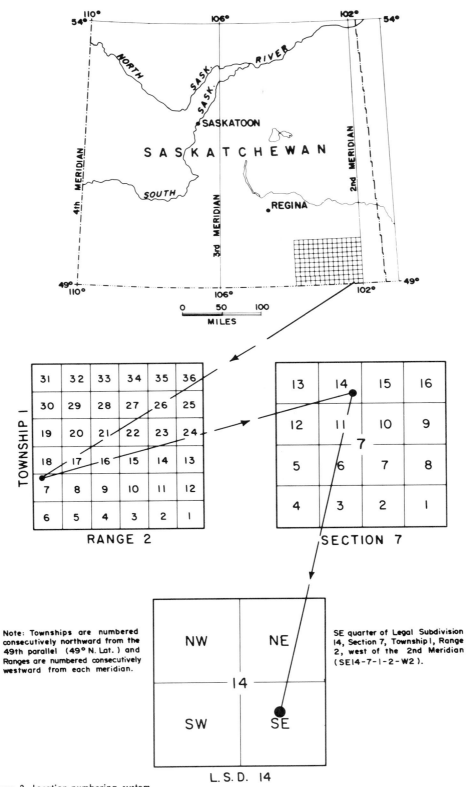

Figure 2. Location-numbering system.

DATA

Field Data

Subsurface data used in this paper were obtained from surface exposures and from cored, rotary testholes. Only one surface exposure was sufficiently complete to be used in the cross-sections, and it required a testhole to establish the base of drift.

The equipment used in the test-drilling program included: (1) a rotary drill capable of drilling 1,000 feet, (2) an electric logger capable of measuring spontaneous potential and resistance (single-point electrode), (3) a geolograph, and (4) a conductivity bridge. Special equipment designed for the program included a sidewall sampler and a mobile field office. Supervisors and drilling and sampling crews were required to live on the drilling site in trailers. Drilling and electric logging were done by private contractors on a footage basis for drilling and electric logging, and on a mileage basis for moving. The sidewall-coring equipment was owned and operated by the Saskatchewan Research Council.

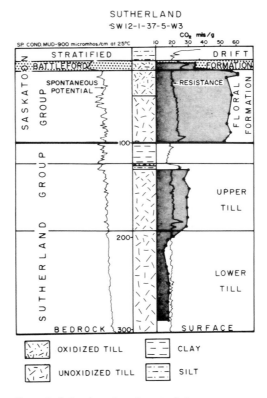

Figure 3. Subsurface data from testhole.

Drilling specifications called for equipment capable of drilling to a depth of 1,000 feet, which is about the maximum thickness of drift encountered in Saskatchewan. Drag bits were used in till, clay, silt, and sand; and rock bits were used in rocks and gravel. Cutting samples were taken by the driller at 10-foot intervals from the circulating drilling fluid and placed in pans with the appropriate depth designation. The samples were dried and described immediately by the geologist. During the course of drilling, exact depths were obtained from the geolograph, on which the history of the testhole was recorded by the driller. When the testhole reached the desired depth, as determined by the geologist, the drilling pipe was removed from the hole and two electric logs were taken.

The electric log (Fig. 3) has two curves, which measure spontaneous potential (SP) and resistance (R). The SP is a measure of the difference in electrical potential between the drilling fluid and the formation water (Dobrin, 1952, p. 378-81), whereas the single-electrode resistance log is a measure of the relative electrical resistance of the strata in the vicinity of the electrode in the testhole (Dobrin, 1952, p. 372-74). If the SP deflects to the left of the till, silt, or clay base, the formation water is more saline than the drilling fluid, and conversely, if it deflects to the right of the base line, the formation water is less saline than the drilling fluid. The conductivity of the drilling fluid is measured by the geologist and recorded on the geological log (Fig. 3). To ensure a good electric log, the water used for drilling must be less mineralized than the formation water. In southern Saskatchewan, drilling water which has a conductivity of less than 1,500 micromhos/cm is specified to ensure a good electric log.

Based on the cutting-sample log, the driller's log, and the electric log, sample depths are chosen for sidewall coring and marked on the electric log to be used by the sidewall-sampling operators. A second electric log is used for geological compilation after the sidewall cores have been described.

The sidewall-coring equipment was designed and built by the Saskatchewan Research Council (Morrison, 1966, 1968). This equipment is capable of taking a one-inch diameter core, two to six inches long, at any desired depth from the wall of the testhole. With this equipment, coring depths can be chosen after the hole has been drilled and after geologic, driller's, and electric logs are available. Sidewall cores are large enough for mineralogical and paleontological analyses, and if carbonaceous material is encountered during drilling, numerous samples can be taken in order to procure sufficient samples for radiocarbon analysis. The number of cores collected within a specific interval varies, depending on the objectives and the logistics of the drilling program. Sidewall sampling is done by a separate crew subsequent to drilling, and proceeds while the drilling rig is being moved to the next site and drilling is being started there. This drilling and coring procedure

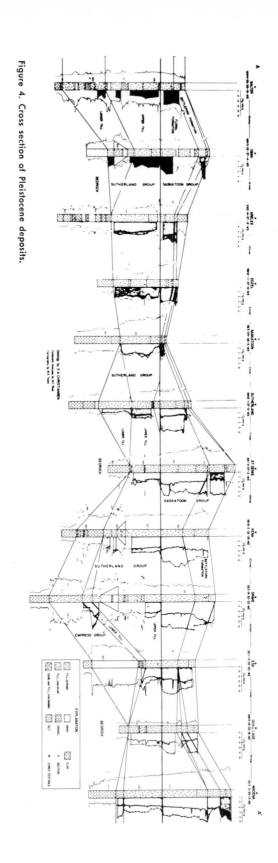

Figure 4. Cross section of Pleistocene deposits.

permits the execution of rapid, cheap, regional subsurface geological investigations at a rate of 2,000 to 3,000 feet per week (5 days, 18 hours/day), a footage which normally represents about 5 to 8 testholes at an average spacing of about 30 miles.

Laboratory Data

The till matrix (that passing a 2.0-mm sieve) was subjected to carbonate, textural, and X-ray diffraction analyses. The carbonate content was recorded as the CO_2 (STP) evolved from 1.0 gram of oven-dried till treated with 20 cc of 6N HCl. The volume of CO_2 was measured by the Chittick apparatus, equipped with mechanical stirrer. These data were plotted against depth and recorded on the geologic log (Fig. 3).

Texture was determined by the pipette method. The silt and clay fraction (<0.05 mm), obtained from the settling column after the textural analysis was completed, was then analyzed by X-ray diffraction.

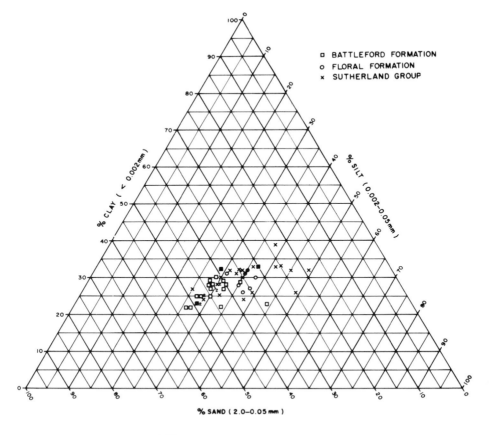

Figure 5. Trilinear plot of texture of tills.

STRATIGRAPHY

Nomenclature

The tills in southern Saskatchewan have been differentiated and correlated on the basis of carbonate content, weathering zones, intertill deposits, firmness, jointing, and electric-log characteristics. The till units have been assigned the stratigraphic rank of "Group" and "Formation" (Christiansen, 1868a, Fig. 3). The lower and upper tills of the Sutherland Group and the Floral and Battleford Formations of the Saskatoon Group have been traced for more than 200 miles, as shown in Figure 4.

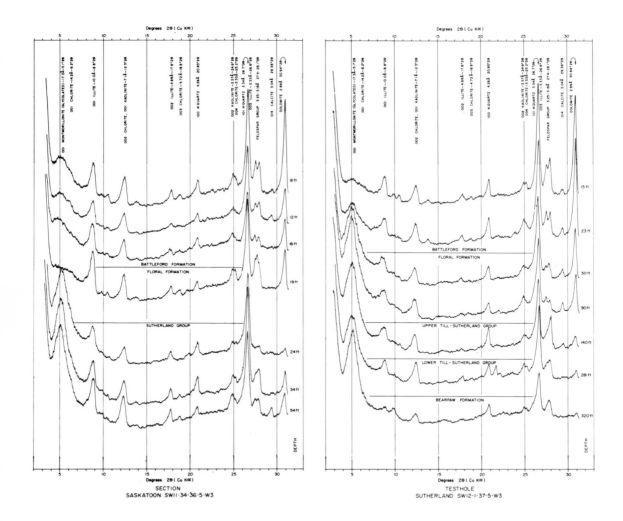

Figure 6. X-ray diffraction patterns of tills.

Lithology

The tills range in thickness from one foot or less (Christiansen, 1968b) to about 250 feet, but are commonly 20 to 100 feet thick (Fig. 4). The tills of the Saskatoon Group contain more carbonate than do the tills of the Sutherland Group. The lower till in the Sutherland Group was also differentiated from the upper till in this group on the basis of carbonate content (Fig. 3).

The tills of the Saskatoon Group are more sandy than those of the Sutherland Group (Fig. 5). The dominant clay minerals of the tills, as shown by X-ray diffraction patterns, are montmorillonite, illite, and kaolinite (Fig. 6), and the 17A° peak on glycolated samples becomes broader and lower as the sand content increases.

The tills of the Sutherland Group are less resistive, electrically, than are the tills of the Saskatoon Group (Figs. 3 and 4). In addition, the Battleford Formation is commonly soft, friable, and massive, whereas the older tills are commonly much firmer, more indurated, and jointed.

WEATHERING OF TILLS

Oxidation of iron and manganese and the translocation of gypsum, iron and manganese oxides, and carbonates are the main results of weathering of tills in southern Saskatchewan. Oxidation is indicated by colors ranging from pale yellow (5Y7/3, dry, Munsell Color Chart) at the top to olive gray (5Y5/2, dry) at the base of the weathered zone. Most buried weathered zones are overlain and underlain by gray, unoxidized till, the upper contact being distinct and the lower contact gradational. Oxidation is the most prevalent indication of weathering and occurs in all weathered zones.

Translocation of gypsum is common in the weathered top of the upper till of the Sutherland Group (Fig. 4, Saskatoon Section) and is less common in the top of the weathered Floral Formation. This translocated gypsum is precipitated in joints to form veinlets of fibrous selenite. Translocation of iron and manganese oxides is indicated by the accumulation of these oxides on joint surfaces both within the zone of oxidation and extending downward into the underlying, unoxidized till. This accumulation of oxides is restricted to tills which pre-date the Battleford Formation and is one of the main criteria for separating it from the older tills.

The translocation of carbonates is restricted to the top of the Sutherland Group. Partial leaching of carbonates to a depth of 40 feet in this weathered zone has been observed in the Saskatoon area (Fig. 7, log 1). Translocation of gypsum and of iron and manganese oxides also occurs in this weathered zone, which is the most highly weathered unit in southern Saskatchewan.

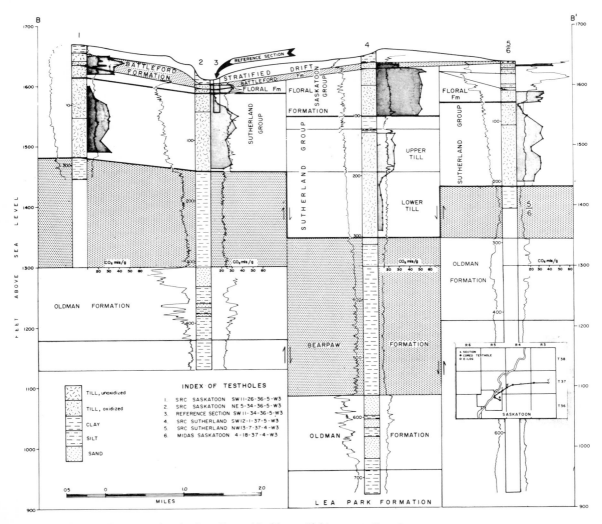

Figure 7. Geologic cross section showing effects of faulting on Pleistocene stratigraphy.

AGES OF TILLS

The ages of tills in southern Saskatchewan and their correlation with tills in Ontario and Illinois are shown in Figure 8. The difference in carbonate content between the lower and upper tills of the Sutherland Group (Figs. 3, 4, 7) suggests that these tills were deposited during different glaciations. The significance of the hiatus between them is precluded, however, because of the absence of a weathered zone in the top of the lower till. In contrast, the great depth of partial leaching of carbonates (Fig. 7, log 1) and the translocation of gypsum and of iron and manganese oxides in the weathered

TABLE 1

RADIOCARBON DATES

	Lab. No.	Location	Stratigraphic Position	Radiocarbon Material	Type of Carbon	Year Dated	Date (Years B.P.)
1)	S-176*‡	SW 2-25-23-25-W3	120 ft. below surface under 2 tills	A-horizon	organic	1963	20,000 ± 850
2)	S-228§	SW 4-20-45-27-W3	Weathering zone under Battleford Formation	carbonaceous silt	organic	1965	21,000 ± 800
2)	S-228A§	SW 4-20-45-27-W3	Weathering zone under Battleford Formation	carbonaceous silt	inorganic	1967	18,000 ± 450
2)	S-228B§	SW 4-20-45-27-W3	Weathering zone under Battleford Formation	carbonaceous silt	organic	1967	19,200 ± 400
2)	S-252§	SE 1-15-42A-1-W3	Organic deposit under 60 ft. of Battleford Formation	carbonaceous silt	organic	1965	33,500 ± 2000
2)	S-267§	SE 3-26-44-27-W3	Weathering zone under Battleford Formation	carbonaceous silt	organic	1965	33,000 ± 2000
2)	S-300A§	NW 12-34-40-28-W3	Surficial, glacial lake silt and clay	carbonaceous silt and clay	inorganic	1967	14,670 ± 240
2)	S-330B§	NW 12-34-40-28-W3	Surficial, glacial lake silt and clay	carbonaceous silt and clay	organic	1967	15,850 ± 225
3)	GSC-1041†‖	SW 11-24-29-3-W3	Between Floral and Battleford Formations	wood	organic	1968	38,000 ± 560

*S   = University of Saskatchewan Radiocarbon Laboratory
†GSC = Geological Survey of Canada Radiocarbon Laboratory
‡McCallum and Wittenberg (1965)
§McCallum and Wittenberg (1968)
‖Blake, Jr. (personal communication)

zone in the top of the Sutherland Group represents the most intensive weathering recorded in southern Saskatchewan.

Radiocarbon dates from carbonaceous material between the Floral and Battleford Formations (Table 1; Fig. 8) suggest that a weathering interval existed from 20,000 to more than 38,000 years ago. During this interval, the upper part of the Floral Formation was oxidized and the gypsum and the iron and manganese oxides were translocated. Translocation of carbonates has not been observed in this weathered zone. Radiocarbon dates (Table 1; Fig. 8) also suggest that the glacier which deposited the Battleford Formation advanced across southern Saskatchewan about 20,000 years ago and was actively retreating from the area about 15,000 years ago.

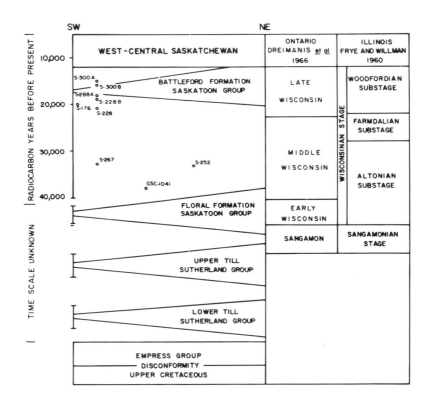

Figure 8. Diagrammatic presentation of Pleistocene chronology.

SOURCE OF TILLS

That the carbonate content increased in the younger tills was recognized by Meneley (1964). He concluded that the change in lithology reflected a change in a single source area as a result of stripping, rather than a change of the source area as a result of a change in glacial movement. He found the hypothesis of a change in glacial movement untenable, because it would require a major shift in the position of the center of the continental ice cap. The fact that the terminus of each glaciation coincides approximately suggests that the center of each ice cap was in approximately the same location, assuming that the ice caps were generally circular.

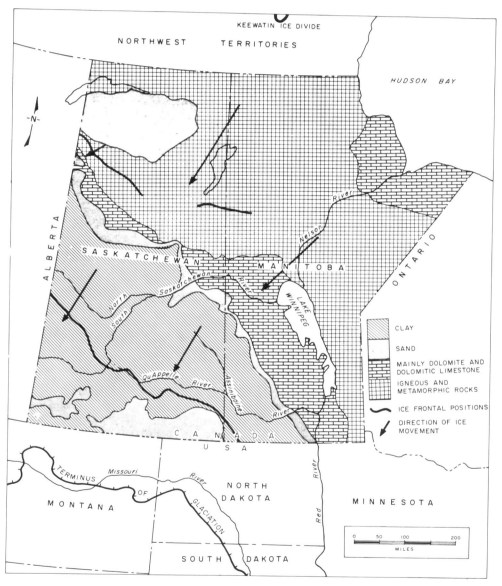

Figure 9. Source of tills.

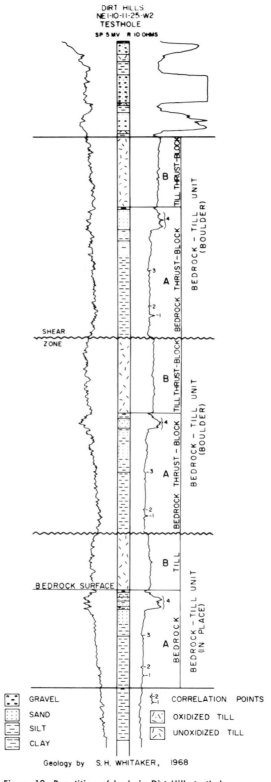

Figure 10. Repetition of beds in Dirt Hills testhole.

Texture and carbonate analyses (Figs. 3, 4, 5, and 7) indicate that the tills of the Sutherland Group have more clay and less carbonate than do the tills of the Saskatoon Group. If one assumes that the Upper Cretaceous clays were more extensive to the northeast (Fig. 9) prior to glaciation, then less carbonate and more clay would be available to the glaciers that deposited the tills of the Sutherland Group. As the older glaciers removed the Cretaceous clays, less clay and more carbonate would have been available to the glaciers that deposited the tills of the Saskatoon Group.

COMPLICATING FACTORS IN TILL CORRELATION

Repetition and omission of beds due to glacial thrusting, the offsetting and preservation of beds by gravity faulting, and major changes of till thickness all occur in the till sequence of southern Saskatchewan. These both confuse identification of individual tills and complicate their correlation.

Glacial thrusting in the zone of compressive flow results in the removal and repetition of beds. A till-bedrock sequence, for example, is repeated three times in the Dirt Hills testhole (Fig. 10) which is located in the Dirt Hills thrust moraine, as described by Byers (1959). Thus, in some areas the stratigraphic sequence is incomplete, whereas in other areas, the sequence is repeated by glacial thrusting. In southern Saskatchewan, the normal stratigraphic sequence is found in the essentially level, gently undulating plains area, where extending flow was presumably operative throughout Pleistocene time.

Gravity faulting (collapse) as a result of removal of salt by groundwater from the Devonian Prairie Evaporite Formation, which occurs several thousand feet below the surface (Christiansen, 1967), has resulted in the local preservation of tills in limited down-dropped areas, in offsetting of marker beds, and in rapid changes in till thickness (Fig. 7). The lower till of the Sutherland Group is preserved only in the graben area, and the Floral Formation is much thicker in the area of this structure. Because of the effect of this collapse structure on Pleistocene stratigraphy, it has been necessary in many places to drill into bedrock to establish structural markers.

Rapid thickening of tills occurs in moraines, in downfaulted areas, and in the fillings of buried valleys. The Battleford Formation changes in thickness from a few feet, north of Sutherland, to 250 feet in the Strawberry Hills Moraine (Fig. 11). Closely spaced changes in till thickness also occur in down-faulted areas (Fig. 7) and in buried valleys, particularly preglacial valleys.

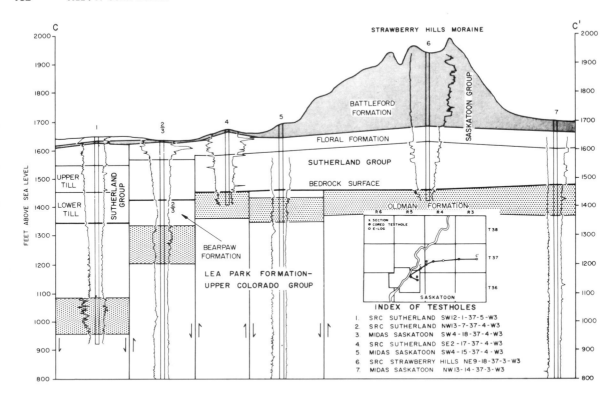

Figure 11. Thickness of Battleford Formation.

ACKNOWLEDGMENTS

Test-drilling programs were financed by the Agricultural and Rural Development Administration (ARDA) and the Saskatchewan Research Council as part of the regional hydrogeology investigation of southern Saskatchewan, and by the National Research Council of Canada in their support of the study of the Physical Environment of Saskatoon. Testholes were drilled by Elk Point Drilling Limited, Edmonton. Laboratory analyses were done by Mr. W. C. Ross, and drafting was done by Mr. W. E. Taylor. Unpublished radiocarbon dates were released by Dr. K. J. McCallum, of the Department of Chemistry and Chemical Engineering, University of Saskatchewan. Discussions with Drs. W. A. Meneley and S. H. Whitaker greatly facilitated this investigation. They also critically read the manuscript.

REFERENCES

Byers, A. R., 1959, Deformation of the Whitemud and Eastend Formations near Claybank, Saskatchewan: Royal Soc. Canada Trans., 3d ser., v. 53, sec. 4, p. 1-11.

Christiansen, E. A., 1967, Collapse structures near Saskatoon, Saskatchewan, Canada: Can. Jour. Earth Sciences, v. 4, p. 757-67.

————, 1968a, Pleistocene stratigraphy of the Saskatoon area, Saskatchewan, Canada: Can. Jour. Earth Sciences, v. 5, p. 1167-73.

————, 1968b, A thin till in west-central Saskatchewan, Canada: Can. Jour. Earth Sciences, v. 5, p. 329-36.

Dobrin, M. B., 1952, Introduction to geophysical prospecting: New York, McGraw-Hill Book Co., Inc., 435 p.

Dreimanis, A., Terasmae, J., and McKenzie, G. D., 1966, The Port Talbot Interstade of the Wisconsin glaciation: Can. Jour. Earth Sciences, v. 3, p. 305-25.

Frye, J. C., and Willman, H. B., 1960, Classification of the Wisconsinan Stage in the Lake Michigan Lobe: Ill. State Geol. Survey Circ. 285, 16 p.

Meneley, W. A., 1964, Geology of the Melfort area (73-A), Saskatchewan: Ph.D. dissertation, Univ. of Illinois (Urbana).

Morrison, B., 1966, A side-wall sampler: Sask. Research Council, Physics Div. Rept. P-66-10, 20 p.

————, 1968, Construction details of the Saskatchewan Research Council side-wall sampler: Sask. Research Council, Physics Div. Rept. P-68-2, 13 p.

Till Stratigraphy of the
Danville Region, East-Central Illinois

**W. Hilton Johnson, David L. Gross,
and Stephen R. Moran**

ABSTRACT

Six tills in the Danville, Illinois, area are established as formal rock stratigraphic units of member rank and two others are correlated to previously
named tills. Three tills of Kansan age are established as the Harmattan, Hillery, and Tilton Till Members of the Banner Formation. Two Illinoian tills,
the Smithboro and Vandalia Till Members of the Glasford Formation, are
identified, and three Wisconsinan tills are established as the Glenburn,
Batestown, and Snider Till Members of the Wedron Formation. The Petersburg Silt and the Mulberry Grove Member of the Glasford Formation, both
of Illinoian age, and the Wisconsinan Richland Loess also occur in the area.
The Yarmouth Soil is preserved locally, where it is developed in the Tilton
Till Member. Correlation between sections in the area and regional correlations with tills of known stratigraphic position elsewhere are based on
texture, carbonate and clay mineralogy, and field characteristics.

INTRODUCTION

A rock-stratigraphic classification is herein established for the Pleistocene
deposits of the Danville, Illinois, area. Six well-exposed sections have been
studied in detail and, from their characteristics, determined in both field
and laboratory, eight till units have been named. Each till unit has distinct
physical and compositional properties that permit it to be treated as a rock-
stratigraphic unit. All units are correlated to previously named units or are
defined as rock-stratigraphic units.

Previous Work

Early reports on the glacial deposits of the Danville area were made by Leverett (1899) and by Campbell and Leverett (1900). Krumbein (1933), one of the first to recognize that individual tills are more or less homogeneous and can be differentiated by their textural and compositional properties, reported the presence of two tills with distinctly different textures in the Danville area.

Our report expands and develops the same principle, meanwhile recognizing many additional tills. The first relatively detailed work was done by Eveland (1952), who described the Pleistocene geology and mapped the surface morphology of the area. He reported several glacial tills, separated by sand and gravel, silt, or weathered materials, and described pebble compositions and color differences between some of the tills. He also was the first to report pre-Illinoian drift in the region and suggested that there was evidence for Nebraskan glaciation. Ekblaw and Willman (1955) redescribed in detail one of Eveland's sections exposed by a newly excavated drainage ditch in Hungry Hollow and reported seven distinct till bodies ranging in age from Illinoian to Wisconsinan (Fig. 1).

Altogether, they distinguished 17 separate Pleistocene units, which they designated by Roman numerals but did not name. Ekblaw and Willman's

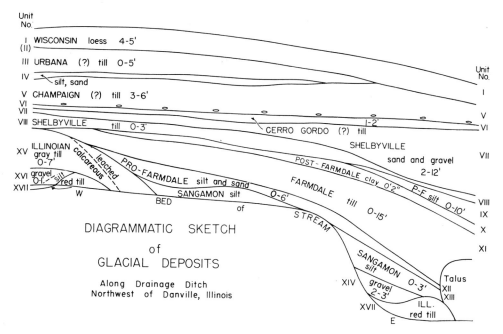

Figure 1. Diagrammatic sketch by Ekblaw and Willman (1955) of glacial deposits exposed along part of the diversion ditch for the Hungry Hollow stream.

work, as well as Eveland's, was based on two assumptions not followed in the present report: (1) the first major weathered zone below the present surface soil is the Sangamon Soil, and (2) the several younger Wisconsinan tills are related to end moraines south of the region. Thus they called the tills Illinoian, Kansan, or Nebraskan, depending on the interpretation of the buried weathered materials, or Shelbyville, Cerro Gordo, West Ridge, Urbana, or Champaign, after the end moraines south of Danville (Eveland, 1952; Ekblaw and Willman, 1955).

The most recent report on the Pleistocene of the Danville area was that of Johnson, Gross, and Moran (1969), which was a preliminary presentation of the work done for this paper.

Location and Laboratory Procedures

Danville is located in east-central Illinois near the Indiana-Illinois state line (Fig. 2). The locations and names of the six described geologic sections are shown in Fig. 2. Descriptions for five sections appear at the end of this report. The sixth section (Drainage Ditch Section) has been well described by Ekblaw and Willman (1955) and is shown diagrammatically in Figures 1 and 7.

Tills in the sections were sampled at intervals of one to two feet. The

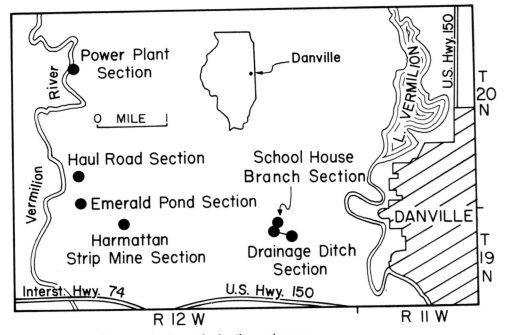

Figure 2. The Danville region, showing section locations and names.

grain-size distribution of all samples was determined by combined hydrometer and sieve analyses (sand 2.0–0.062 mm., silt 0.062–0.0039 mm., and clay < 0.0039 mm.). The amount of calcite and dolomite in the minus-200-mesh grade (less than 74 microns) was determined by use of a Chittick gasometric apparatus (Dreimanis, 1962). The clay-mineral composition of the tills was determined from oriented aggregates of the clay-size fraction by standard X-ray diffraction procedures. The terms illite, kaolinite, and chlorite are used with their generally accepted meaning. The term *expandable clay minerals* refers to montmorillonite, mixed-layered, and degraded clay minerals that expand to about 17 Angstroms after treatment with ethylene glycol.

The results of the laboratory analyses are given in Tables 1 and 2 and are shown in Figures 3, 4, and 5. Table 1 gives the average and standard deviation of the laboratory data for each stratigraphic unit, and Table 2 gives the data values for all samples.

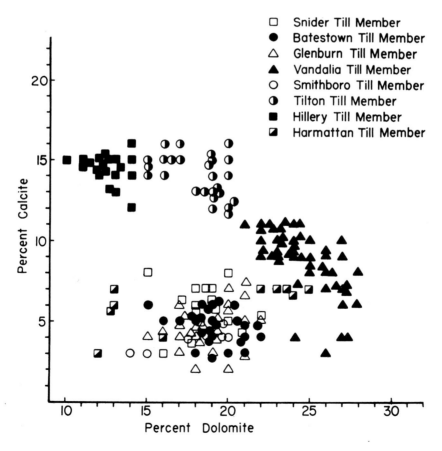

Figure 3. Carbonate composition of the < 74µ fraction of the tills.

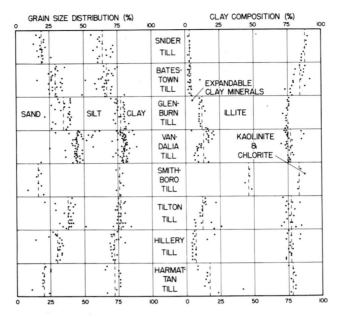

Figure 4. Grain-size distribution and clay-mineral composition of the till units. Each sample is plotted relative to its position within a unit in a particular geologic section. The dashed lines show the mean percentages of sand, silt, clay, and of expandable clay minerals, illite, and kaolinite plus chlorite for each unit.

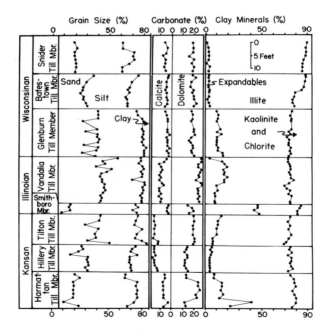

Figure 5. Composite section showing the grain size and the carbonate and clay-mineral composition of the till units from typical sections. Snider Till, Batestown Till, and Glenburn Till Members from Emerald Pond Section; Vandalia Till, Tilton Till, Hillery Till, and Harmattan Till Members from the Strip Mine Section; and Smithboro Till Member from the School House Branch Section. Stratified drift and paleosols are not shown.

TABLE 1

MEAN GRAIN SIZE, CARBONATE CONTENT, AND CLAY MINERALOGY OF TILLS

	°	Grain Size (%)			Carbonates (%)		Clay Minerals (%)		
		Sand	Silt	Clay	Calcite	Dolomite	Expand-ables	Illite	Kaolinite & Chlorite
Snider Till Member	X	19	45	36	6	19	3	85	12
	s	2.9	5.0	5.0	1.4	1.9	0.6	1.8	1.5
	N	16	16	16	16	16	8	8	8
Batestown Till Member	X	29	38	33	5	19	4	80	16
	s	5.7	4.3	5.6	1.3	1.7	2.2	3.8	5.0
	N	27	27	27	27	27	18	18	18
Glenburn Till Member	X	35	42	23	5	19	12	63	25
	s	5.6	7.0	3.6	1.4	1.8	2.0	2.3	1.1
	N	26	26	26	26	26	15	15	15
Vandalia Till Member	X	45	34	21	9	25	13	64	23
	s	6.9	7.4	3.6	2.0	1.9	4.6	5.1	3.6
	N	50	50	50	50	50	34	34	34
Smithboro Till Member	X	16	60	24	4	18	46	37	17
	s	3.6	2.8	1.8	0.6	2.4	1.9	1.8	0.8
	N	7	7	7	7	7	4	4	4
Tilton Till Member	X	38	39	23	14	18	13	64	23
	s	5.7	5.9	2.7	1.3	1.8	5.3	3.5	4.4
	N	26	26	26	26	26	18	18	18
Hillery Till Member	X	30	42	28	14	12	7	70	23
	s	4.3	5.1	2.9	0.9	1.2	4.9	2.4	3.2
	N	20	20	20	20	20	18	18	18
Harmattan Till Member	X	20	52	28	6	19	17	61	22
	s	3.9	7.0	4.4	1.4	5.5	9.8	9.2	1.6
	N	10	10	10	10	10	9	9	9

°Note: X = mean; s = standard deviation; N = number of samples.

TABLE 2

GRAIN SIZE, CARBONATE, AND CLAY MINERAL ANALYSES OF SAMPLES

Sample No.	Stratigraphic Unit	Ft. below Top of Unit	Grain Size (%)			Carbonates (%)		Clay Minerals (%)		
			Sand	Silt	Clay	Calcite	Dolomite	Expand-ables	Illite	Kaolinite & Chlorite
1-1	Batestown	0.5	36	39	25	3	18	3	85	12
1-2	Batestown	1.5				4	21	3	84	13
1-3	Batestown	3.0	29	39	32	5	17	3	84	13
1-4	Batestown	5.0	26	38	36	5	16	3	83	14
1-5	Batestown	6.0	26	42	32	6	19	3	82	15
1-6	Batestown	7.0	24	43	33	5	18	3	76	21
1-7	Batestown	8.5	27	37	36	5	19	3	75	22
1-8	Batestown	9.5	27	38	35	4	19	2	75	23
1-9	Batestown	11.0	29	37	34	4	22	3	74	23
1-10	Batestown	12.0	30	36	34	6	19	4	74	22
1-11	Glenburn	1.0	40	34	26	6	17	10	66	24
1-12	Glenburn	2.5	40	35	25	5	18	12	63	25
1-13	Glenburn	4.0	40	39	21	4	18	11	65	24
1-14	Glenburn	5.0	26	54	20	4	17	10	66	24
1-15	Glenburn	8.0	39	40	21	4	19	13	63	24
1-16	Glenburn	9.0	29	51	20	4	19	10	64	26
1-17	Glenburn	10.0	39	41	20	2	20	14	61	25
1-18	Glenburn	11.0	39	40	21	5	17	10	62	28

TABLE 2 — *Continued*

Sample No.	Stratigraphic Unit	Ft. below Top of Unit	Grain Size (%)			Carbonates (%)		Clay Minerals (%)		
			Sand	Silt	Clay	Calcite	Dolomite	Expand-ables	Illite	Kaolinite & Chlorite
1-19	Glenburn	12.0	41	39	20	4	15	10	64	26
1-20	Glenburn	13.0	40	39	21	4	16	10	64	26
1-21	Glenburn	14.0	27	53	20	4	18	10	64	26
1-22	Glenburn	15.0	40	43	17	2	18	11	64	25
1-23	Glenburn	16.0	38	43	19	5	17	11	63	26
1-24	Glenburn	17.0	26	55	19	3	17	15	60	25
1-25	Glenburn	18.0	37	44	19	3	19	16	57	27
1-26	Tilton	0.5	28	47	25	12	20	13	65	22
1-27	Tilton	1.5	39	35	26	15	16	12	68	20
1-28	Tilton	3.0	40	36	24	15	17	12	67	21
1-29	Tilton	4.0	42	34	24	13	19	11	70	19
1-30	Snider	3.0	18	44	38	6	18	2	88	10
1-31	Snider	5.0	21	41	38	4	18	3	86	11
1-32	Snider	6.5	18	54	28	5	18	3	86	11
1-33	Snider	8.0	18	53	29	8	15	3	85	12
1-34	Snider	9.5	19	53	28	6	19	3	84	13
1-35	Snider	11.0	20	47	33	6	17	2	84	14
1-36	Snider	12.5	20	41	39	7	19	4	82	14
1-37	Snider	14.0	19	42	39	6	18	3	83	14
2-1	Snider	1.0	15	41	44	7	18			
2-2	Snider	2.5	21	40	39	8	20			
2-3	Snider	4.0	22	38	40	7	19			
2-4	Snider	5.5	19	48	33	4	21			
2-5	Snider	7.0	26	44	30	5	20			
2-6	Batestown	0.5	29	44	27	4	21			
2-7	Batestown	2.0	24	47	29	4	19			
2-8	Batestown	4.5	26	37	37	5	18			
2-9	Batestown	6.0	34	34	42	4	18			
2-10	Batestown	7.5	24	35	41	3	19			
2-11	Batestown	10.5	25	45	30	4	19			
2-12	Glenburn	1.5	31	39	30	7	20			
2-13	Glenburn	2.0	32	41	27	7	21			
2-14	Glenburn	3.0	36	38	26	5	22			
2-15	Glenburn	4.5	36	39	25	7	20			
2-16	Glenburn	6.0	33	41	26	7	21			
2-17	Glenburn	7.5	34	40	26	6	20			
2-18	Glenburn	9.0	39	34	27	6	20			
2-19	Glenburn	10.5	38	38	24	5	21			
2-20	Glenburn	12.0	21	61	18	5	19			
2-21	Glenburn	13.5	40	32	28	4	17			
2-22	Glenburn	14.0	28	45	27	4	18			
3-1	Vandalia	1.0	47	35	18	8	26	14	62	24
3-2	Vandalia	2.5	47	34	19	9	27	17	59	24
3-3	Tilton	0.5	41	34	25	13	19	14	58	28
3-4	Tilton	1.5	21	59	20	13	19	14	60	26
3-5	Tilton	0.5	43	33	24	15	19	14	61	25
3-6	Tilton	2.0	42	32	26	15	20	12	61	27
3-7	Tilton	3.5	29	44	27	14	19	12	63	25
3-8	Tilton	5.0	42	38	20	15	19	10	63	27
3-9	Tilton	7.0	38	37	25	15	19	10	66	24

TABLE 2 — *Continued*

Sample No.	Stratigraphic Unit	Ft. below Top of Unit	Grain Size (%)			Carbonates (%)		Clay Minerals (%)		
			Sand	Silt	Clay	Calcite	Dolomite	Expand-ables	Illite	Kaolinite & Chlorite
3-10	Tilton	8.5	39	35	26	16	20	6	68	26
3-11	Tilton	0.5	32	48	20	12	20	7	65	28
3-12	Tilton	1.5	51	33	16	12	19	7	63	30
3-13	Hillery	0.5	27	47	26	15	14	6	68	25
3-14	Hillery	1.5	28	43	29	15	13	5	69	26
3-15	Hillery	2.5	24	46	30	15	11	6	71	23
3-16	Hillery	3.5	16	60	24	14	14	5	68	27
3-17	Hillery	4.5	33	36	31	16	14	4	72	24
3-18	Hillery	5.5	32	37	31	15	12	4	72	24
3-19	Hillery	6.5	32	37	31	15	12	4	71	25
3-20	Hillery	7.5	33	37	30	14	12	4	72	24
3-21	Hillery	8.5	30	40	30	13	13	4	73	23
3-22	Hillery	9.5	32	39	29	12	14	4	69	27
3-23	Harmattan	0.5	21	45	34	6	13	8	67	25
3-24	Harmattan	2.0	25	40	35	6	13	8	70	22
3-25	Harmattan	0.5	19	57	24	7	24	15	64	21
3-26	Harmattan	2.0	20	55	25	7	24	15	65	20
3-27	Harmattan	3.5	19	57	24	7	25	14	63	23
3-28	Harmattan	5.0	19	57	24	7	22	16	62	22
3-29	Harmattan	6.5	19	57	24	7	23	13	67	20
3-30	Harmattan	1.0	11	62	27	3	12	42	38	20
3-31	Harmattan	2.5	24	43	33	7	13	23	54	23
4-1	Batestown	0.5	30	35	35	5	19	3	81	16
4-2	Batestown	1.5	31	38	31	6	20	2	79	19
4-3	Batestown	2.5	27	45	28	5	21	2	78	20
4-4	Batestown	3.5	33	41	26	5	21	2	78	20
4-5	Vandalia	0.5	58	20	22	4	27	4	69	27
4-6	Vandalia	1.5	45	33	22	7	27	15	58	27
4-7	Vandalia	2.5	47	30	23	11	23	17	57	26
4-8	Vandalia	3.5	39	41	20	10	23	19	56	25
4-9	Vandalia	4.5	53	28	19	8	26	16	59	25
4-10	Vandalia	5.5	48	30	22	10	25	17	57	26
4-11	Vandalia	6.5	44	36	20	11	24	17	58	25
4-12	Vandalia	7.5	43	26	21	10	24	19	57	24
4-13	Vandalia	8.5	44	27	29	11	22	17	61	22
4-14	Vandalia	9.5	44	38	18	8	26	17	57	26
4-15	Vandalia	0.5	46	36	18	7	27	10	60	30
4-16	Vandalia	1.5	46	34	20	11	23	11	62	27
4-17	Vandalia	2.5	47	29	24	11	23	11	64	25
4-18	Vandalia	3.5	45	35	20	10	23	8	66	26
4-19	Vandalia	4.5	43	35	22	11	24	8	68	24
4-20	Vandalia	5.5	44	35	21	9	23	—	—	—
4-21	Vandalia	6.5	43	36	21	10	24	8	67	25
4-22	Vandalia	7.5	42	37	21	9	26	8	67	25
5-1	Vandalia	1.0	48	32	20	11	22	15	64	21
5-2	Vandalia	2.0	41	40	19	9	24	14	64	22
5-3	Vandalia	3.0	44	36	20	9	24	10	68	22
5-4	Vandalia	4.0	44	35	21	8	25	8	70	22
5-5	Vandalia	5.0	36	51	13	9	26	11	69	20
5-6	Vandalia	5.5	18	63	19	7	27	10	69	21

TABLE 2 — *Continued*

Sample No.	Stratigraphic Unit	Ft. below Top of Unit	Grain Size (%)			Carbonates (%)		Clay Minerals (%)		
			Sand	Silt	Clay	Calcite	Dolomite	Expand-ables	Illite	Kaolinite & Chlorite
5-7	Smithboro	0.5	17	57	26	3	14	47	37	16
5-8	Smithboro	1.5	15	61	24	4	18	47	36	17
5-9	Smithboro	2.5	16	61	23	4	19	43	39	18
5-10	Smithboro	3.5	8	66	26	3	15	48	34	18
5-11	Tilton	7.8	39	32	29	0	0	22	61	17
5-12	Hillery	0.5	35	40	25	15	10	17	67	16
5-13	Hillery	1.5	33	42	25	15	12	7	71	22
5-14	Batestown	0.5	16	36	48	0	0	11	81	8
5-15	Mulberry Grove	0.2	12	57	31	2	13	62	24	14
5-16	Mulberry Grove	0.5	2	79	19	1	21	—		—
5-17	Mulberry Grove	0.8	0	79	21	0	0	54	12	34
5-18	Mulberry Grove	1.4	16	42	42	0	0	80	9	11
5-19	Mulberry Grove	2.3	16	56	28	0	0	79	11	10
5-20	Mulberry Grove	2.6	28	33	39	0	0	52	38	10
5-21	Mulberry Grove	3.1	28	35	37	0	0	69	21	10
5-22	Mulberry Grove	3.6	28	29	43	0	0	57	31	12
5-23	Petersburg	0.2	25	46	29	0	0	59	23	18
5-24	Petersburg	0.7	18	55	27	0	0	61	20	19
5-25	Petersburg	1.2	14	68	18	0	0	58	21	21
5-26	Tilton	0.5	21	41	38	0	0	63	19	18
5-27	Tilton	1.0	27	35	38	0	0	60	25	15
5-28	Tilton	1.5	31	35	34	0	0	64	20	16
5-29	Tilton	2.0	32	31	37	0	0	61	25	14
5-30	Tilton	2.5	36	32	32	0	0	60	23	17
5-31	Tilton	3.3	51	07	42	0	0	61	24	15
5-32	Tilton	3.8	36	34	30	0	0	61	25	14
5-33	Tilton	4.5	48	26	26	0	0	61	24	15
5-34	Tilton	5.3	48	25	27	0	0	63	26	11
5-35	Tilton	6.0	35	31	34	0	0	29	54	17
5-36	Tilton	6.5	38	32	30	0	0	22	57	21
5-37	Tilton	7.0	38	37	25	15	15	26	59	15
5-38	Tilton	7.5	38	36	26	14	15	20	60	20
11-1	Batestown	0.3	26	39	35	6	15	3	82	15
11-2	Batestown	1.0	30	35	35	5	18	3	85	12
11-3	Vandalia	0.5	47	30	23	9	22	20	65	15
11-4	Vandalia	2.0	46	31	23	10	23	21	66	13
11-5	Vandalia	3.5	44	32	24	7	24	15	70	15
11-6	Vandalia	5.0	42	34	24	9	23	11	68	21
11-7	Vandalia	0.5	57	32	11	8	28	8	70	22
11-8	Vandalia	2.0	44	40	16	7	27	9	73	18
11-9	Vandalia	3.5	43	40	17	7	27	7	71	22
11-10	Vandalia	5.0	47	35	18	7	26	6	73	21
11-11	Vandalia	6.5	45	37	18	6	28	7	69	24
11-12	Tilton	0.5	40	39	21	13	19	22	62	16
11-13	Tilton	2.5	39	40	21	15	16	18	63	19
11-14	Hillery	0.5	29	41	30	15	11	21	63	16
11-15	Hillery	2.0	33	41	26	14	13	15	68	17
11-16	Hillery	3.5	32	40	28	15	12	6	72	22
11-17	Hillery	5.0	31	40	29	15	11	7	71	22
11-18	Hillery	6.0	33	41	26	15	12	7	71	23
11-19	Hillery	7.0	29	42	29	13	13	6	72	22
11-20	Vandalia	3.0	52	28	20	10	22			
11-21	Vandalia	4.0	49	29	22	11	22			
11-22	Vandalia	5.5	56	25	19	10	24			
11-23	Vandalia	2.0	57	30	13	3	26			
11-24	Vandalia	1.5	28	55	17	4	27			
11-25	Vandalia	0.5	62	24	14	4	24			

TABLE 2 — *Continued*

Sample No.	Stratigraphic Unit	Ft. below Top of Unit	Grain Size (%)			Carbonates (%)		Clay Minerals (%)		
			Sand	Silt	Clay	Calcite	Dolomite	Expand-ables	Illite	Kaolinite & Chlorite
11-26	Snider	5.0	13	51	36	3	16			
15-1	Batestown	4.0	34	36	30	5	22			
15-2	Hillery	1.5	36	44	20	14	12			
15-3	Hillery	1.0	32	44	24	14	13			
15-4	Tilton	4.0	35	42	23	14	20			
15-5	Tilton	1.5	38	39	23	12	20			
15-6	Tilton	6.0	40	40	20	16	17			
15-7	Tilton	3.0	40	37	23	14	16			
15-8	Vandalia	4.0	43	35	22	8	25			
15-9	Snider	1.5	16	41	43	6	19			
15-10	Batestown	1.5	34	27	39	3	21			
15-11	Batestown	3.0	29	38	33	6	19	8	84	8
15-12	Vandalia	1.5	49	24	27	9	23			
15-13	Vandalia	4.0	45	27	28	10	27			
15-14	Tilton	1.0	37	38	25	15	16			
15-15	Tilton	3.0	40	34	26	13	18			
15-16	Batestown	9.0	33	38	29	6	18			
15-17	Vandalia	1.5	49	26	25	9	22			
15-18	Snider	6.0	17	45	38	5	22			
15-19	Harmattan	2.0	25	50	25	4	16			
15-20	Smithboro	2.0	18	60	22	4	19			
15-21	Smithboro	4.0	18	61	21	4	20			
15-22	Smithboro	6.0	20	57	23	5	20			
15-23	Vandalia	4.0	44	31	25	9	24			
15-24	Tilton	1.0	38	36	25	13	19			
15-25	Vandalia	4.0	44	34	22	9	25			
15-26	Tilton	1.0	34	44	22	15	15			
15-28	Vandalia	1.5	37	40	23	9	24			
15-29	Vandalia	3.5	42	33	25	9	23			
15-30	Batestown	6.5	49	31	20	3	20			
15-31	Vandalia	2.5	49	29	22	11	21			

STRATIGRAPHY

Principles

All units are defined as rock-stratigraphic units and are distinguished by their physical characteristics. Where possible, till units in the area have been correlated to previously named tills and the names of the previously named units are used in this report. These correlations to named tills are based on physical characteristics and stratigraphic position.

New units are defined in accordance with the Code of Stratigraphic Nomenclature (Am. Comm. Strat. Nomenclature, 1961). The tills of Woodfordian age are defined as members of the Wedron Formation (Frye, Willman, Rubin, and Black, 1968). The tills of Illinoian age are correlated to units first named by Jacobs and Lineback (1969) and later formalized as members of the Glasford Formation by Willman and Frye (1970). The tills of Kansan age are defined as members of the Banner Formation (Willman and Frye, 1970).

Kansan Stage

The three oldest till units in the Danville area and their related deposits are tentatively assigned to the Kansan Stage. The age designation is based on physical correlation of the youngest of these tills to till that Johnson (1964) and Jacobs and Lineback (1969) considered to be Kansan in Illinois. A thick paleosol, believed to be the Yarmouth Soil, is developed in this till in the School House Branch and Drainage Ditch sections. The age interpretation is also supported by the occurrence above the paleosol of a silt that is correlated with the Petersburg Silt, the oldest rock-stratigraphic unit in the Illinoian Stage in Illinois. As there is no indication of a significant time interval between any of the tills, all are assigned to the Kansan Stage. However, it is possible that one or more of these units could be of Nebraskan age.

Banner Formation

The Banner Formation is defined by Willman and Frye (1970) as including all the tills of the Kansan Stage represented in the type section in a borrow pit in SW¼ NE¼ sec. 31, T. 7 N., R. 6 E., Peoria County, Illinois. The three oldest tills of the Danville region are defined as members of the Banner Formation.

Harmattan Till Member

The oldest of the three tills is here named the Harmattan Till Member for exposures in the Harmattan Strip Mine of the Ayrshire Coal Company. The till is at present best exposed in the strip mine located in the NE¼ sec. 4, T. 19 N., R. 12 W. The Strip Mine Section is designated the type section, but, because of the nature of strip-mining operations, this section is no longer exposed and the strip mine is designated the type reference area. The till is also present in the easternmost portion of the Drainage Ditch Section.

The Harmattan Till Member is gray, calcareous, and quite dense and hard. The upper few feet of the till are pebbly, and gravelly sand is present in local channels in the upper part of the unit. The till becomes less pebbly with depth and more olive-gray in color. Wood is present throughout the till, but is more abundant in the upper portion.

The Harmattan Till Member appears to have three compositional zones. Because of deformation within the unit, the internal stratigraphy has not been definitely determined. The lower and upper portions of the unit contain about 7 percent calcite and 13 percent dolomite (Fig. 5). In general, the middle carbonate zone (7 percent calcite and 23 percent dolomite) is more silty and contains less clay and sand than do the zones above and below it (Fig. 5), although exceptions have been observed in other sections in the strip mine. The clay-mineral composition of the three zones also is different. Although the kaolinite and chlorite contents are uniform, the amount of expandable clay minerals in the clay fraction varies from about 30 percent in the lower zone to 15 percent in the middle zone and to 7 percent in the upper zone (Fig. 5). The illite content also increases upward in the section.

The Harmattan Till Member rests on Pennsylvanian bedrock, residual soil, or alluvial and colluvial conglomerate, silt, or clay. There has been much deformation near the lower contact, and the subjacent materials are often sheared up into the basal portion of the till. The till is overlain by the Hillery Till Member, a distinct reddish brown till, or by reddish brown stratified drift related to the Hillery Till Member.

The Harmattan Till Member is thickest where it occurs in a valley cut into the bedrock; however, it does not appear to be entirely confined to the valley. It is 10 to 15 feet thick in most places, but it is as much as 22 feet thick in the deeper part of the bedrock valley.

The carbonate composition of this member, characterized by considerably more dolomite than calcite, is typical of tills deposited by glaciers from the Lake Michigan basin (Willman, Glass, and Frye, 1963). Thus, the Harmattan Till Member is thought to have been deposited by the Lake Michigan Lobe.

The upper portion of the Harmattan Till Member is tentatively correlated with the Kansan till (part of Wayne's (1965) Cloverdale Till Member of the Jessup Formation) exposed at the Cataract Lake Spillway Section in Indiana. The texture, clay-mineral, and carbonate compositions of the two tills are similar, although there is some slight variation.

The lower portion of the Harmattan Till Member may be correlative with Kansan till described in central Illinois by Johnson (1964) and tentatively related to the Lake Michigan Lobe. The Kansan till in central Illinois and the lower portion of the Harmattan both contain over 20 percent expandable clay minerals, and have similar textures and carbonate compositions.

Hillery Till Member

The Hillery Till Member is here named from its occurrence in sections near the town of Hillery, Illinois. The Power Plant Section (NW¼ SW¼ SW¼ sec. 21, T. 20 N., R. 12 W., Vermilion County) is designated the type section. The unit is well exposed in the Harmattan Strip Mine, and is also present in the School House Branch and Drainage Ditch Sections of Hungry Hollow. In the Drainage Ditch Section, this till was interpreted as Kansan by Eveland (1952), and as Illinoian or possibly Kansan (Unit XVII in Fig. 1) by Ekblaw and Willman (1955).

The Hillery Till Member is reddish brown, calcareous, massive, and quite hard. The lower portion is slightly darker red than the upper. The unit contains reddish brown silt, fine sand, and pinkish gray to brown gravelly sand. The silt occurs most commonly as stringers in the upper portion of the till, and the coarser stratified drift is common near both the lower and upper contacts.

The Hillery Till Member is underlain by either the gray Harmattan Till Member or by Pennsylvanian bedrock and is overlain by the gray Tilton Till Member. Both contacts are abrupt and easily identified in the field. The unit varies from about five to ten feet thick.

The Hillery Till Member is medium-textured and contains a mean of 30 percent sand, 42 percent silt, and 28 percent clay (Fig. 4). It has slightly more calcite than dolomite (Fig. 3) and a total carbonate content of about 26 percent. Except in parts of two sections, the till contains large amounts of illite, moderate amounts of kaolinite and chlorite, and small amounts of expandable clay minerals (Fig. 4). In the type section and School House Hollow Section, the upper portion of the till contains more expandable clay minerals. The high calcite content relative to that in Lake Michigan Lobe tills suggests that the Hillery Till Member was deposited by ice entering Illinois from the east or northeast, probably out of the Lake Erie basin (Willman, Glass, and Frye, 1963).

The Hillery Till Member has been observed and sampled 115 miles west of Danville at the Kincaid dam site near Springfield, Illinois, by the senior author. An equivalent till has been reported south of Shelbyville, Illinois, by Bleuer (1967). It is present in the subsurface near Urbana, Illinois, and has been observed and sampled 45 miles east of Danville near Crawfordsville, Indiana (units 5 and 6 in the Liberty School Geologic Section of Wayne, 1965). It is therefore an extensive, easily recognized rock unit and serves as a useful marker among the older tills in this general area. Its absence in many exposures and borings suggests that it is discontinuous and is preserved only locally. The origin and significance of the red color is still a matter for continuing study.

Tilton Till Member

The till overlying the Hillery Till is here named the Tilton Till Member from the town of Tilton, Illinois. The School House Branch Section (SE¼ NE¼ NE¼ sec. 2, T. 19 N., R. 12 W., Vermilion County) is designated the type locality. The unit is also well exposed in the Harmattan Strip Mine Section and is present in the Emerald Pond, Power Plant, and Drainage Ditch Sections. In the Drainage Ditch Section, this till was interpreted as Illinoian (Unit XV in Fig. 1) by Ekblaw and Willman (1955). In his School House Branch Section, Eveland (1952) called this till Kansan.

Where unoxidized, the Tilton Till Member is dark gray, calcareous, and quite hard. The unit contains considerable interbedded silt, and sand and gravel, particularly near the upper and lower contacts. This till is silty and sandy, having about 38 percent sand, 39 percent silt, and 23 percent clay (Fig. 4). It contains about 14 percent calcite and 18 percent dolomite (Fig. 3), and has a mean clay-mineral composition of 64 percent illite, 24 percent kaolinite and chlorite, and 12 percent expandable clay minerals (Fig. 4).

The upper part of the Tilton Till Member in many places contains either a thick weathered zone or, where the soil is truncated, the oxidized zone below the weathering profile. In the School House Branch Section, the unit has been leached 6.5 feet. The soil is complex and does not show typical profile characteristics. It is developed in partly sorted material as well as till, and the dark gray to greenish brown soil color reflects development under poorly drained conditions. The clay fraction contains large amounts of expandable clay minerals of the type that are common in the B-zones of weathered till in Illinois (Willman, Glass, and Frye, 1966). A leached zone also occurs at the top of the Tilton Till Member in the Drainage Ditch Section.

The Tilton Till Member lies on the reddish brown Hillery Till Member, with a lower contact that is obvious in the field. The unit is variously overlain by the Petersburg Silt, Vandalia Till Member, or Glenburn Till Member. Where the weathered zone and/or oxidized portion of the till is preserved, the contact is easily identified in the field (Power Plant, Emerald Pond, Drainage Ditch, and School House Branch Sections). Where only unoxidized till is present, the upper contact is marked by an abrupt but subtle change in color from dark gray in the Tilton to brownish gray in the younger till units (Strip Mine Section). Laboratory analyses of samples from above and below the contact can be used to check the field contact. The unit is about 5 to 10 feet thick in these sections.

The Tilton Till Member is correlated to a till called "eastern Kansan" by Johnson (1964) in central Illinois and "Kansan" by Jacobs and Lineback (1969) in south-central Illinois, and with the lowermost till exposed at the Lake Bloomington Spillway Section (Leonard and Frye, 1960) near Bloomington, Illinois. It is also probably present above till equivalent to the Hillery Till Member in the Liberty School Section of Wayne (1965) near Crawfordsville, Indiana. In recent years, this till has been called eastern Kansan in Illinois and has been interpreted as having been deposited by a glacier entering Illinois from the northeast, probably out of the Lake Erie basin (Willman, Glass, and Frye, 1963).

Illinoian Stage

The following silt and till units are assigned to the Illinoian Stage because of physical correlation to stratigraphic units of definite Illinoian age in Illinois. Little evidence of the Sangamon Soil is present in this area.

Petersburg Silt

Silt lying above a weathered zone and occurring below till correlated to the Smithboro Till Member (Jacobs and Lineback, 1969) of south-central Illinois is tentatively assigned to the Petersburg Silt (Willman, Glass, and Frye, 1963). Eveland (1952) referred to this silt as Loveland Loess, and Ekblaw and Willman (1955) called it Sangamon silt (Unit XIII in Fig. 1).

The silt, which occurs in two sections (School House Branch and Drainage Ditch Sections), is massive, noncalcareous, sparsely fossiliferous (a few mollusks), and dark brown to gray. It contains abundant montmorillonite (approximately 60 percent in the clay fraction), as does the Petersburg Silt in western Illinois (Willman, Glass and Frye, 1963).

Glasford Formation

The Glasford Formation is defined by Willman and Frye (1970) as including all the tills of the Illinoian Stage represented in the type section, a borrow pit in SW¼ NE¼ sec. 31, T. 7 N., R 6 E., Peoria County, Illinois. Two tills and a slit unit in the Danville region are correlated to named members of the Glasford Formation.

Smithboro Till Member

The oldest Illinoian till of the Danville region is correlated to the Smithboro Till Member on the basis of strong textural and compositional similarity. Smithboro Till Member was first named by Jacobs and Lineback (1969) in the Vandalia region, and later was formalized as a member of the Glasford Formation by Willman and Frye (1970). The Smithboro Till Member in the type area contains an average of 25 percent sand, 49 percent silt, and 26 percent clay, plus large amounts of expandable clay minerals (Jacobs and Lineback, 1969). Five samples of the Smithboro Till Member from near the type area average 4 percent calcite and 16 percent dolomite. Till in central Illinois, correlative to the Smithboro and called Jacksonville-Mendon by Johnson (1964), is also silty, and has a mean clay-mineral composition of 34 percent montmorillonite, 46 percent illite, and 20 percent kaolinite and chlorite; it contains a mean of 3 percent calcite and 17 percent dolomite.

In the Drainage Ditch Section, Eveland (1952) called this till Illinoian, and Ekblaw and Willman (1955) called the till Farmdale (Unit XI in Fig. 1). In recent years, this till has been referred to as the "Danville till" of Altonian age (Frye and Willman, 1960; Willman, Glass, and Frye, 1963).

In the Danville area, the Smithboro Till Member is quite silty and contains abundant wood, carbonaceous material, and molluscan fossils. It is gray-brown where unoxidized and upon exposure oxidizes rapidly to dark brown. The unit has a mean of 16 percent sand, 60 percent silt, and 24 percent clay (Fig. 4); it contains a mean of 4 percent calcite and 18 percent dolomite (Fig. 3), and about 46 percent expandable clay minerals, 37 percent illite, and 17 percent kaolinite and chlorite (Fig. 4).

The unit occurs in only two of the sections (School House Branch and Drainage Ditch Sections) and is easily distinguished from both underlying and overlying tills. It is up to 10 feet thick in the Drainage Ditch Section. The silty character, the presence of wood and molluscan fossils, and the high montmorillonite content suggest that the Smithboro glacier incorporated large amounts of the underlying Petersburg Silt in the Danville area. The predominance of dolomite over calcite indicates that the Smithboro Till Member was deposited by Lake Michigan Lobe ice.

Mulberry Grove Member

Silt, clay, and colluvial materials overlying the Smithboro Till Member are correlated with the Mulberry Grove silt, which was named by Jacobs and Lineback (1969). The unit was later formalized as a member of the Glasford Formation by Willman and Frye (1970). The correlation is based primarily on the stratigraphic position between the Smithboro and Vandalia Till Members. These materials were called Farmdale Loess and Illinoian gumbotil by Eveland (1952), and post-Farmdale silt and clay (Units X and IX in Fig. 1) by Ekblaw and Willman (1955).

In the School House Branch Section, the Mulberry Grove Member consists of an upper carbonaceous and calcareous silt, a middle noncalcareous clayey silt, and a lower noncalcareous colluvial zone. The unit fills a depression in the underlying materials, and a thin peat mat locally covers the upper surface of the unit. All three zones contain large amounts of expandable clay minerals, and in the middle zone these minerals have the same X-ray diffraction characteristics as the well-crystallized montmorillonite in accretion-gleys in Illinois (Willman, Glass, and Frye, 1966).

In the type area near Vandalia, Illinois, the Mulberry Grove Member is gray and calcareous, and contains an organic-rich silt cap (Jacobs and Lineback, 1969). The existence of the unit 125 miles northeast of its type area and the presence of noncalcareous colluvial materials indicate a significant withdrawal of the ice after deposition of the Smithboro Till Member.

Because the underlying till was thought in recent years to be Altonian (Ekblaw and Willman, 1955; Willman, Glass, and Frye, 1963), the Mulberry Grove Member was in the proper stratigraphic position to be the Robein Silt (Willman and Frye, 1970). Consequently, wood from the silt was submitted for radiocarbon dating. The resulting date (> 40,000 radiocarbon years B.P., ISGS 23) is clearly older than Farmdalian and is consistent with our interpretation. However, the date alone does not rule out the possibility of an early Altonian age for both this deposit and the underlying till.

Vandalia Till Member

The Vandalia Till Member was first named by Jacobs and Lineback (1969) and later was formalized as a member of the Glasford Formation by Willman and Frye (1970). The Vandalia is extensive throughout south-central Illinois. Till in the same stratigraphic position was called Jacksonville in central Illinois by Johnson (1964) after the Jacksonville Moraine (Ball, 1937; Ekblaw, 1957). The rock-stratigraphic name, Vandalia Till Member, is preferable to the morphostratigraphic name, Jacksonville till, and is used for correlative tills in the Danville area. This till in the Drainage Ditch Section was called Tazewell by Eveland (1952) and Shelbyville (Unit VII in Fig. 1) by Ekblaw and Willman (1955).

In south-central Illinois, the Vandalia Till Member is sandy (mean of 43 percent sand, 38 percent silt, and 19 percent clay) and the clay-mineral fraction contains about 10 to 20 percent expandable clay minerals, 60 to 70 percent illite, and 20 to 25 percent kaolinite and chlorite (Jacobs and Lineback, 1969). Two samples of the Vandalia Till Member from the type section average 10 percent calcite and 22 percent dolomite.

In the Danville area, sandy till, occurring stratigraphically above the Tilton and Smithboro Till Members and below the Glenburn and Batestown Till Members, is included in the Vandalia Till Member. The till is gray to brownish gray where unoxidized, and becomes yellowish brown to brown upon oxidation. The unit contains a significant amount of interbedded sand and gravel and some silt.

The Vandalia Till Member contains an average of 45 percent sand, 34 percent silt, and 21 percent clay (Fig. 4). The till contains 9 percent calcite and 25 percent dolomite (Fig. 3), and the clay fraction has a mean of 13 percent expandable clay minerals, 64 percent illite, and 23 percent kaolinite and chlorite (Fig. 4). However, as is evident in Figures 4 and 5, the lower portion of the unit usually contains 5 to 10 percent less expandable clay minerals than does the upper portion.

The upper samples in the Power Plant Section (samples 11-23 to 11-25) and in the Strip Mine Section (sample 4-5) contain significantly less calcite than the rest of the unit (Fig. 3). The sample from the Strip Mine also has more illite and less expandable clay minerals (Fig. 4). These samples probably represent another stratigraphic unit that as yet has not been well documented in the area, which may be equivalent to the Buffalo Hart till of Johnson (1964), the youngest Illinoian till in Illinois. The samples are compositionally similar to this till and are in the proper stratigraphic position.

The Vandalia Till Member rests on the Mulberry Grove Member or the oxidized Tilton Till Member, and, in most places, the lower contact is easy to recognize in the field. It is overlain by either the pinkish to pinkish gray Glenburn Till Member or the silty, gray Batestown Till Member. The color and textural differences are distinctive. The Vandalia is one of the thicker tills in the area and is over 20 feet thick in the Power Plant Section.

The only evidence of weathering of the Vandalia Till Member is the oxidation in the upper portion below the unoxidized Wisconsinan till in the Drainage Ditch Section, and the degradation of chlorite indicated by the lower kaolinite and chlorite contents for samples 11-3 to 11-5 in the Power Plant Section. Evidently the Sangamon Soil in this area was almost completely eroded by Wisconsinan glaciers. As there was no evidence of pre-Wisconsinan weathering, this till was called Wisconsinan and related to the Shelbyville Moraine by Eveland (1952) and by Ekblaw and Willman (1955).

The Vandalia Till Member in the Danville area and the few samples analyzed from the type area contain about 3 to 6 percent more calcite than does till of the Lake Michigan Lobe in the same stratigraphic position in central Illinois (Johnson, 1964). This variation supports the idea that the Vandalia Till Member was deposited by a glacier from the east or northeast, as suggested by Willman, Glass, and Frye (1963), and not by ice from the Lake Michigan Lobe, as suggested by Jacobs and Lineback (1969) on the basis of clay-mineral composition and fabric studies. Comparison of the carbonate composition of the Vandalia Till Member in the Danville area with tills probably derived from Lake Erie (the Hillery and Tilton Till Members) and with tills deposited by the Lake Michigan Lobe (the Harmattan, Smithboro, Glenburn, Batestown, and Snider Till Members) also supports a generally eastern source for the Vandalia (Fig. 3). Its calcite content lies between that of the two groups of tills and suggests that it may have been deposited by ice of the Saginaw Bay Lobe, as postulated by Willman, Glass, and Frye (1963).

Wisconsinan Stage

Altonian and Farmdalian Substages

Materials of the Altonian and Farmdalian Substages (Frye and Willman, 1960) are not recognized in the Danville sections. Till and silt in the Drainage Ditch Section, which had been interpreted to be Altonian and Farmdalian (Ekblaw and Willman, 1955; Willman, Glass, and Frye, 1963) on the basis of regional correlations, are now considered to be Illinoian.

Woodfordian Substage

Wedron Formation. — The Wedron Formation was defined by Frye, Willman, Rubin, and Black (1968) as including the units below the Richland Loess and above the Farmdale Silt, as found in the type section in Pit No. 1

of the Wedron Silica Company. The youngest tills in the Danville region are defined as members of the Wedron Formation because they occur below the Richland Loess and have been correlated to tills to the west that rest on the Farmdale Silt.

Glenburn Till Member. — The Glenburn Till Member is here named for the town of Glenburn, and the Emerald Pond Section (NE¼ SW¼ SW¼ sec. 33, T. 20 N., R. 12 W., Vermilion County) is designated the type section. The unit is also present in the Haul Road Section and is the oldest Woodfordian till known in the area.

The Glenburn Till Member is brownish gray, but when oxidized has a distinct pink to brown color. It has a rather blocky structure, and brown staining is common on the joint surfaces. Wood and thin interbedded sand stringers are present in the unit.

The unit typically contains about 35 percent sand, 42 percent silt, and 23 percent clay (Fig. 4). However, the texture is variable, and a number of the samples are finer grained. These finer-grained samples have a rather uniform texture (Fig. 5), so Moran (1971) has suggested that they may represent a finer-grained element sheared into more typical Glenburn Till deposits. This till has a mean carbonate composition of 5 percent calcite and 19 percent dolomite (Fig. 3), and the clay-size fraction contains about 12 percent expandable clay minerals, 63 percent illite, and 25 percent kaolinite and chlorite (Fig. 4).

The Glenburn Till Member is overlain by the gray, silty Batestown Till Member. The lower contact has been observed only in the type section where the unit rests on the Tilton Till Member. In that section, the Glenburn is 18.5 feet thick. The Glenburn has a carbonate composition indicating a Lake Michigan Lobe source.

The Glenburn Till Member is correlated with till Unit 4 in the Champaign-Urbana area, which was described by Kempton, DuMontelle, and Glass (1971, this volume). The correlation is based on stratigraphic position, color, and clay mineral composition.

Wood collected near the base of the Glenburn Till Member in the Emerald Pond Section was dated as >38,000 radiocarbon years B. P. (ISGS 15). The date is anomalous to our interpretation, but we believe that the wood is probably pre-Wisconsinan wood that was picked up by the ice and incorporated into the Woodfordian drift. This interpretation is based on correlation of the Glenburn Till to the Woodfordian till that rests on Robein Silt to the west (Kempton, DuMontelle, and Glass, 1971).

Batestown Till Member. — The Batestown Till Member is a distinct gray till that occurs in most sections and is easily recognized by its texture, structure, and color. It is here named for the town of Batestown, Illinois, and the Emerald Pond Section (NE¼ SW¼ SW¼ sec. 33, T. 20 N., R. 12 W., Vermilion County) is named the type section. In Eveland's (1952) Drainage Ditch Section, this till was called Tazewell; Ekblaw and Willman (1955) described it as two units, the Cerro Gordo and the Champaign (Units V and VI in Fig. 1).

The Batestown Till Member is dark gray and oxidizes to a light olive-brown. It is calcareous, silty, and clayey, and contains abundant shale fragments. It has a characteristic platy to small, irregular blocky structure, and in most sections is quite friable. The upper portion of the unit commonly contains interbedded silt, sand, and till that appear to have been partly sorted and worked by water. Inclusions and zones of stratified drift are common in the till.

The Batestown Till Member has a mean texture of 29 percent sand, 38 percent silt, and 33 percent clay (Fig. 4). It contains about 5 percent calcite and 19 percent dolomite (Fig. 3), and contains more illite than do the older tills in the area. The characteristic clay-mineral composition of the unoxidized till is 2 percent expandable clay minerals, 75 percent illite, and 23 percent kaolinite and chlorite (Fig. 4). The higher illite content and the lower kaolinite and chlorite content shown in Figure 4 for the upper portion of this till are the result of surface weathering.

A thin till, called Cerro Gordo by Ekblaw and Willman (1955), occurs in the base of the Batestown Till Member in the Drainage Ditch Section (Fig. 7). It is sandier, contains more boulders than does the Batestown Till above, and has a slightly pinkish cast. A somewhat similar zone has been observed at the base of the Batestown Till Member in other nearby sec-

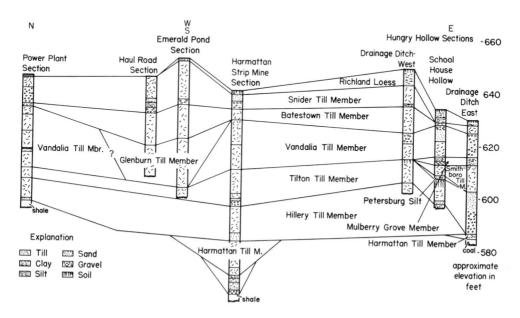

Figure 6. Correlation of stratigraphic units in the sections. Length of section approximately five miles.

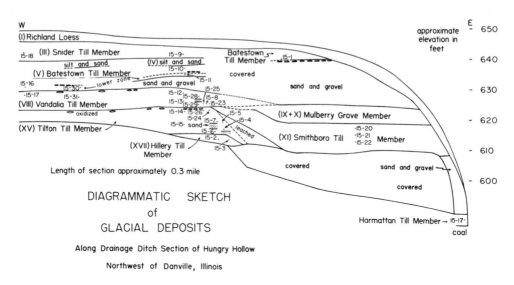

Figure 7. The Drainage Ditch Section of Hungry Hollow showing locations of samples.

tions. However, it does not have the same composition as the Glenburn Till Member, and may be equivalent to the till referred to as Unit 3 by Kempton, DuMontelle, and Glass (1971).

The Batestown Till Member commonly overlies either the Glenburn or the Vandalia Till Member, but it has been observed on units as old as the Tilton Till Member. It is either overlain by the Snider Till Member or is the surface till in areas below an elevation of about 640 feet. The unit is about 12 feet thick in both the Emerald Pond and Haul Road Sections. The Batestown Till Member is correlated to till Unit 2 in the Champaign-Urbana area (Kempton, DuMontelle, and Glass, 1971) on the basis of field characteristics, texture, and clay-mineral composition.

Snider Till Member. — The youngest till in the Danville area is separated from the Batestown Till Member by stratified drift and is here named for the town of Snider, Illinois. The Emerald Pond Section (NE¼ SW¼ SW¼ sec. 33, T. 20 N., R. 12 W., Vermilion County) is designated the type section. In the Drainage Ditch Section, this till was called Tazewell by Eveland (1952) and was described as the Urbana till and a silt-sand unit (Units III and IV in fig. 1) by Ekblaw and Willman (1955).

The Snider Till Member is gray-brown to light olive-brown where oxidized, and dark gray where unoxidized. It is calcareous and clayey, and contains very few pebbles. The unit has a characteristic medium-to-coarse blocky structure, and manganese staining and concentration of secondary calcite are common along joint surfaces. Locally the unit contains interbedded sand or silt.

The Snider Till Member is the finest textured till in the Danville area and averages 19 percent sand, 45 percent silt, and 36 percent clay (Fig. 4). In composition it is similar to the Batestown Till Member, although the calcite content is often 1 to 2 percent higher (Fig. 3).

The Snider Till Member is the surface drift in those areas above an elevation of about 640 feet. Stratified sands, gravels, and silts that appear to be continuous stratigraphically below the till are included in the unit, and it is overlain by a thin cover of Richland Loess (Frye and Willman, 1960). The Snider Till Member is about 15 feet thick in the type section.

Richland Loess. — Most of the Danville area is covered by a thin veneer of Richland Loess (Frye and Willman, 1960). The loess and related silts reach a maximum thickness of about 5 feet in the area, though locally they are completely missing as a result of a post-glacial erosion. The loess is yellow-brown to brown in color and is commonly completely leached. In it is the surface soil that developed under the forest vegetation of this area. This loess was not described by Eveland (1952), and was simply called Wisconsin loess by Ekblaw and Willman (1955).

DISCUSSION

Use of Laboratory Data

The till units in the Danville area are, for the most part, distinct in the field. In some localities, however, it is difficult to make a positive field identification. As a result, the use of laboratory data on texture and composition has greatly aided the development of the stratigraphy.

Each stratigraphic unit has one or more textural or compositional characteristics that, when used in conjunction with field data, allow it to be positively identified. These compositional and textural variations are evident in Table 1 and show particularly well on the composite vertical section of the units (Fig. 5).

Carbonate composition has been the most useful characteristic for distinguishing the pre-Woodfordian tills. The oldest Kansan till, the Harmattan Till Member, contains 8 percent less calcite than do the younger Hillery and Tilton Till Members. The two last-named tills both contain 14 percent calcite, but the Tilton contains 6 percent more dolomite than does the Hillery. There are also slight textural variations between these older units, as shown by the sand contents.

The Smithboro Till Member, the oldest Illinoian till in the area, is distinguished by its low carbonate content and high silt content, and the high percentage of expandable clay minerals in its clay fraction. The overlying Vandalia Till Member contains more calcite and dolomite, more illite, and much more sand. Its carbonate composition can be used to distinguish it from the overlying Woodfordian tills and also from the underlying Kansan units.

All the Woodfordian tills have similar carbonate compositions. However, the Glenburn Till Member contains about 12 percent less illite and is coarser textured than the unoxidized Batestown Till Member. The Batestown Till Member and the Snider Till Member have similar compositions, but the Snider contains about 10 percent less sand.

Correlation of Sections

Figure 6 shows the correlation of stratigraphic units in the six sections. The Harmattan Till Member appears to be preserved only locally in pre-glacial bedrock valleys. The Hillery and Tilton Till Members are persistent units and are usually overlain by the Vandalia Till Member. However, the Smithboro Till Member, as interpreted by Ekblaw and Willman (1955), is preserved in a valley cut into the Kansan drift. The oldest Woodfordian drift, the Glenburn, also is preserved locally only in a valley position, the valley being cut into and through the Vandalia Till Member. The youngest tills, the Batestown and Snider, are relatively continuous in the sections in the southwestern part of the area.

The importance of filled valleys cut in till (Norris and White, 1961) is obvious in the development of a complete stratigraphic section. Major erosional unconformities are present at the base of the Hillery, Vandalia, and Batestown units, and the Harmattan, Smithboro, and Glenburn units are preserved only because they locally filled valleys incised into the bed-rock or older drift. Other stratigraphic units may be preserved in the area in similar form.

Regional Correlation

Because the stratigraphic units could be distinguished on the basis of their textural and compositional characteristics, meaningful regional correlations could be established with other tills of known stratigraphic position. These correlations have necessitated reinterpretation of the Pleistocene succession in the area.

Because the paleosol developed in the Tilton Till Member is the only major soil preserved below the modern soil, it has previously been interpreted as the Sangamon Soil (Ekblaw and Willman, 1955). However, the tills below the paleosol are correlative with tills considered to be Kansan in Illinois, and the overlying tills are correlative with tills of the Illinoian drift in Illinois. Therefore, the soil developed on the Tilton represents weathering during Yarmouth time. The Sangamon Soil must have been almost completely eroded away.

Obviously more data on the physical characteristics of tills throughout the glaciated area of central United States are needed. Correlation is possible in Illinois for areas where information is available; but there are large gaps between these areas, and essentially no information is available eastward into Indiana. Subsurface drilling and sampling will be necessary to collect the needed information and samples for many areas. Work in Illinois (Hackett, 1960; Kempton, 1963, 1966; Kempton and Hackett, 1968a, 1968b; Frye, Glass, Kempton, and Willman, 1969; Kempton, DuMontelle, and Glass, 1971; this volume) and in Saskatchewan (Christiansen, 1968a, 1968b, 1971) has demonstrated that it is possible to develop a meaningful subsurface stratigraphy and to relate it to surface sections. Until physical data are available from both surface sections and subsurface borings, regional correlations between areas will be subject to uncertainties.

Relation of Buried Soils

Buried soils provide the major basis for subdivision of the Pleistocene Epoch into time-stratigraphic divisions and provide valuable information for the development of a geologic history. However, because of the nature of the glacial record, their usefulness as key beds in defining rock-stratigraphic units is limited. For example, in the Danville area both the Sangamon Soil and Yarmouth Soil are completely missing in all but two sections, making a rock-stratigraphy based on these soils unworkable. However, where these soils are preserved, they do provide valuable information, are helpful in correlation, and allow rock-stratigraphic units to be fitted into a time-stratigraphic classification.

Relation to Morphologic Features

As yet, the younger till units in the Danville area have not been related to local morainal ridges. Before that is possible, it will be necessary to demonstrate that a particular till is responsible for a particular morphologic feature. For example, in the Danville area (except for the Power Plant Section), the sections are located at the frontal margin of a ridge that has been called the Newtown Moraine (Willman and Frye, 1970). The Snider Till Member does not appear to extend beyond these sections and is present only in topographic locations above an elevation of about 640 feet. As this is also the approximate elevation of the drift plain to the south, it suggests that the till could thicken and be responsible for the topographic ridge just north of the sections. Conversely, the Snider Till could have about the same thickness to the north, veneering an older topographic feature. The relations of the Glenburn and Batestown Till Members to end moraines to the south are not known.

CONCLUSIONS

Eight till units have been recognized in the Danville area in the development of a local rock-stratigraphic classification.

The three oldest units are named (from oldest to youngest) the Harmattan Till Member, the Hillery Till Member, and the Tilton Till Member. They are assigned to the Banner Formation of the Kansan Stage on the basis of correlations of the Tilton Till Member to till in Illinois that is considered to be Kansan (Johnson, 1964; Jacobs and Lineback, 1969; and Willman, Glass, and Frye, 1963). The Harmattan Till Member was deposited by ice from the Lake Michigan Lobe, and the Hillery and Tilton Till Members were deposited by ice from the Lake Erie Lobe. A well-developed Yarmouth Soil is preserved in the Tilton Till Member in two sections.

Two till and two silt units are correlated to known stratigraphic units of Illinoian age. The Petersburg Silt (Willman, Glass, and Frye, 1963) occurs above the Yarmouth Soil and is overlain by the Smithboro Till Member (Jacobs and Lineback, 1969). Colluvium and peaty silt that are assigned to the Mulberry Grove Member (Jacobs and Lineback, 1969) occur above the Smithboro Till Member and indicate a significant withdrawal of the ice before the deposition of the overlying Vandalia Till Member (Jacobs and Lineback, 1969). The Smithboro, Mulberry Grove, and Vandalia are members of the Glasford Formation (Willman and Frye, 1970). The Smithboro was deposited by ice from the Lake Michigan Lobe, but the Vandalia apparently had a Saginaw Bay source. Evidence of the Sangamon Soil has been almost completely eroded from the area.

The three youngest tills and the overlying loess are included in the Woodfordian Substage of the Wisconsinan Stage on the basis of their correlation to known stratigraphic units of Woodfordian age in Illinois (Kempton, DuMontelle, and Glass, 1971; this volume; Frye and Willman, 1960). These tills, which are part of the Wedron Formation, are named (from oldest to youngest) the Glenburn, Batestown, and Snider Till Members, and were deposited by the Lake Michigan Lobe. The Richland Loess forms a surficial mantle over most of the area.

The till units have distinct characteristics that can be noted in the field or determined in the laboratory, and these criteria allow them to be treated as rock-stratigraphic units. Detailed sampling and data on texture, carbonate content, and clay minerals have greatly aided the development of the stratigraphy and have made meaningful correlations possible.

ACKNOWLEDGMENTS

This study has been supported in a variety of ways by the Illinois State Geological Survey. H. D. Glass analyzed the clay minerals and assisted in the interpretation of the clay-mineral data. The grain-size analyses were performed in the laboratories of the Survey under the supervision of W. A. White, and two samples of wood were dated in the Survey's Radiocarbon Dating Laboratory by S. M. Kim. A. M. Jacobs and J. P. Kempton furnished samples from the Vandalia and Champaign-Urbana areas, respectively, for use in regional correlations. J. C. Frye and H. B. Willman examined the sections and discussed stratigraphic problems with us, as did Kempton.

Professor G. W. White, University of Illinois, originally encouraged us to undertake the study. His advice during several field trips to the study area and his reading of the manuscript are most gratefully acknowledged.

REFERENCES

American Commission on Stratigraphic Nomenclature, 1961, Code of Stratigraphic Nomenclature: Am. Assoc. Petroleum Geologists Bull., v. 45, p. 645-60.

Ball, J. R., 1937, The physiography and surficial geology of the Carlinville Quadrangle, Illinois: Illinois Acad. Sci. Trans., v. 30, p. 219-33.

Bleuer, N. K., 1967, Geology of the southeast quarter of the Shelbyville, Illinois, Quadrangle: M. Sc. thesis, Univ. Illinois, 139 p.

Campbell, M. R., and Leverett, Frank, 1900, Description of the Danville Quadrangle, Illinois-Indiana: U. S. Geol. Survey Geol. Atlas. Folio 67, 10 p.

Christiansen, E. A., 1968a, A thin till in west-central Saskatchewan, Canada: Can. Jour. Earth Sciences, v. 5, p. 329-36.

———, 1968b, Pleistocene stratigraphy of the Saskatoon area, Saskatchewan, Canada: Can. Jour. Earth Sciences, v. 5, p. 1167-1173.

———, 1971, Tills in southern Saskatchewan, Canada; in this volume.

Dreimanis, Aleksis, 1962, Quantitative gasometric determination of calcite and dolomite by using Chittick apparatus: Jour. Sed. Petrology, v. 32, p. 520-29.

Ekblaw, G. E., 1957, Glacial drift of western Illinois (map), in Wanless, H. R., 1957, Geology and mineral resources of the Beardstown, Glasford, Havana, and Vermont Quadrangles: Illinois Geol. Survey Bull. 82, 233 p.

Ekblaw, G. E., and Willman, H. B., 1955, Farmdale drift near Danville, Illinois: Illinois Acad. Sci. Trans., v. 47, p. 129-38.

Eveland, H. E., 1952, Pleistocene geology of the Danville region: Illinois Geol. Survey Rept. Inv. 159, 32 p.

Flint, R. F., Colton, R. B., Goldthwait, R. P., and Willman, H. B., 1959, Glacial map of the United States east of the Rocky Mountains: Geol. Soc. America.

Frye, J. C., Glass, H. D., Kempton, J. P., and Willman, H. B., 1969, Glacial tills of northwestern Illinois: Illinois Geol. Survey Circ. 437, 47 p.

Frye, J. C., and Willman, H. B., 1960, Classification of the Wisconsinan Stage in the Lake Michigan glacial lobe: Illinois Geol. Survey Circ. 285, 16 p.

Frye, J. C., Willman, H. B., Rubin, Meyer, and Black, R. F., 1968, Definition of Wisconsinan Stage: U. S. Geol. Survey Bull. 1274-E, 22 p.

Hackett, J. E., 1960, Ground-water geology of Winnebago County, Illinois: Illinois Geol. Survey Rept. Inv. 213, 63 p.

Jacobs, A. M., and Lineback, J. A., 1969, Glacial geology of the Vandalia, Illinois, region: Illinois Geol. Survey Circ. 442, 23 p.

Johnson, W. H., 1964, Stratigraphy and petrography of Illinoian and Kansan drift in central Illinois: Illinois Geol. Survey. Circ. 378, 38 p.

Johnson, W. H., Gross, D. L., and Moran, S. R., 1969, Till stratigraphy of the Danville region, east-central Illinois: Geol. Soc. America Abs. with Programs for 1969, pt. 6, p. 23-24.

Kempton, J. P., 1963, Subsurface stratigraphy of the Pleistocene deposits of central northern Illinois: Illinois Geol. Survey Circ. 356, 43 p.

————, 1966, Radiocarbon dates from Altonian and Twocreekan deposits at Sycamore, Illinois: Illinois Acad. Sci. Trans., v. 59, p. 39-42.

Kempton, J. P., DuMontelle, P. B., and Glass, H. D., 1969, Stratigraphic implications of lithologic and clay mineral data from Wisconsinan tills in east-central Illinois: Geol. Soc. America Abs. with Programs for 1969, pt. 6, p. 25.

————, 1971, Subsurface stratigraphy of Woodfordian tills in the McLean County region, Illinois, *in* this volume.

Kempton, J. P., and Hackett, J. E., 1968a, The Late-Altonian (Wisconsinan) glacial sequence in northern Illinois, *in* Means of correlation of Quaternary succession: Proc. VII Congress of INQUA, Univ. Utah Press, v. 8, p. 535-546.

————, 1968b, Stratigraphy of the Woodfordian and Altonian drifts of central northern Illinois, *in* The Quaternary of Illinois: Univ. Illinois College Agriculture Spec. Pub. 14, p. 27-34.

Krumbein, W. C., 1933, Textural and lithological variations in glacial till: Jour. Geology, v. 41, p. 382-408.

Leonard, A. B., and Frye, J. C., 1960, Wisconsinan molluscan faunas of the Illinois Valley region: Illinois Geol. Survey Circ. 304, 32 p.

Leverett, Frank, 1899, The Illinois glacial lobe: U. S. Geol. Survey Mon. 38, 817 p.

MacClintock, Paul, 1929, I. Physiographic divisions of the area covered by the Illinois drift sheet in southern Illinois. II. Recent discoveries of pre-Illinoian drift in southern Illinois: Illinois Geol. Survey Rept. Inv. 19, 57 p.

Moran, S. R., 1971, Glacio-tectonic structures in till; *in* this volume.

Norris, S. E., and White, G. W., 1961, Hydrologic significance of buried valleys in glacial drift: U. S. Geol. Survey Prof. Paper 424-B, Art. 17, p. B34-35.

Wayne, W. J., 1965, Western and central Indiana, *in* Schultz, C. B., and Smith, H. T. U., eds., Great Lakes-Ohio River Valley: Guidebook for Field Conf. G., VIIth Cong. Internat. Assoc. Quaternary Research; Lincoln, Nebraska Acad. Sci., p. 27-39.

Willman, H. B., and Frye, J. C., 1970. Pleistocene stratigraphy of Illinois: Illinois Geol. Survey Bull. 94, 204 p.

Willman, H. B., Glass, H. D., and Frye, J. C., 1963, Mineralogy of glacial tills and their weathering profiles in Illinois. Pt. I, Glacial tills: Illinois Geol. Survey Circ. 347, 55 p.

————, 1966, Mineralogy of glacial tills and their weathering profiles in Illinois. Pt. II, Weathering profiles: Illinois Geol. Survey Circ. 400, 76 p.

GEOLOGIC SECTIONS

The sections are arranged alphabetically. Color notations and color names refer to those of the Munsell color charts and are for moist samples. Numbers in parentheses at the ends of descriptions are sample numbers.

Drainage Ditch Section of Hungry Hollow

Section in a drainage ditch in the NE¼ sec. 2 and the NW¼ sec. 1, T. 19 N., R. 12 W. This section was described by Eveland (1952) and by Ekblaw and Willman (1955). Much of the section is no longer well exposed, but it is unusual in that it contains all of the till units in the area except the Glenburn Till Member. Figure 7 is a diagrammatic sketch modified from the Eveland and Ekblaw and Willman figures and shows sample locations. The Roman numerals refer to numbers used by Ekblaw and Willman to refer to the various units.

Emerald Pond Section

Section along gravel road parallel to east valley bluff of the Middle Fork, Vermilion River; upper unit best exposed and described 0.25 mile north of east-west black-top road; NE¼ SW¼ SW¼ sec. 33, T. 20 N., R. 12 W. Type section for the Snider Till Member, the Batestown Till Member, and the Glenburn Till Member.

	Thickness (feet)
Pleistocene Series	
Wisconsinan Stage	
Woodfordian Substage	
Richland Loess	
Silt, thin, upper part of surface soil	0.5-1.0
Wedron Formation	
Snider Till Member	
Till, upper 2 feet leached and part of B-zone of surface soil; till, calcareous, 2.5Y 5/3 light olive-brown grading to 2.5 5/2 gray-brown at base of unit, clayey, not many pebbles; shale fragments common; jointed; coarse to medium irregular blocky structure; manganese and iron staining and accumulation of secondary $CaCO_3$ common on joint surfaces; locally some interbedded sand and silt (samples 1-30 top to 1-37 base)	14.5
Silt, calcareous, yellow-brown, laminated	0.1
Sand and gravel, calcareous, yellow to dark gray-brown, mostly well-sorted coarse sand; locally, upper 1 foot is coarse sandy gravel	2.0
Batestown Till Member	
Till, calcareous, 2.5Y 5/4 light olive-brown grading to 5Y 4/1 dark gray at 6.5 feet below top; upper portion contains many sand and silt stringers with no preferred orientation and a boulder concentration, with thin silt bed 4 feet below top; lower portion is silty, soft, with small irregular blocky to platy structure (1-1 top to 1-10 base)	13.0
Glenburn Till Member	
Till, calcareous, 7.5YR 4/2 brown to dark brown at top to 10YR 3/3 dark brown at base; has pinkish cast; joints prominent with iron staining and oxidation of till along joint surfaces; sand stringers locally in unit (1-11 top to 1-25 base). Radiocarbon date of wood from base > 38,000 radiocarbon years B. P. (ISGS 15)	18.5

Thickness
(feet)

Kansan Stage
  Banner Formation
    Tilton Till Member
      Till, calcareous, slightly oxidized, 10 YR 3.5/3 dark brown to 10 YR 4/2 dark gray-brown at base; numerous 1- to 3-inch horizontal and irregular sand zones; concentration of white sandstone fragments locally at upper contact; oxidized joints less prominent (1-26 top to 1-29 base)   10.0

                                                         Total   58.6

Harmattan Strip Mine Section

Section measured in two lifts along active high wall of the mine located in the E½ NW¼ NE¼ sec. 4,T. 19 N., R. 12 W. Type section for the Harmattan Till Member.

Upper lift

Pleistocene Series
  Wisconsinan Stage
    Woodfordian Substage
      Richland Loess
        Silt, noncalcareous; contains surface soil   1.5
      Wedron Formation
        Snider Till Member
          Sand, noncalcareous, rusty orange-brown, massive, well-sorted; 3-inch clay band near base   2.5
          Sand and gravel, noncalcareous, poorly sorted; upper 12 inches dark reddish brown; lower portion yellow to olive-brown   3.0
        Batestown Till Member
          Till, calcareous, 5Y 4/3 olive to 5Y 5/2 olive-gray in upper 1.5 feet; lower portion 5Y 4/1 dark gray; pebble concentration at upper contact; breaks into small irregular blocky fracture (4-1 top to 4-4 base)   4.0
  Illinoian Stage
    Glasford Formation
      Vandalia Till Member
        Till, calcareous, 5Y 4/1.5 dark gray to olive-gray in upper 2 feet, massive, soft; pebble concentration at top of unit; lower portion 10YR 3.5/2 dark gray-brown to very dark gray-brown; blocky structure; locally discontinuous sand and silt stringers (4-5 top to 4-14 base)   9.5
        Till, calcareous, 10YR 4/2 dark gray-brown at top to 10YR 4/1 dark gray at base, harder, more massive (4-15 top to 4-22 base)   7.5

                                                         Total   28.0

Lower lift

Illinoian Stage
  Glasford Formation
    Vandalia Till Member
      Till, calcareous, 2.5YR 5/2 dark gray-brown, massive with only a few joints, hard; thin sand zones in the till and a thin sand at the base; sands contain small wood fragments (3-1 top to 3-2 base)   3.5

Thickness
(feet)

Kansan Stage

    Banner Formation

        Tilton Till Member

Till, calcareous, 10YR 3/1 very dark gray, massive; locally contains thin, persistent gray sand zones or inclusions of contorted yellow sand and gravel (3-3 top to 3-10 base)   9.0

Till, calcareous, 10YR 4/1 dark gray, massive; fewer pebbles and more sand zones (3-11 top to 3-12 base)   2.0

Sand, calcareous, gray, medium-textured, well-sorted, moderately cross-bedded   1.0

Gravel, calcareous, gray to yellow-brown, poorly sorted; some interbedded till; piping common   2.0

        Hillery Till Member

Till, calcadeous, 7.5YR 3.4/2 dark brown in upper 5 feet, 5YR 3/3 dark reddish brown in lower 5 feet, very hard; numerous silty streaks in upper 3 feet of till; fewer pebbles than in till above (3-13 top to 3-22 base)   10.0

Sand, calcareous, pink to pinkish gray, well-sorted; some interbedded gravel and silt; coarser units cross-bedded   2.5

        Harmattan Till Member

Till, calcareous, 5Y 4/1.5 dark gray to olive-gray, hard, pebbly; contains much woody material, brownish in color; local channels of gray to olive sand at upper contact (3-23 top to 3-24 base)   3.0

Till, calcareous, same color as above, less pebbly; not as much wood; thin sand at upper contact (3-25 to 3-29 base)   7.0

Till, calcareous, same color as above, more pebbly; local bedrock common (3-30 top to 3-31 base)   3.0

Kansan Stage or older deposits

Clay, calcareous, brown to greenish gray; irregular fracture; many slickensides; more laminated at base   1.0

Silt, calcareous, 10YR 3/2 very dark gray-brown, carbonaceous; peat mat at top; fossiliferous, gradational to unit below   1.5

Silt, noncalcareous, 5Y 3.5/2 dark olive-gray, 5GY 3/1 dark greenish gray; where reduced, massive; contains numerous sandstone and siltstone pebbles and some large red chert pebbles   3.0

Gravel, noncalcareous, olive-yellow to dark brown, moderately well-cemented, primarily siltstone and sandstone pebbles, poorly sorted   1.5

Pennsylvanian System

Shale, gray; green residual soil developed on the bedrock locally   ——

                                                  Total   50.0

Haul Road Section

Section north of strip-mine haul road and on south bank of small tributary to the Vermilion River in the SW¼ SW¼ NW¼ sec. 33, T. 20 N., R. 12 W.

Pleistocene Series

  Wisconsinan Stage

    Woodfordian Substage

      Wedron Formation

        Snider Till Member

Thickness
(feet)

Till, calcareous, 5Y 4.5/3 olive to 2.5Y 5/3 gray-brown to light olive-brown at base; irregular blocky structure; secondary $CaCo_3$ in upper 2 feet; irregular oxidation on fracture surfaces (2-1 top to 2-5 base) ... 7.5

Sand and interbedded silt, calcareous, brown; silt at base; coarser at top ... 1.5

Silt, calcareous, olive in upper 4 inches, blue-gray for 12 inches, olive to yellow-brown in lower 8 inches; laminated ... 2.0

Sand and gravel, fine sand, and interbedded till, calcareous; coarser units yellow-brown, fine units gray to gray-brown; sand zones locally well-cemented ... 2.0

Batestown Till Member

Till, calcareous, 2.5Y 4.5/4 light olive-brown to olive brown at top to 5Y 4/1 dark gray at base; pebble concentration 1.5 feet below top; silt stringers and sand inclusions common about 2 feet below top; lower portion of till harder and more blocky (2-6 top to 2-11 base) ... 12.0

Glenburn Till Member

Till, calcareous, 10 YR 4/2 dark gray-brown at top to 10 YR 4/1 dark gray 5 feet below top; locally sand and/or pebble concentration at upper contact; some wood and carbonaceous material present near base; till oxidized along joints (2-12 top to 2-22 base) ... 14.0

Total   39.0

Power Plant Section

Located on the east bank of the Middle Vermilion River just north of bridge in NW¼ SW¼ SW¼ sec. 21, T. 20 N., R. 12 W. Approximately 6 feet of the Snider Till is exposed 20 yards north of the section above the gravels (sample 11-26). Type section of the Hillery Till Member.

Pleistocene Series

Wisconsinan Stage

Woodfordian Substage

Wedron Formation

Snider Till Member

Gravel, calcareous, rusty brown; most pebbles less than 1 inch in diameter, well-rounded, flat, slabby shape; bedding obscure ... 6.0

Batestown Till Member

Till, calcareous, 2.5 5/2 gray-brown, silty, clayey; irregular blocky structure; slight gravel concentration on upper contact (11-1 top to 11-2 base) ... 1.2

Illinoian Stage

Glasford Formation

Vandalia Till Member

Till, silt, and interbedded sand, calcareous; beds vary from 0.3 to 1.0 feet thick; till, 10YR 5/4 yellowish brown to 10YR 6/3 pale brown; sand, yellow-brown; maroon streaks near top (11-25 top to 11-23 base) ... 3.0

Till, sand, and interbedded silt, calcareous; beds vary from 0.5 to 2.0 feet thick; till, 10YR 4/3 brown to dark brown and contains sand stringers; 4-inch sand at base highly ozidized (11-20 top to 11-22 base) ... 6.0

Till, calcareous, 10YR 4/2.5 dark gray-brown to brown; irregular blocky structure; a few small wood fragments; 10YR 4/1 dark gray at base (11-3 top to 11-6 base) ... 6.5

	Thickness (feet)

Silt, calcareous, light gray — 0.5

Gravel, calcareous, rusty brown, sandy — 1.0

Till, calcareous, 10YR 4/1 dark gray; much irregular interbedded oxidized sand in upper portion; iron stains on joints (11-7 top to 11-11 base) — 7.5

Kansan Stage

  Banner Formation

    Tilton Till Member

Till, calcareous, 10YR 4/2 dark gray-brown, hard, blocky; iron stains on joints; boulders 4 to 6 inches in diameter concentrated near base; 9 inches of sand about 3 inches below top, light gray, medium to fine with a few pebbles (11-12 top to 11-13 base) — 4.0

    Hillery Till Member

Till, calcareous, 5YR 3/2 dark reddish brown; massive to coarse blocky structure; very hard (11-14 top to 11-19 base) — 7.5

Gravel, calcareous, clayey; dark gray-brown to olive-gray; pebbles are well-rounded; contains some till-like material; red sand at base — 1.0

Pennsylvanian System

  Shale, gray

Total   44.2

School House Branch Section of Hungry Hollow

Located along the east bank of a meander in the SE¼ NE¼ NE¼ sec. 2, T. 19 N., R. 12 W. Type section of the Tilton Till Member.

Pleistocene Series

  Wisconsinan Stage

    Woodfordian Substage

      Richland Loess

Silt, noncalcareous, massive, yellow-brown to dark brown; contains surface soil — 3.0

Silt, calcareous, stratified, yellow-brown with gray mottling, not present on north portion of exposure — 1.5

      Wedron Formation

        Batestown Till Member

Colluvium, noncalcareous, silt, sand, and tabular rock fragments; north end of exposure only — 0.5

Till, noncalcareous, 2.5Y 5/4, light olive-brown, clayey; few pebbles; dark brown clay skins; only 5 feet exposed laterally at north end of exposure; probably slumped to some degree (5-14 center) — 1.0

  Illinoian Stage

    Glasford Formation

      Vandalia Till Member

Gravel and interbedded sand, calcareous, gray to gray-brown; high-angle cross-bedding dipping to the south; upper 6 inches colluviated and poorly sorted — 2.0

	Thickness (feet)

Sand, calcareous, yellow-brown to gray-brown, well sorted; medium size; angular to subangular; lower 6 inches gravelly — 4.0

Till, calcareous, 10YR 3.5/3 dark brown in upper 6 inches; grades to 10YR 4/1 dark gray at base; hard; coarse blocky structure; stains on joints common; lower portion of till contains streaks of sand and silt (5-1 top to 5-6 base) — 6.0

Mulberry Grove Member

Silt, calcareous, carbonaceous, very dark brownish gray; contains wood fragments; lower 3 inches gray, not calcareous or carbonaceous; peat mat locally at top of unit (5-15 top to 5-17 base) — 1.0

Clay, noncalcareous, gray to brownish gray; contains a few pebbles; locally carbonaceous in upper part; cracks on surface when dry (5-18 and 5-19) — 1.0-1.5

Colluvium, noncalcareous, sandy, pebbly clay, yellow to olive-brown, faintly laminated; lower portion gray and till-like (5-20 top to 5-22 base) — 0.5-1.5

Smithboro Till Member

Till, calcareous, 10YR 3/3 dark brown, very silty with only a few pebbles; contains abundant wood fragments and a few molluscan fossil fragments (5-7 top to 5-10 bottom) — 2.0-4.0

Petersburg Silt:

Silt, noncalcareous; gray, gray-brown, and olive-brown; massive except for color banding; contains a few pebbles; sparsely fossiliferous (5-23 top to 5-25 base) — 1.5

Kansan Stage

Banner Formation

Tilton Till Member (contains Yarmouth Soil)

Till, noncalcareous, 2.5Y 3/2 very dark gray-brown in upper 2 feet; iron stains on joints; next 2 feet more gravelly and sandy, dark gray with iron staining and mottled appearance; next 1.5 feet sandy, olive-brown with iron stains and pebble concentration at base; lower 1 foot till, 10YR 3/3 dark brown (5-26 top to 5-36 base) — 6.5

Till, ealcareous, 10YR 3.5/1 dark gray (5-37 and 5-38) — 1.5

Sand, calcareous, yellow-brown to brown; fine gravel at top and grades to fine sand at base — 1.5

Hillery Till Member

Till, calcareous, 5YR 3/3 dark reddish brown, very hard; base not exposed (5-12 top to 5-13 base) — 1.5

Total   35.0

*Subsurface Stratigraphy of the
Woodfordian Tills in the
McLean County Region, Illinois*

**J. P. Kempton, P. B. DuMontelle,
and H. D. Glass**

ABSTRACT

Five till units within the Wedron Formation (Woodfordian Substage) have been substantiated in the McLean County area of Illinois by subsurface data, principally from closely spaced split-spoon samples of more than 60 borings. The Farmdale Silt (Farmdalian Substage), widely distributed throughout the area, provides an excellent stratigraphic datum at the base of these tills.

The lower two till units are generally rather pinkish or brownish gray, whereas the upper three units are generally gray or brownish gray. The lower two tills also contain less illite than do the upper three. Each of the units can be distinguished from the others by grain size, as well as by overall texture, clay-mineral composition, color, and stratigraphic position.

Repetitive till sequences and sheared peats that have been observed locally within the Le Roy and Bloomington Moraines indicate that shear stacking was one mechanism responsible for development of the moraines. Stratigraphic evidence in this region suggests that morainal ridges locally reflect the pre-existing topography. In the southeastern part of the area, east of an interlobate reentrant, three of the five Woodfordian till units are correlated across the interlobate reentrant into the western part of the region.

INTRODUCTION

McLean County and adjacent areas of central Illinois contain a sequence of glacial deposits that averages nearly 300 feet thick and is locally more than 400 feet thick. Deposits of the Wisconsinan Stage locally compose more than half of the total drift present, ranging up to nearly 200 feet thick, but averaging about 60 feet thick regionally. Figure 1 shows the area of the state covered by drift of Wisconsinan age, the principal Woodfordian moraines, and the outline of the region discussed in this report. The area studied includes most of McLean County and parts of De Witt, Champaign, Tazewell, Piatt, Woodford, and Logan Counties (Fig. 2).

The work reported here is part of a broader subsurface study made to define the stratigraphic position and distribution of sand and gravel aquifers

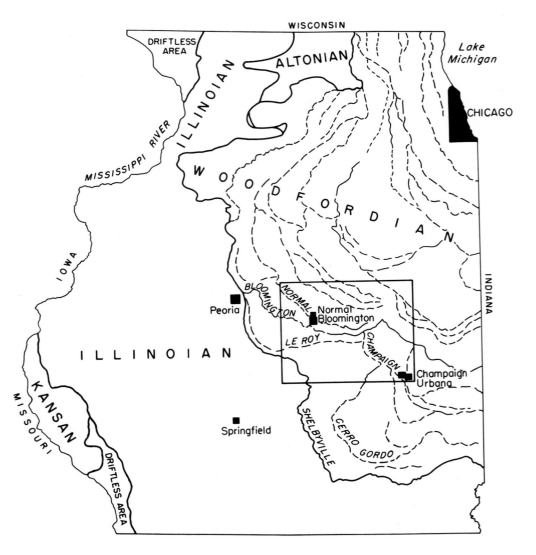

Figure 1. Outline of study area and outer margin of principal Woodfordian moraines.

and the engineering properties of the various drift units in the McLean County region. Representative split-spoon samples from more than 60 foundation borings, many of which have penetrated the entire Wisconsinan drift sequence, provided the basic stratigraphic data for the study. Our project was begun when initial examination of the cores and exposures from central and southern McLean County revealed structural features, stratigraphic relations, and till-distribution patterns that appeared unrelated to morainal features. This report summarizes the results of the study dealing with Wisconsinan deposits and suggests possible relations between Woodfordian till units and surficial and structural features.

The glacial geology of McLean County has received considerable attention, owing in part to the availability of exposures where multiple till units can be viewed in stratigraphic sequence and in part to the numerous moraines that cross the area. The massive Bloomington Moraine has been particularly significant in the development of a geologic history of the region.

One of the earliest descriptions of the glacial geology of the McLean County region, and probably the most comprehensive, is in Leverett's study (1899) of the Illinois Glacial Lobe. Leverett described in some detail the

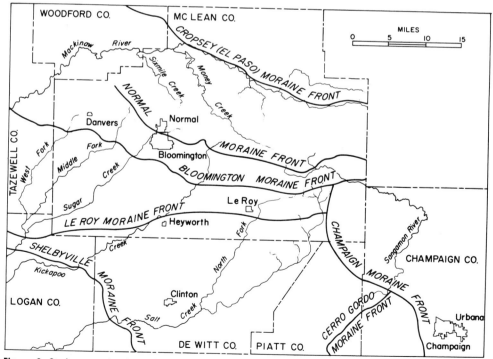

Figure 2. Study area showing principal moraine fronts. (Cropsey Moraine renamed El Paso Moraine by Willman and Frye, 1970.)

morainic systems, and the occurrences of a "buried soil" (probably in the Farmdale Silt) and its general elevation throughout the region. Later, Ek-blaw (1946) described a multiple-till exposure in a cut for the Lake Bloomington spillway, suggesting that these tills were related to the Wisconsinan moraines to the south. Horberg (1953) described the pre-Wisconsinan deposits in detail, principally from study of well samples, but also from many exposures in which Wisconsinan deposits also were found. He noted differences in colors of till along the Bloomington Moraine east of the city of Bloomington, and also presented a contour map of the "buried Sangamon plain," which in present terminology is the top of the Farmdale Silt, thus defining the base of the sequence of deposits of Woodfordian age.

More recent regional studies (Frye, Glass, and Willman, 1968; Kempton and Hackett, 1968; McComas, 1969) have shown that the Woodfordian tills of central and northern Illinois can be differentiated into two distinct groupings, or families: (1) a lower family of reddish brown, pinkish gray to brownish gray, generally sandy, silty tills, of which the Shelbyville, Le Roy, Bloomington, and Metamora Moraines are composed; and (2) an overlying family of gray, generally silty, clayey tills of the Lake Michigan Lobe, that form the younger moraines of the area, such as the Normal and Crop-sey* Moraines. The lower family of tills is distinguished mineralogically from the upper by a distinctly lower illite content. The southwestern boundary of the upper family of tills has been considered to be the outer margin of the Normal Moraine (Frye, Glass, and Willman, 1968). Both general families of lithologies can be recognized throughout the McLean County region.

Our purpose in this project was to define the subsurface sequence, characteristics, and regional correlations of the various Wisconsinan units in the McLean County region. The distribution patterns and boundaries of the tills suggested in this study are subject to modification when future field mapping can be coordinated with the subsurface data. However, the stratigraphic units recognized and described should provide a basis for mapping the surface tills.

STRATIGRAPHY

The sequence and character of the tills of the Wedron Formation (Frye, Willman, Rubin, and Black, 1968) in the McLean County Region, as well as Wisconsinan Stage deposits here, are shown in Table 1. Sand and gravel deposits related to one or more of the till units are not included in the table. The Wisconsinan deposits overlie a relatively consistent sequence of Illinoian deposits in which the Sangamon Soil is frequently preserved at the

*Renamed El Paso in Willman and Frye (1970).

top. Numerous borings have penetrated the uppermost Illinoian till, which is also exposed in at least two sections in McLean County — the Danvers Section (Frye, Glass, and Willman, 1962, p. 50-51) and the Lake Bloomington Spillway Section (Leonard and Frye, 1960, p. 29).

Roxana Silt

The oldest deposit of Wisconsinan age recognized in the region is a silt, probably loess, that rests on an accretion-gley of probable Sangamonian age, underlies the Farmdale Silt, and has been correlated with the Roxana Silt (Frye and Willman, 1960). It has not been identified in any of the samples taken from borings used in this study, but has been noted in the Danvers Section (Frye, Glass, and Willman, 1962), where it is 1.5 feet thick and is a noncalcareous, very dark gray, clayey silt, with very little sand and only an occasional small pebble.

Farmdale Silt

The Farmdale Silt (Frye and Willman, 1960) is a widespread marker bed throughout the McLean County region, and can easily be recognized from

TABLE 1—STRATIGRAPHIC UNITS OF WISCONSINAN AGE IN STUDY AREA

STAGE	SUBSTAGE	UNIT		TYPICAL TILL COLOR and RADIOCARBON DATES B.P.		Samples	Grain size (%)			Clay minerals (%)			Carbonates (counts/sec)	
							Sand	Silt	Clay	Montmorillonite	Illite	Chlorite & kaolinite	Calcite	Dolomite
WISCONSINAN	WOODFORDIAN	WEDRON FORMATION	RICHLAND LOESS											
			1	Oxidized: tan, yellowish-brown Unoxidized: gray, olive-gray	Devonian black shale, spores	30	15	49	36	2	82	16	10	16
			2	Oxidized: tan, yellowish-brown Unoxidized: gray, brownish-gray		150	27	45	28	3	79	18	14	21
			3	Oxidized: yellowish-brown, tan-brown Unoxidized: brownish-gray		50	39	38	23	6	76	18	16	18
			4	Oxidized: pink, brown, pinkish-tan Unoxidized: reddish-brown, pinkish-gray, brownish-gray		250	30	40	30	8	70	22	22	33
			5	Oxidized: not observed Unoxidized: brown, dark-brown		30	24	54	22	20	59	21	10	15
		MORTON LOESS		(20,500 ±700 from moss at top) Danvers Section										
	FARMDALIAN	FARMDALE SILT		(21,950 ±500, youngest date in area) (27,200 ±900, oldest date in area)										
	ALTONIAN	ROXANA SILT												
SANGAMONIAN		SANGAMON SOIL												
(ILLINOIAN and older drift)														

water-well samples and drillers' descriptions. It is usually a very dark brown organic silt, in which the organic component ranges from minor amounts of silt-sized wood fibers and other plant debris to a true peat that locally includes abundant pieces of wood. It averages about three feet thick and is commonly as much as five feet thick in McLean County. Its position is rather well defined by drilling records for the area. In addition, the elevation of the top of the unit has been mapped in some detail by Horberg (1953, Fig. 8, p. 26), although he placed the unit within the Sangamonian Stage. Recent drilling has only slightly modified Horberg's mapping of this unit in McLean County. Six radiocarbon dates determined from samples taken at various intervals within the unit at five localities in the area range from 21,950 ± 500 (I-2517) to 27,200 + 1,000, − 900 (I-2220) radiocarbon years B.P. (Table 2).

Morton Loess

The Morton Loess (Frye and Willman, 1960; Frye, Glass, and Willman, 1968) occurs locally above the Farmdale Silt and below the Woodfordian tills of the Wedron Formation. It is exposed only in the Danvers Section and has not been recognized in the subsurface. It is a yellow-tan to gray, massive silt, about 2½ feet thick. Locally, it contains a few molluscan fossils. A radiocarbon date of 20,500 ± 600 radiocarbon years B.P. (W-483) was determined from a sample of moss taken from the top of the unit and directly below the overlying till.

TABLE 2—RADIOCARBON DATES FROM FARMDALIAN AND
WOODFORDIAN DEPOSITS IN STUDY AREA

Locality	Unit	Radiocarbon Date	Lab No.
(1) Danvers Section	Morton Loess	20,500 ±600	W-483
	Farmdale Silt	25,150 ±700	W-406
	Farmdale Silt	23,900 ±200	ISGS-12
(2) Farmer City	Farmdale Silt	21,950 ±500	I-2517
(3) Le Roy	Farmdale Silt	27,200 +1000 −900	I-2220
(4) Normal	Farmdale Silt in Unit 4	22,450 ±500	I-2518
(5) Damsite test boring on Six Mile Creek	Farmdale Silt	26,500 +900 −800	I-2218
(6) Lake Bloomington spillway	Illinoian silt	>40,000	ISGS-16

Wedron Formation

The tills and related deposits stratigraphically framed by the underlying Morton Loess and the overlying Richland Loess have been assigned to the Wedron Formation (Frye, Willman, Rubin, and Black, 1968). These tills, in general, form a series of onlapping layers in northeastern Illinois, with the lowermost till reaching farthest south and west. In the McLean County region, two principal families of tills can be distinguished within the Wedron Formation. Within these two families, five informal subsurface rock-stratigraphic units are differentiated — Units 1, 2, and 3 within the upper family and Units 4 and 5 within the lower family (Table 1). Grain size and clay-mineral composition serve to distinguish each unit (Fig. 3), together with color, texture, and stratigraphic position.

Unit 5, the basal till unit, is a dark brown, silty till that locally contains abundant organic fibers. Mineralogically, it is characterized by the highest content of expandable clay minerals (referred to as montmorillonite) and the lowest content of illite of all the tills in the Wedron Formation in this

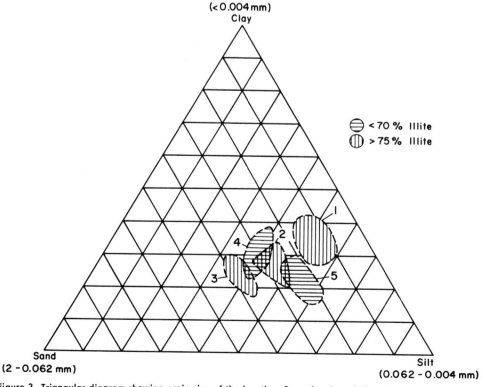

Figure 3. Triangular diagram showing grain size of the less-than-2-mm fraction of till units and their relation to clay-mineral composition.

area. It averages 20 percent montmorillonite, 59 percent illite, and 21 percent chlorite plus kaolinite. Grain-size analyses of the unit showed an average of 24 percent sand, 54 percent silt, and 22 percent clay. Unit 5 has not been found exposed in the area. It is generally less than 15 feet thick where encountered in the subsurface and averages about 5 feet thick. Unit 5 is much siltier than is the overlying Unit 4 (Fig. 3), and can be distinguished from all other units on the basis of clay-mineral composition and color.

Till Unit 4 is, on the average, the thickest unit of the Wedron Formation in the McLean County region. Along the crest of Bloomington Moraine, in the vicinity of Danvers, Unit 4 locally may be as much as 180 feet thick, thinning slightly eastward to less than 120 feet east of Bloomington (Figs. 4, 5). Both north and south of the Bloomington Moraine, the average thickness of this till is about 50 feet, although in some places only 3 to 5 feet of till is found. It is normally a reddish brown to pinkish gray till, though locally it grades into brownish gray. It can nearly always be differentiated from the upper three units in the region on the basis of color and of clay-mineral composition, particularly when one or more of the upper units are found in sequence with Unit 4. Unit 4 is slightly silty, averaging 30 percent

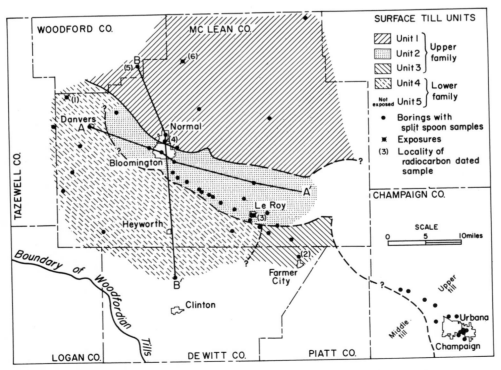

Figure 4. Distribution of till units based on subsurface data. A-A' and B-B' are lines of cross section.

sand, 40 percent silt, and 30 percent clay. The average clay-mineral content is 8 percent montmorillonite, 70 percent illite, and 22 percent chlorite plus kaolinite. No significant change in clay-mineral composition or grain size of the unit occurs with the changes in color. Unit 4 contains the greatest amount of carbonate of all five units (Table 1).

In the upper family of tills, three units have been differentiated in the region. All contain some Devonian spores and black shale, which apparently increase in abundance upward through each succeeding unit.

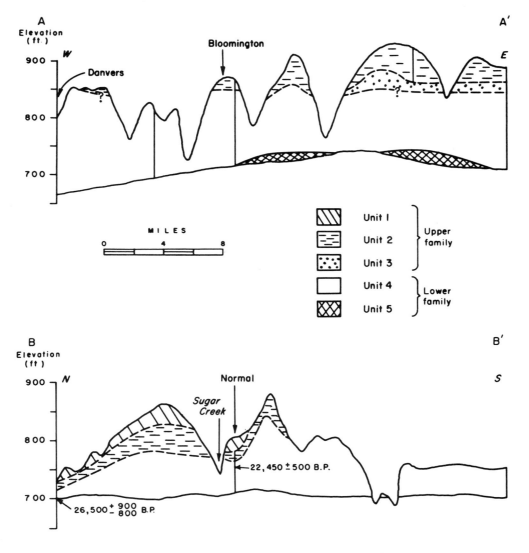

Figure 5. Generalized cross sections of the Woodfordian tills west-east (A-A') along, and north-south (B-B') across, the Bloomington Moraine.

Unit 3 has been encountered only in the southeastern part of the McLean County region (Fig. 4), where it occurs as the surface drift above Unit 4. Its position in sequence between Units 2 and 4 in the subsurface, although inferred, has not been established. This till is the sandiest of the five composing the Wedron Formation of this region and can generally be distinguished on this basis (Fig. 3). It also can be distinguished from the overlying units in the subsurface by a slight brownish tint, as opposed to the more nearly gray color of the upper two units. Devonian black shale and spores are present, but are usually sparse. Unit 3 is characterized by a clay-mineral composition averaging 6 percent montmorillonite, 76 percent illite, and 18 percent chlorite plus kaolinite, and by an average grain size of 39 percent sand, 38 percent silt, and 23 percent clay. It generally averages about 25 feet thick, but shows considerable local variation.

Unit 2 is typically a gray, silty till, containing locally abundant black shale pebbles (principally of Devonian age) and significant numbers of Devonian spores. A pinkish tint has been noted in some samples along the outer margin of the till. A maximum thickness of 60 feet of this unit has been encountered east of Bloomington. The clay-mineral composition of the till averages 3 percent montmorillonite, 79 percent illite, and 18 percent chlorite plus kaolinite, while the grain size averages 27 percent sand, 45 percent silt, and 28 percent clay (Table 1). Unit 2 is locally similar in grain size (Fig. 3) and color to Unit 4, particularly along its southern margin. However, the clay-mineral composition readily distinguishes the two units (Fig. 3).

Unit 1, the youngest till unit, covers much of the northern half of the region, although it is relatively thin and patchy. The maximum thickness encountered is 30 feet. The unit is gray or yellowish gray to olive-gray, has the highest clay content of any till in the region, and contains the lowest percentage of sand (Table 1). It has an average clay-mineral composition of 2 percent montmorillonite, 82 percent illite, and 16 percent chlorite plus kaolinite, and an average grain size of 15 percent sand, 49 percent silt, and 36 percent clay (Table 1).

DISTRIBUTION OF TILL UNITS

A sufficient coverage of the McLean County region by subsurface data is now available to suggest a general distributional pattern of the till units identified in the region. The distribution of each unit shown on Figure 4 is based principally on the identification of that unit both in boring samples and in a few sampled exposures. Surficial features such as moraines or streams are used as boundaries of these units where such features fall near the outermost known occurrence of a given till unit. More precise boundaries await detailed field mapping.

McLean and De Witt Counties

All till units except Unit 5 occur at the surface in some part of McLean County or northern De Witt County. Unit 4, the uppermost of the reddish brown family of tills, characterized by an illite content of 70 percent in the clay fraction, occurs in an area generally southwest of a line extending from north of Danvers southeastward to south of Le Roy (Fig. 4). Correlation of Unit 4 throughout the area is based on color, clay-mineral composition, and grain-size data. Unit 4 is the till that composes the bulk of the material in the Bloomington Moraine, even where it is not the surface drift (Fig. 5). Thicknesses in excess of 100 feet are common for this unit along the crest of the moraine, except where drainageways have cut through the ridge.

Unit 3, the sandiest of the tills and the lowermost of the tills that contain Devonian black shale fragments and spores, is found in a small area in the southeastern part of McLean County and northeastern part of De Witt County. Sparse data from water wells and outcrops suggest that this unit continues some distance both to the southwest and to the east. As Unit 3 has not been found in the subsurface between Units 2 and 4 in the vicinity of Bloomington and Normal, its western subsurface boundary must lie to the east of these cities (Fig. 4). Where Unit 3 is the top till, south of the boundary of Unit 2 and in the vicinity of Le Roy (Fig. 4), it has only been found east of North Fork Salt Creek (Fig. 3). Therefore, the western boundary of Unit 3 in this area has been placed along North Fork. The general land surface and the surface of the Farmdale Silt east of North Fork are approximately 20 to 40 feet lower in elevation than they are west of North Fork.

Unit 2 is well documented as the surface till along the Bloomington Moraine just southeast of Bloomington where it overlies Unit 4. Throughout the area north and east of Bloomington, it occurs in the subsurface below Unit 1 (Figs. 4, 5) and is distinct in color and texture from both Units 1 and 4. However, along the outer margins of the area mapped, between Danvers and Bloomington and in the vicinity of Le Roy, Unit 2 is generally thin and tends to incorporate some of the characteristics of Unit 4, making it difficult to distinguish. However, the clay-mineral composition of the unaltered tills can be used to separate the two units.

The boundary of Unit 2 also is difficult to define because this unit overlaps the major surficial feature of the region, the Bloomington Moraine, just west of Bloomington (Figs. 2, 4, and 5). To the south of the Bloomington Moraine, in the vicinity of Le Roy, a few borings suggest that Unit 2 is present there. The extension of the Le Roy Moraine east of Le Roy would appear to mark the maximum extent of this unit in that area, but the position of the boundary of the unit to the northwest has not been definitely established.

Unit 1, the uppermost of the till units within the Wedron Formation in McLean County, is characterized by low sand and relatively high clay contents. It has been found only north of the outer margin of the Normal Moraine (Figs. 2 and 4), so this boundary has been used to delimit Unit 1.

Champaign and Northern Piatt Counties

Subsurface data from the vicinity of Champaign-Urbana, southeast of McLean County (Fig. 2), have shown the presence of three distinct till units that may be correlated with till Units 2, 3, and 4 of McLean County. These data were obtained from borings made mainly during foundation investigations for a number of new buildings on the University of Illinois campus, for new bridges in Champaign-Urbana for the interstate highway system, and for municipal well sites immediately west of Champaign (Kempton, Sherman, and Cartwright, in preparation).

Color, grain size, and illite content of the Woodfordian tills in the Champaign-Urbana area are summarized in Table 3, and are correlated with units within the McLean County region. A gray silty till very similar to Unit 2 in color, grain size, and clay-mineral composition forms the surface till in the Champaign-Urbana area and probably continues toward the northwest into Pitt County as the surface till on the Champaign Moraine (Figs. 2, 4). The outer margin of this till appears, from the stratigraphic relations found in the study of the Champaign-Urbana well borings, to be at the front of the Champaign Moraine in this area.

Below this upper till is a gray to brownish gray, sandy till, very similar in all characteristics to Unit 3, which occurs in the subsurface throughout the Champaign-Urbana area. This till appears to be the surface till imme-

TABLE 3—SUGGESTED CORRELATION OF WOODFORDIAN TILLS
IN MC LEAN COUNTY WITH THOSE NEAR CHAMPAIGN–URBANA, ILLINOIS

McLean County Region							Champaign-Urbana Vicinity					
Till	Color	Sand (%)	Silt (%)	Clay (%)	Illite (%)		Till	Color	Sand (%)	Silt (%)	Clay (%)	Illite (%)
1	Gray	15	49	36	82	Devonian black shale, spores						
2	Gray Brownish gray	27	45	28	79		Upper	Gray	27	44	29	80
3	Brownish gray	39	38	23	76		Middle	Brownish gray	37	37	26	76
4	Reddish brown, reddish gray, grayish brown	30	40	30	70		Lower	Reddish brown, reddish gray	33	35	32	71
5	Brown	24	54	22	59							

Farmdale Silt, Accretion-gley,
or Sangamon Soil

diately west and south of the boundary of the upper till in west-central Champaign County and northern Piatt County.

Both of these tills in the Champaign-Urbana area contain Devonian black shale fragments and spores, and both are characterized by an illite content of more than 75 percent in the clay fraction. The stratigraphic position of both units and their geographic location provide striking continuity with the distributions of till Units 2 and 3 in the McLean County region, even though moraine configurations suggest that these two areas are in separate glacial sublobes.

Below the upper two tills in the Champaign-Urbana area and directly above the Farmdale Silt or Sangamon Soil is a reddish brown to reddish gray till that is similar to Unit 4 in McLean County. It also is like Unit 4 in that it contains no Devonian shale or spores.

The three Woodfordian tills identified in east-central Champaign County, therefore, are tentatively correlated with Units 2, 3, and 4 in McLean County. Further subsurface and field studies are necessary to substantiate these correlations and the continuity of occurrence of these units between Champaign and McLean Counties.

SPECIAL FEATURES OF THE TILLS

Several features of the tills in the McLean County region are significant to the interpretation and understanding of their distributional pattern and the glacial history inferred from them. These features include vertical and lateral color variations within till units, and shear zones that cause repetitive till sequences and thickness variations. Shear zones and other glaciotectonic structures are discussed by Moran (1971; this volume).

At two localities, extensive lenses of organic silt have been encountered within Unit 4 at least 18 feet above the position of the Farmdale Silt. In borings for an addition to Brokaw Hospital in Normal, such an organic silt zone ranging from 3 to 8 feet thick was penetrated within 10 feet of the top of Unit 4. A radiocarbon date from this zone was reported as 22,450 ± 500 (I-2518) radiocarbon years B.P. The silt thus falls within the time interval of the Farmdalian Substage (Frye and Willman, 1960), although the base of this silt zone is 47 feet above the mapped position of the Farmdale Silt (Horberg, 1953, Fig. 8, p. 26).

An excavation in the village of Le Roy exposed another dark brown, highly organic silt lens, 6 to 12 inches thick, within 2 to 5 feet of the top of Unit 4. Borings at the site encountered the Farmdale Silt 18 feet below this silt lens, with only about 3 feet of Unit 5 intervening.

At both of these localities, Unit 2 directly overlies Unit 4. Thus, without regional control on the top of the Farmdale Silt, the stratigraphy at these sites could be quite misleading, possibly suggesting a higher position for the Farmdale Silt and therefore a thinner sequence of Woodfordian tills, or additional tills within the Wedron Formation. A more likely explanation, however, is that shear zones within the tills, particularly within Unit 4, explain these anomalous positions of the Farmdale Silt, as well as repetitive sequences and excessive local thicknesses of till.

A more striking example of repetitive till sequences is illustrated in a boring drilled at the crest of the Le Roy Moraine about 3 miles southeast of Le Roy (Fig. 4). Regional data suggest that this moraine marks the southern limit of Unit 2. Units 2, 3, and 4 and the Farmdale Silt were recognized from the samples taken from the boring, and five repetitions of two or more of these units can be distinguished.

The other suggested effect of shearing in the McLean County region is the local thickening of Unit 4, where it forms the backbone of the Bloomington Moraine. The uniform character of Unit 4, wherever the unit is encountered within the moraine, and the strictly local occurrence of lenses of sand and gravel suggest that the sorting action of water at the ice margin, normally associated with stagnant ice or melting glaciers, was essentially absent. As the ice flowed up the gentle backslope of a low, broad ridge, developed during previous glaciations (Horberg, 1953, Fig. 8, p. 26), it sheared over less mobile or stagnating ice to the south. This resulted in the stacking of two or more debris-laden wedges of ice, which, upon melting, deposited excessive thicknesses of till. The wedges may have locally carried up with them some of the Farmdale Silt, frozen into their bases, thus explaining the presence of the silt at one or more positions in the till sequence of these areas, considerably above its normal position.

One other significant feature of at least two of the till units is the local variation in color and in other properties. Color variations, both lateral and vertical, are frequently noted in Units 2 and 4, where no appreciable change has taken place in other properties, such as grain size or clay-mineral composition. Borings encountering Unit 4 reveal that the color may vary locally, both vertically and horizontally within a few tens of feet. Color gradations in Unit 4 are predictable — reddish browns and reddish grays occur in the western third of the county, while grayish browns and brownish grays are more common in the east. Locally, and most noticeably in the west, a thin, brownish gray to gray zone has been found at the base of Unit 4. However, the other properties of Unit 4 remain essentially the same, especially the clay mineral composition, so that, in spite of color changes, Unit 4 can always be distinguished from other units.

Unit 2, the unit locally most similar in grain size to Unit 4, may also be quite similar to it in color, particularly along the outer margin of Unit 2 just west and northwest of Bloomington, and along and just south of the Bloomington Moraine southeast of Bloomington. In these areas, Unit 2 has a slightly pinkish cast and is slightly less silty than normal, suggesting that the ice that deposited the till locally incorporated some material from the underlying Unit 4.

DISCUSSION

The rock-stratigraphic identification and classification of the tills of the McLean County region has suggested a somewhat different interpretation of the glacial history of the region from the interpretation based on a morphostratigraphic classification. This difference is mainly due to the quantity of representative subsurface samples available throughout the region, and to the distinctions made between rock-stratigraphic units and physiographic features. In the McLean County region, the distributional patterns, thicknesses, and stratigraphic relations of the various till units suggest that the principal moraines, at least in part, are a composite of two or more tills, and do not necessarily record the history of the most recent glacier that covered them. Similar occurrences have been reported in northern Illinois (Kempton, 1963; Kempton and Hackett, 1968), and others are well documented in northeastern Ohio and northwestern Pennsylvania (White, 1962; Totten, 1969).

The regional stratigraphy proposed also suggests the need for a re-evaluation of the significance of the lobate pattern of the moraines now mapped in this region. The correlation of the three units found in the Champaign-Urbana area with three units present in McLean County (Table 3) is surprising, in light of the lobate pattern separating the two areas (Fig. 1). If these separate lobes have derived material from different sources, it is remarkable that tills in the same stratigraphic positions should be so similar in their properties. However, regardless of the history that might be inferred from such correlations, these units should be mapped as the same rock-stratigraphic units throughout the region if the correlations are confirmed through field examination and additional subsurface data.

SUMMARY

For this study, based principally on data from boring samples, a rock-stratigraphic classification of the deposits has provided a basic stratigraphic framework of the Wisconsinan glacial deposits of the region. Within the Wedron Formation, two previously recognized till families were substantiated for the McLean County region. Within these two families, five subsurface till units were recognized and correlated throughout the region. Identification and definition of each unit were made in the areas where they were most typically developed and stratigraphically well framed. Correlations were then made on the basis of stratigraphic relations by using all available data on the properties of the units.

Glaciotectonic structures, such as shear zones and repetitive till sequences, may explain the excessive local thickness of the till units and may therefore be mechanisms for morainal development. The stratigraphy and distribution of the till units in the region indicate that the moraines are, in part, a composite of two or more tills and that the surface till may be only a veneer over an earlier buried moraine, composed of an older till. Regionally, three of the till units may be correlated across an interlobate reentrant, from the McLean County area to the Urbana-Champaign area.

Detailed field mapping and sampling are now needed for a better definition of the surface boundaries of the till units, variations in physical properties, outcrop patterns, and the precise relation of the subsurface and surface characteristics and features. Additional subsurface data also will help to refine the mapping of the subsurface units. As additional data become available, it should be possible to draw isopach and contour maps of the tops of each unit to make clearer the glacial history of the region. This type of information would also provide the basic Pleistocene stratigraphic framework of the region so vital to land-use planning.

ACKNOWLEDGMENTS

The boring samples used in this study were provided by the Illinois Division of Highways and foundation engineers in the region. Particular thanks are due District 8, Illinois Division of Highways, Ottawa, for drilling and sampling seven sites particularly for this project, and to Robert K. Morse, Engineering Geologist, El Paso, Illinois, for the considerable effort and time he expended in providing information and samples for the study. Radiocarbon dates were provided by the Washington Laboratory of the United States Geological Survey (W), Isotopes Incorporated (I), and the Illinois State Geological Survey Radiocarbon Dating Laboratory (ISGS).

REFERENCES

Ekblaw, George E., 1946, Significant exposure of our Tazewell tills (Abs.): Geol. Soc. America Bull., v. 57, p. 1189-90.

Frye, J. C., Glass, H. D., and Willman, H. B., 1962, Stratigraphy and mineralogy of the Wisconsinan loesses of Illinois: Illinois Geol. Survey Circ. 334, 55 p.

———, 1968, Mineral zonation of Woodfordian loesses of Illinois: Illinois Geol. Survey Circ. 427, 44 p.

Frye, J. C., and Willman, H. B., 1960, Classification of the Wisconsinan Stage in the Lake Michigan glacial lobe: Illinois Geol. Survey Circ. 285, 16 p.

Frye, J. C., Willman, H. B., Rubin, Meyer, and Black, R. F., 1968, Definition of Wisconsinan Stage: U. S. Geol. Survey Bull. 1274-E, p. E1-E22.

Horberg, Leland, 1953, Pleistocene deposits below the Wisconsin drift in northeastern Illinois: Illinois Geol. Survey Rept. Inv. 165, 61 p.

Kempton, J. P., 1963, Subsurface stratigraphy of the Pleistocene deposits of central northern Illinois: Illinois Geol. Survey Circ. 356, 43 p.

Kempton, J. P., and Hackett, J. E., 1968, Stratigraphy of the Woodfordian and Altonian drifts of central northern Illinois, in The Quaternary of Illinois: Univ. Illinois College Agr. Spec. Pub. 14, p. 27-34.

Kempton, J. P., Sherman, F. B., and Cartwright, Keros (in preparation), Hydrogeology of the Champaign-Urbana, Illinois, west well field: Illinois Geol. Survey Environmental Geol. Note.

Leonard, A. B., and Frye, J. C., 1960, Wisconsinan molluscan faunas of the Illinois Valley region: Illinois Geol. Survey Circ. 304, 32 p.

Leverett, Frank, 1899, The Illinois glacial lobe: U.S. Geol. Survey Mon. 38, 817 p.

McComas, M. R., 1969, Pleistocene geology of the Middle Illinois Valley: Ph.D. dissertation, Univ. Illinois (Urbana).

Moran, S. R., 1971, Glaciotectonic structures in drifts, in this volume.

Totten, S. M., 1969, Overridden recessional moraines of north-central Ohio: Geol. Soc. America Bull., v. 80, p. 1931-46.

White, G. W., 1962, Multiple tills of end moraines: U.S. Geol. Survey Prof. Paper 424-C, p. 71-73.

Willman, H. B., and Frye, J. C., 1970, Pleistocene stratigraphy of Illinois; Illinois Geol. Survey Bull. 94, 204 p.

5

Composition

Bimodal Distribution of Rock and Mineral Fragments in Basal Tills

A. Dreimanis and U. J. Vagners

ABSTRACT

Every lithologic component of basal till, fragments of both rocks and their constituent materials, has a bimodal particle-size distribution, if the rock is monomineralic or consists of minerals of similar physical properties. One of the modes is in the clast-size group, the other reflects the mineral fragments in the till matrix. Several such modes develop in the till matrix, if the rock comminuted by glacial transport consists of minerals with differing physical properties.

The clast-size mode is relatively larger than is that of the matrix mode near the source, where the glacier picked up the rock fragments. With increasing distance of glacial transport from the source, the matrix modes become larger and the clast-size modes are reduced or may even disappear.

The matrix modes are restricted to certain particle-size grades, which are typical for each mineral. These mineral grades are called "terminal grades," because they are the final product of the glacial comminution. The particle-size grade produced depends upon the original sizes and shapes of the mineral grains in the rocks, and upon the resistance of each mineral to comminution during glacial transport. Predominance of "terminal grades" over clast-size modes indicates a higher degree of maturity of till, as compared with shorter-transported, less mature tills which consist mainly of clast-size particles.

INTRODUCTION

The conclusions on rules which govern the textural and the lithologic composition of till, as presented in this paper, are based mainly upon inves-

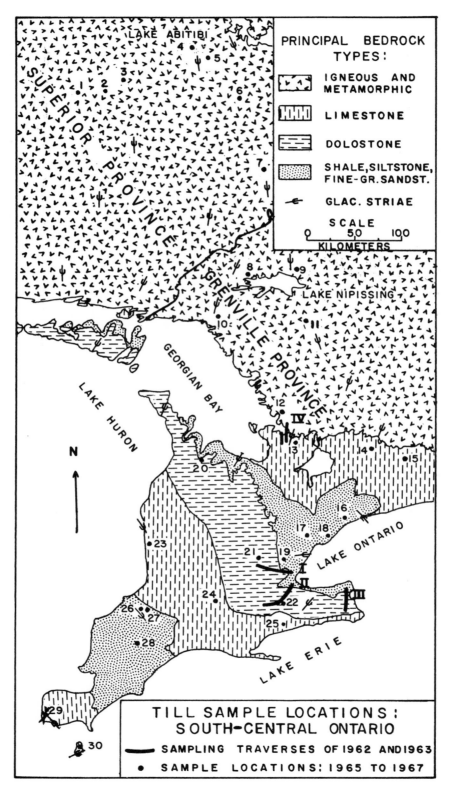

Figure 1. Central and southern Ontario: till sample locations, major bedrock types, and directions of glacial movements as indicated in striae.

tigations of Wisconsin-age tills in Ontario (Fig. 1). These tills were deposited by a continental ice sheet; therefore all their clastic components were derived from drift material which was incorporated in the ice probably exclusively at the base of the glacier. Absence of any evidence of nunataks rising above the Wisconsin ice sheet precluded addition of true superglacial drift. After incorporation of the drift material (fragments of bedrock and of older Quaternary sediments), the drift was transported in the ice either near the base of the glacier ("basal transport") or higher up within the glacial ice ("englacial transport"). Both these modes of transport caused comminution of the rock fragments and of the lumps of sediments carried by glacier. The most probable mechanism by which comminution was accomplished was by crushing and abrasion (Dreimanis and Vagners, 1965), though the disruptive effect of the growth of ice crystals and solution by water should also be considered.

Some of the englacially transported drift material became superglacial during surface melting of the ice sheet, thus producing ablation till. The effect of the superglacial environment upon the initially englacially transported material will not be discussed here, as all the samples investigated were taken from dense basal till, deposited most probably from englacially or basally transported drift (Fig. 2). The composition of the deformed till which has been subjected to hardly any transport, and the waterlaid till, which has resulted from a combination of glacial and lacustrine or marine sedimentation (Fig. 2), will also not be discussed here. It is only the basal till which will be considered.

Various investigations of basal till (for a summary, see Flint, 1957, p. 122-29) point out that its composition depends mainly upon the following two factors: (1) the multilithologic source material, consisting of bedrock and of older Quaternary sediments, and (2) comminution of this material

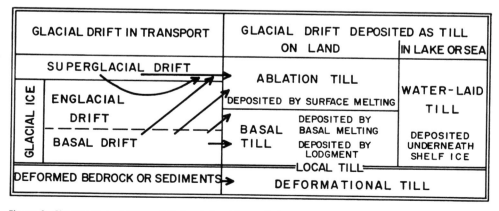

Figure 2. Classification of tills and their relationship to glacial drift in transport. (After Dreimanis, 1967a.)

during glacial erosion and transport, which in turn depends upon the durability of the rocks, mode and position of transport (basal or englacial), and distance of transport.

Investigations of the compositions of tills in Ontario were begun by the senior author in 1950, with their principal objective to develop quantitative criteria for (a) stratigraphic correlations, (b) differentiation of various till units, and (c) indicator tracing in a search for ore deposits. The first phase of these investigations dealt mainly with application of various quantitative laboratory and field methods and modification of them (Dreimanis and Reavely, 1953; Dreimanis and others, 1957; Dreimanis, 1961 and 1962).

Following a standard practice of most till investigators, arbitrary particle-size boundaries were used at the beginning, for both the textural and the lithologic studies. The first step toward selecting a natural particle-size range for quantitative determination of carbonates in the till matrix was stimulated by finding that carbonates had a bimodal distribution in tills and that the carbonate mineral modes were always in the silt- and clay-size fractions (Dreimanis and Reavely, 1953). In order to investigate the distribution of noncarbonate minerals and rocks in tills, and to select the particle-size grades most suitable for their investigation, the next step was to study the factors which determine the composition of till. These problems have been investigated since 1962, particularly by the junior author. By analysing large till samples, an effort was to relate their lithologic composition to (1) the distance from the most probable bedrock source, (2) the mode of transport, and (3) the mode of deposition. Several progress reports have resulted from these studies (Dreimanis and Vagners, 1965; Vagners, 1966; Dreimanis, 1969; Vagners, 1969). As both authors are still continuing these till investigations — one of them on Ontario, and the other in the Atlantic Provinces of Canada — this paper is essentially another progress report.

METHODS OF INVESTIGATION

In the initial study, twenty-five till samples, 0.5-1.5 cubic meters large, collected over limestone and dolostone bedrock areas, were investigated (Vagners, 1966). Main objective of this study was to determine the mode of incorporation of these two rock types in the associated, overlying till and the distribution of their fragments throughout the particle-size range, from cobbles to clay-size, along four selected transects parallel to glacial movement (I through IV in Fig. 1). The samples were split into 18 size fractions according to Wentworth's (1922) scale. The quantitative lithologic composition of each fraction was then investigated separately. In the 0.062-mm-to-256-mm grades, all the rock and mineral grains were identified visually, but in the less-than-0.062 mm grades, only the percentages of calcite and dolomite were determined, using the gasometric method of Dreimanis (1962) (for details, see Vagners, 1966).

After the rule of the bimodal distribution for several rock groups investigated in basal tills was etablished (Dreimanis and Vagners, 1965), more detailed studies were conducted, dealing with the distribution of individual common minerals throughout various particle-size fractions in tills. In this study, begun in 1965, the junior author investigated 30 till samples, randomly collected over various bedrock terrains — Precambrian igneous and metamorphic bedrock, and Ordovician, Silurian, and Devonian limestones, dolostones, and fine-grained clastics in central and southern Ontario (Fig. 1). The samples were taken along transects 300 to 750 km long and parallel to the predominantly north-south to northeast-southwest-oriented glacial movements. The main objective of the study was to determine the results of comminution produced by glacial transport on such common minerals as quartz, the plagioclases, the potassium feldspars, amphiboles and pyroxenes as a group, garnets, heavy minerals, calcite, and dolomite. As in the previous study, each till sample was split according to the Wentworth scale, but this time in the size range from 8 to 0.001 mm. Staining methods were used for identification of several mineral groups in the particle-size fractions coarser than 0.062 mm, but X-ray-diffraction techniques were applied for quantitative determinations in the less-than-0.062-mm particle-size fractions (for details, see Vagners, 1969).

RESULTS AND CONCLUSIONS

Some results of the above investigations have been reported by Dreimanis and Vagners before (1965, 1969). Only the main conclusions will be given here, each of them illustrated by typical examples (Figs. 3-8).

Conclusion I

When, during glacial erosion and transport, rocks become comminuted to their constituent minerals, abrupt transition from rocks to their minerals takes place (Figs. 3 and 4). As a result, at least two modes or two groups of modes develop for each lithic component of the till; one is in the clast size, consisting predominantly of rock fragments, and the other is in the till matrix, consisting mainly of mineral fragments. Several mineral modes may develop in the till matrix, depending on the different physical properties of the minerals present (see Conclusion III for elaboration on the mineral modes).

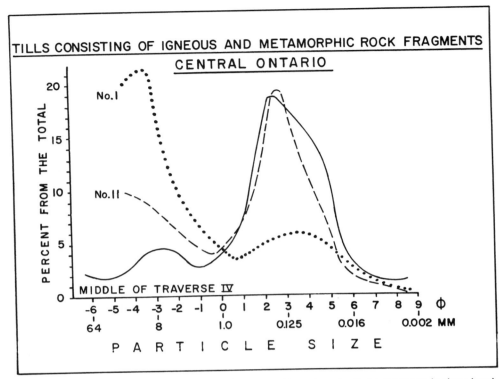

Figure 3. Frequency distribution of igneous and metamorphic rock and mineral fragments in three basal tills from the Canadian Shield, central Ontario. Maximum transport distances of igneous and metamorphic rock fragments prior to the deposition of the tills: No. I: 220 km; No. II and the Middle of Traverse IV: 1,200 km (probably less); see Figure 1 for location of samples. (After Dreimanis, 1967a.)

Conclusion II

Near the source, where the glacier picked up the rock fragments, the clast-size mode is always relatively larger than the matrix mode. For instance, in the local till at Niagara (0-3 km from bedrock source in Fig. 4), the fragments consist mostly of dolostone clasts.

With increasing distance of glacial transport away from the source, the matrix modes grow larger, recording increasing comminution of clast-size particles (Figs. 3 and 4), while the clast-size mode becomes relatively smaller. For instance (Fig. 4), the clast-size dolostone mode has nearly disappeared after the 300-to-500-km transport from the Beekmantown Dolostone bedrock area in the St. Lawrence Lowland down to the Hamilton area at the northeastern end of traverse II in Figure 1.

Conclusion III

The matrix mode, or modes, resulting from multi-mineral rocks, is restricted to certain particle-size grades, typical for each mineral. These

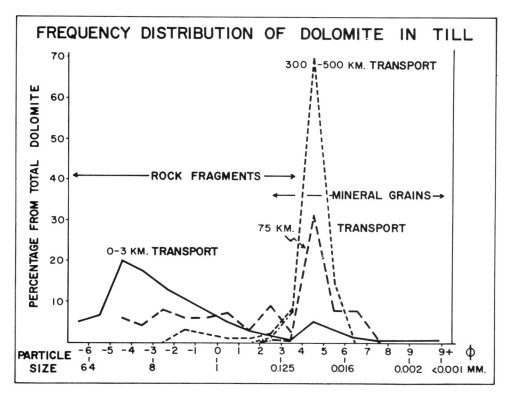

Figure 4. Frequency distribution of dolostone-dolomite in three selected till samples from the Hamilton-Niagara area (Traverses II and III, see Figure 1).

mineral grades are called the "terminal grades," because they are the final product of the glacial comminution. They depend upon the original sizes and shapes of the mineral grains in the rocks, and upon the resistance of each mineral to comminution during glacial transport.

The "terminal grades" of some minerals common in tills are shown in Figure 5. Their gradual development and their final restriction to a certain particle-size range typical for each mineral is demonstrated by two examples: dolomite (Fig. 4), representing a medium-hard mineral with excellent cleavage, and garnet (Fig. 6), a much harder mineral with very poor cleavage.

For each mineral investigated (Fig. 5), the particle-size range of the "terminal grade" appears to be relatively constant, as concluded already by Dreimanis and Vagners (1965). Thus this "terminal grade" is between 0.031 and 0.062 mm for dolomite in Ontario, independent of the distance of its transport. The "terminal grade" of dolomite apparently represents its equilibrium size, beyond which the dolomite grains do not become crushed or abraded any more by lengthy glacial transport.

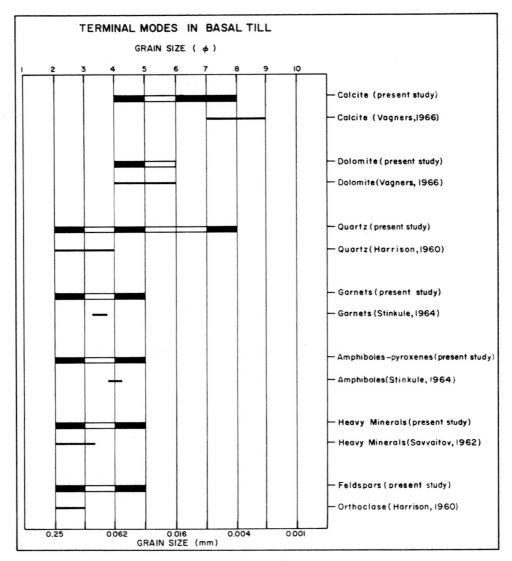

Figure 5. "Terminal grades" (both white and black bars) and major modes in them (black bars) of selected minerals and their groups in basal tills. (After Vagners, 1969.)

Several minerals (Fig. 5) have bi- or tri-modal "terminal grades." These may be explained in several ways. For chemically and mechanically very resistant minerals, such as quartz and garnet, the main cause for the several modes in their "terminal grades" may be the different original sizes of these mineral grains in their source rocks. Thus Figure 6 suggests predominance of finer grained garnet in the Precambrian rocks of the Superior province, while garnets from the Grenville province, which is the main source of garnets in Ontario, show a bimodal distribution, containing both coarse and fine grains. Though the tills of southern Ontario contain greater quantities of garnets from the Grenville than from the Superior province bedrock (Dreimanis and others, 1957), the finer grained mode (maximum between 0.031 and 0.062 mm.) predominates in the tills 300-600 km down-glacier from the Grenville province. Apparently, some of the coarse-grained garnets (0.125-to-0.25-mm mode) have become comminuted during a lengthy transport of hundreds of kilometers from the Grenville province to southwestern Ontario. Therefore the 0.031-to-0.062-mm grade should be considered the true terminal modal class for garnets.

Quartz has a tri-modal distribution in its terminal grade, because it has derived from rocks containing quartz in particle sizes ranging from pebbles to silt size. As quartz has been picked up by the glacier not only from Precambrian rocks but also from belts of the various Paleozoic clastic sediments, it is difficult to determine how much comminution of quartz grains occurred during glacial transport.

Besides the original particle size of minerals and the effect of crushing or abrasion, solution during or after glacial transport must be also considered as one of the factors affecting the "terminal grade." This would apply particularly to the more soluble minerals, such as calcite. This is especially

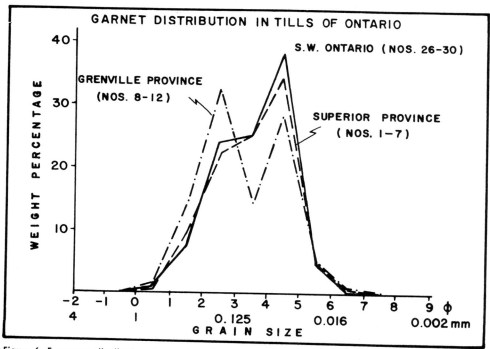

Figure 6. Frequency distribution of garnets in three groups of tills in Ontario. The bedrock source areas of garnets are the Grenville and the Superior provinces of the Canadian Shield. See Figure 1 for locations. (After Vagners, 1969.)

true if the content of the mineral in the till is low, and if the waters circulating through the till are not saturated with calcium bicarbonate. Vagners (1969) noticed that, in those tills which were low in calcite, the fine-grained side of its "terminal grade" is reduced.

The "terminal grades" of all the minerals listed in Figure 5 have been discussed in more detail by Vagners (1969). Because of the variety of factors involved in producing the "terminal grades," these investigations are still far from complete.

Conclusion IV

Predominance of "terminal grades" over clast-size modes indicates a higher degree of maturity of such tills, as compared with short-transported tills, for instance in local moraines, which consist mainly of clast-size particles. The highest degree of maturity of a till is attained when all of its rocks have become comminuted to the terminal grades of all of their constituent minerals. In Ontario, none of the tills have reached this degree of maturity.

SIMILAR INVESTIGATIONS IN OTHER AREAS

The bimodal distribution of rocks and their constituent mineral fragments in tills have been overlooked by most till investigators. However, some have noted either both the rock and the mineral modes, or merely the concentration of minerals in certain particle-size ranges in till matrix. Examples include Harrison (1960), who related till composition in Indiana to its probable source; various workers in the Baltic area, for instance Raukas (1961) in Estonia, Savvaitov (1962) and Stinkule (1964) in Latvia, and Gaigalas (1964) in Lithuania; and in Ohio, Smeck and others (1968) and Wilding and others (1971). Their accounts (see summary in Fig. 6) are merely descriptive; they do not discuss the dynamic aspects of the transition from the clast-size modes to the mineral modes during glacial transport. Nevertheless, their findings on the modes of various minerals in tills are in agreement with those discussed in this paper. Apparently the "terminal grades" which have been found in Ontario are not merely local phenomena. Nevertheless, many other areas should be investigated in order to establish possible variances in the "terminal grades" for each mineral or its varieties, for instance quartz derived from sandstone, siltstones, shales, intrusives, extrusives, metamorphics, etc. Kalinko (1948) and Sindowski (1949) have noticed that also in non-glacial clastic sediments each mineral has its maximum abundance in a specific particle-size range. Apparently "terminal grades" develop not only in tills, but also in other clastic sediments.

Experimental simulation of glacial comminution of rocks and their minerals is difficult, because most of the artificial grinding, milling, or crushing equipment does not work in the same manner as does comminution by a glacier. However, it is interesting to note that artificial grinding in ball mills and pebble mills (Gaudin, 1926) also produced bimodal distribution of rock and mineral fragments in most cases (as in Fig. 7). However, by increasing the time of grinding, the fine-grained mode also becomes progressively finer. The cumulative curve of artificially crushed gneiss (Elson, 1961) also suggests a bimodal distribution. Crushing of hard shale (Elson, 1961), however, did not produce a bimodal distribution of its fragments, because the shale had been comminuted only to approximately 0.05 mm, which is considerably coarser than the individual mineral grains usually present in shale.

The straight-line cumulative curves of the granulometric composition of tills, if plotted on Rosin and Rammler's "law of crushing" paper, mentioned by Elson (1961) as being characteristic for tills, may develop because of the multi-lithologic composition of most tills. If several bimodal distribution curves are superimposed one upon the other, the resulting curve becomes more or less straight. However, if a monomineralic rock type predominates in till, the cumulative curve of its granulometric composition does not approach a straight line (Fig. 8; see also Figs. 5 and 6 in Dreimanis and Vagners, 1969).

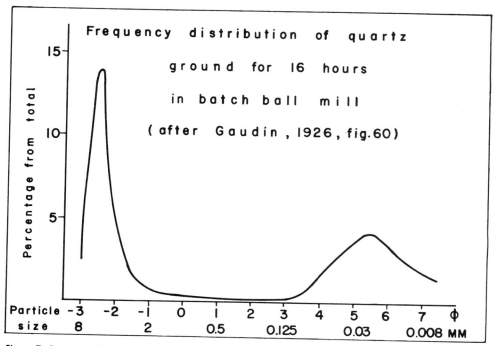

Figure 7. Frequency distribution of artificially ground quartz. (After Gaudin, 1926.)

DOLOMITIC TILLS OF S.W. ONTARIO, CANADA

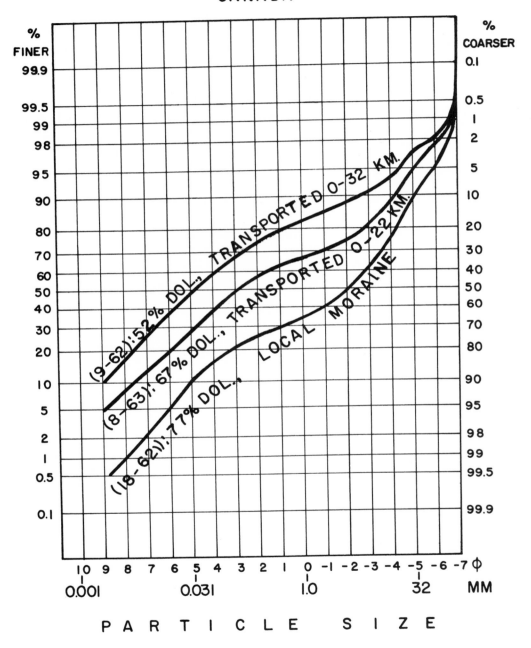

Figure 8. Cumulative granulometric curves of three dolostone-dolomite-rich tills from the traverses I (8-63), II (9-62), and III (18-62). See Figure 1 for locations. (After Dreimanis, 1967b.)

IMPLICATIONS OF THE BIMODAL DISTRIBUTION OF THE ROCK-MINERAL COMPONENTS OF TILL

The bimodal or even multimodal distribution of fragment sizes of each rock and its constituent minerals in till requires that at least two particle-size groups should be investigated for description of lithologic composition of a till (Dreimanis, 1969):

A. the clasts consisting predominantly of rock fragments;
B. the till matrix, consisting predominantly of minerals. Preference should be given to their "terminal grades."

The bimodal distribution should also be considered when separating the till matrix from clasts, for instance when determining the granulometric composition of till matrix. The granulometric composition of tills is strongly governed by their lithologic composition and by the inter-relationship of the clast-size modes and the "terminal modes." Another factor, not discussed here, is the admixture of incorporated nonconsolidated sediments.

ACKNOWLEDGMENTS

The authors gratefully acknowledge support for the till investigations provided by the Geology Department of the University of Western Ontario, the Ontario Research Foundation, the Department of the University Affairs of Ontario, and the National Research Council of Canada, and are thankful to J. L. Forsyth for critical reading of the manuscript.

REFERENCES

Dreimanis, A., 1961, Tills of Southern Ontario: *in* Legget, R. F., (ed.), Soils in Canada, Royal Soc. Canada Spec. Public. No. 3, p. 80-94.

————, 1962, Quantitative gasometric determination of calcite and dolomite by using Chittick apparatus: Jour. Sed. Petrology, v. 32, p. 520-29.

————, 1969, Selection of genetically significant parameters for investigation of tills: Zesz. Nauk. Univ. A. Mickiewicza, Geografia No. 8, Poznan, Poland, p. 15-29.

Dreimanis, A., and Reavely, G. H., 1953, Differentiation of the lower and the upper till along the north shore of Lake Erie: Jour. Sed. Petrology, v. 23, p. 238-59.

Dreimanis, A., Reavely, G. H., Cook, R. J. B., Knox, K. S., and Moretti, F. J., 1957, Heavy mineral studies in tills of Ontario and adjacent areas: Jour. Sed. Petrology, v. 27, p. 148-61.

Dreimanis, A., and Vagners, U. J., 1965, Till-bedrock lithologic relationship: Abstracts, INQUA VII Internat. Congr. Gener. Sess., p. 110-11.

————, 1969, Lithologic relation of till to bedrock, *in* Wright, H. E., Jr. (ed.), Quaternary geology and climate: Washington, D. C., Nat. Acad. Sci., p. 93-98.

Elson, J. A., 1961, The geology of tills: Proceed. 14th Can. Soil Mech. Confer., Nat. Res. Counc. Can. Assoc. Com. Soil and Snow Mech. Techn. Mem. no. 69, p. 5-36.

Flint, R. F., 1957, Glacial and Pleistocene geology: New York, J. Wiley and Sons, 553 p.

Gaigalas, A. I., 1964, Mineralo-petrograficheskii sostav moren pleistocena iugo-vostochnoi Litvy: Lietuvos TSR Mokslu Akad. Darbai, ser. B, v. 4 (39), p. 185-211.

Gaudin, A. M., 1926, An investigation of crushing phenomena: Transact. Amer. Inst. Min. and Metallurg. Engineers, v. 73, p. 253-316.

Harrison, P. W., 1960, Original bedrock composition of Wisconsin till in central Indiana: Jour. Sed. Petrology, v. 30, p. 432-446.

Raukas, A., 1961, Mineralogiia moren Estonii: Eesti NSV Teaduste Akad. Toimetised X Koide, Füüsik.-Matemat. ja Tehn. Teaduste Ser. Nr. 3, p. 244-58.

Savvaitov, A. S., 1962a, O sostave melkozema morennykh otlozhenii v baseine r. Salaca, *in* I. Danilans (ed.), Questions on Quaternary geology, I: Akad. Sci. Latvian SSR, Instit. geol., Riga, p. 115-22.

————, 1962b, O soderzhanii tiazhelykh mineralov v morennykh suglinkakh, *in* I. Danilans (ed.), Questions on Quaternary geology, I: Akad. Sci. Latvian SSR, Inst. geol., Riga, p. 123-28.

Sindowski, F. K. H., 1949, Results and problems of heavy mineral analysis in Germany: a review of sedimentary-petrological papers, 1936-1948: Jour. Sed. Petrology, v. 19, p. 3-25.

Smeck, N. E., Wilding, L. P., and Holowaychuk, N. 1968, Properties of argillic horizons derived from Wisconsin-age till deposits of varying physical and chemical composition: Proc. Soil Sci. Soc. Amer., v. 32, p. 550-56.

Stinkule, A. V., 1964, O raspredelenii khimicheskikh elementov v melkozeme moreny, *in* I. Danilans (ed.), Questions on Quaternary geology, III: State Com. Geol. USSR, Instit. geol., Riga, p. 311-20.

Vagners, U. J., 1966, Lithologic relationships of till to carbonate bedrock in southern Ontario: M.Sc. thesis, Univ. West. Ontario, London, Canada. 154 pp.

————, 1969, Mineral distribution in tills, south-central Ontario: Ph.D. dissertation, Univ. West. Ontario, London, Canada, 270 p.

Wentworth, C. K., 1922, A scale of grade and class terms for clastic sediments: Jour. Geology, v. 30, p. 377-92.

Wilding, L. P., Dress, L. R., Smeck, N. E., and Hall, G. F., 1971, Mineral and elemental composition of Wisconsin-age till deposits in west-central Ohio, *in* this volume.

Till / a Symposium

Grain-Size and Mineralogical Gradations within Tills of the Allegheny Plateau

David L. Gross and Stephen R. Moran

ABSTRACT

Feldspar content of fine sand of the Lavery, Kent, and Titusville Tills (Wisconsinan) decreased regularly in the direction of ice movement. A corresponding increase in sand content in the same direction is especially well shown in the Titusville Till. The feldspar content also increases upward through the Titusville Till. In stratigraphic sequence, the tills of the Allegheny Plateau grade progressively, older to younger, from a coarse-grained and feldspar-poor (potassium feldspar dominant) composition to a fine-grained and feldspar-rich (plagioclase dominant) composition.

Gradations in composition are the result of a progressive dilution of a fine-textured, feldspar-rich parent material by coarse-textured, quartz-rich debris derived from upper Paleozoic clastic rocks. Areal gradations in till composition reflect greater incorporation of bedrock material with increased distance of transport; vertical gradations in till composition reflect greater incorporation of bedrock material near the base of the till sheet. Differences in mean composition among the various till sheets reflect decreased influence of local bedrock as it was progressively mantled by successive drift sheets.

Gradations in feldspar content indicate that more than 50 percent of the Titusville Till was derived from within 20 miles of the site of deposition. Because the compositional changes occur in the direction of transport, compositional isopleths may be considered equipotential lines for reconstruction of glacial flow lines.

INTRODUCTION

Purpose

Although the problem of ice-erosion has been studied continuously and from many angles, relatively few facts have been tendered in proof or disproof. The literature is largely a repetition of unfounded assertions or of indirect arguments of doubtful validity, conclusions varying with the predilections and field experiences of the disputant (Charlesworth, 1957, p. 218).

Several analytical techniques have been used in attempts to determine the source areas of glacial till materials. Holmes (1952, 1960) in New York State and Gillberg (1965, 1967) in Sweden have used large numbers of pebble counts. Anderson (1957) studied pebble and sand lithologies of the six Wisconsinan glacial lobes of the Central Lowland. Krumbein (1933) studied texture and pebble lithology of till from the Michigan Lobe. Harrison (1960) studied all size fractions of a few till samples from Marion County, Indiana. All of these authors based their conclusions on abundant quantitative data for the tills they studied, so that the opening quotation by Charlesworth does not apply to their papers. The quotation does, however, remain as a warning against making ill-founded conclusions about the origin of tills.

This study examines the grain size and light minerals of the tills of the Allegheny Plateau. The large number of samples used made it possible to detect and quantify regional gradations in these compositional parameters. This evidence permits formulation of hypotheses as to the source areas and glacial mechanisms that contributed to the formation of this drift.

Location

The area under consideration includes most of the glaciated portion of the Allegheny Plateau in northwestern Pennsylvania and northeastern Ohio, in the area of the Grand River and Killbuck Lobes, as shown in Figure 1 (after White, 1969). The hypotheses presented in this paper were formulated principally from work in the Grand River Lobe, which includes the glaciated portions of Erie, Crawford, Mercer, Warren, Verango, Butler, Lawrence, and Beaver counties in northwestern Pennsylvania, and Lake, Geauga, Portage, Stark, Columbiana, Mahoning, Trumbull, and Ashtabula counties in northeastern Ohio.

Stratigraphy

The Pleistocene stratigraphy of the Allegheny Plateau in Pennsylvania is described in considerable detail by White, Totten, and Gross (1969), and in a more general way by White (1969). Tills in this area are classified with a dual system of nomenclature, which uses independent rock units and time units, and are named as formal rock-stratigraphic units of formation rank (Fig. 1). The time-stratigraphic classification uses the terminology of Frye and Willman (1960).

The Ashtabula, Hiram, Lavery, and Kent Tills of the Woodfordian Substage (Wisconsinan) are the surficial tills present throughout most of the Grand River Lobe. Each of these till units is lithologically distinctive (White, Totten, and Gross, 1969) and may be identified in hand specimen by its color, texture, pebble lithology, and degree of compaction, and also by its stratigraphic and geographic position. Although these tills blanket large areas, they are extremely thin (White, 1971; this volume, p. 149).

The Titusville Till of the Altonian Substage (Wisconsinan) is the thickest of the plateau-area tills and often composes the bulk of the drift. Lithologically, it is particularly distinctive. The Titusville Till was chosen for detailed study because it is the most widely traced till unit in the area, extending from the New York-Pennsylvania state line west through all of the Grand River Lobe in Pennsylvania and northeastern Ohio.

The Mapledale Till (Illinoian) is somewhat discontinuous across the area, and is generally thickest in filled valleys. It crops out in a belt a few miles wide adjacent to the southern glacial boundary of the area.

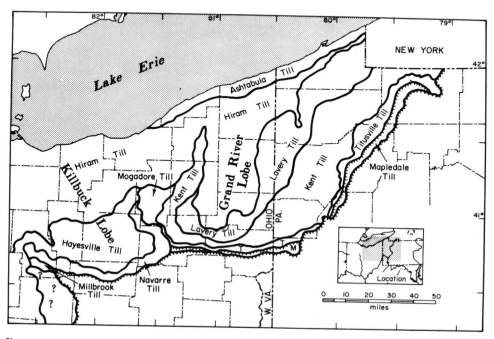

Figure 1. Glacial map of northwestern Pennsylvania and northeastern Ohio. (After White, 1969.)

Procedure

Each of the till units of the Allegheny Plateau can be identified on the basis of characteristics observable in the field. They can be recognized both in their outcrop area and in cuts north of their type areas, where they may be covered by one or more younger tills. Interstate highway construction and numerous strip mines have provided many miles of continuous Pleistocene exposures in the plateau area. These excellent exposures have made it possible to describe and to sample each of the till units in a wide geographic area, both in their outcrop areas and in the subsurface farther north. More than 50 of these continuous exposures are described and illustrated by White, Totten, and Gross (1969).

All till samples reported were taken from calcareous till, and most were collected from unleached and unoxidized till. Some of the samples were collected from unleached and oxidized till, but such samples were always taken from more than one foot below the depth of leaching. Stratigraphic identifications were made on the basis of field evidence. All analyzed samples have been included, so that the reported degree of variability is an accurate estimate of the relative degree of homogeneity of each till unit.

Grain-size analyses were performed by the pipette method and are presented as percentages of sand, silt, and clay in Table 1. The light-mineral (minerals with low specific gravity, principally quartz and feldspar) analyses were performed on the 0.125-0.177-mm-size fraction by a staining procedure. The feldspar percentage values are the percentages of feldspar in the light minerals of that size fraction. The potassium-feldspar value is the potassium-feldspar percentage of the total feldspar. This light-mineral laboratory technique is the subject of a separate paper (Gross and Moran, 1971).

TABLE 1

MEAN GRAIN SIZE AND FELDSPAR CONTENT OF THE TILLS

	Sand %			% Silt			% Clay			% Total Feldspar			% Potassium Feldspar		
	°X	S	N	X	S	N	X	S	N	X	S	N	X	S	N
Ashtabula Till	28	7.8	65	46	5.0	65	26	7.0	65	32	7.7	37	39	12.2	37
Hiram Till	▲20	9.5	180	45	6.6	180	▲35	11.3	180	▲27	6.8	158	▲44	10.2	158
Lavery Till	▲30	9.2	85	45	5.7	85	▲25	6.9	85	▲21	9.3	51	45	13.8	51
Kent Till	▲41	8.3	304	▲41	7.0	304	▲18	5.8	304	17	6.8	141	▲49	12.9	141
Titusville Till	▲45	7.1	389	▲37	6.0	389	18	4.5	389	▲13	7.0	326	50	17.6	326
Mapledale Till	▲42	7.8	44	36	5.5	44	▲22	5.7	44	▲6	3.6	32	55	19.0	32

°X = mean; S = standard deviation; N = number of samples.
▲Means proved significantly different at a 95 percent level of confidence.

Previous Work

Aside from incidental comments in the various bedrock-geology reports, the first studies specifically concerned with glacial geology of the plateau region were by Lewis (1882), Chamberlin (1883), and the classic monograph by Leverett (1902). Later studies included those of Leverett (1934), DeWolf (1929), and Preston (1950).

In the last 20 years, extensive work in the Allegheny Plateau region has been done by Professor G. W. White and his students. General studies include White (1949, 1951, 1953a, 1953b, 1953c, 1957, 1960, 1961, 1962, 1968, 1969, and 1971, this volume, p. 144); White and Totten (1965, 1967); White, Totten, and Gross (1969); DeLong and White (1963); Goldthwait, Dreimanis, Forsyth, Karrow, and White (1965); Goldthwait, White, and Forsyth (1961); Moran (1967b); Shepps and others (1959); and Totten, Moran, and Gross (1969).

The clay mineralogy has been studied by Droste (1956a, 1956b), Droste and Doehler (1957), and Droste and Tharin (1958). The till matrix has been studied by Droste, White, and Vatter (1958), and by Sitler and Chapman (1955). Textural studies have been reported by Shepps (1953, 1958), and by Tharin (1958).

The present study is an outgrowth and extension of masters' theses by the authors (Gross, 1967; Moran, 1967a) at the University of Illinois. An earlier version of this report was given by Gross and Moran (1969).

RESULTS

General Statement

From more than 1000 grain-size analyses and 700 light-mineral analyses, three types of compositional gradations have been recognized in the tills of the Allegheny Plateau:

1. Progressive differences in mean values between the various stratigraphic units

2. Areal gradations within individual till units

3. Vertical gradations within individual till units

The compositional gradations described are based on grain-size and light-mineral data. The authors chose these two compositional parameters from the many that could have been used to describe glacial tills, because they were particularly well suited for working with large numbers of samples.

Totten (1960) and Heath (1963) first demonstrated the presence of differences in feldspar content between the various tills of northeastern Ohio. The usefulness of grain-size analyses in describing the tills of northwestern Pennsylvania was first demonstrated by Shepps (1953) and Tharin (1958). These two parameters have served as the means by which several glacial processes (involving erosion and deposition) could be described. The same glacial processes could possibly have been examined by other analytical techniques (heavy minerals, clay minerals, magnetic susceptibility, etc.), but grain-size and light-mineral analyses have proved to be particularly useful for working with large numbers of samples (Gross and Moran, 1970) because samples can be analyzed in large batches.

Differences among Units

Differences in values for grain-size and light-mineral data among the various stratigraphic units are illustrated in Table 1, which summarizes the data presented by Gross (1967), Moran (1967a), Heath (1963), Totten (1960), and Tharin (1958). Statistically significant differences exist among all adjacent stratigraphic units in three or more of the five measured parameters. Each of the five parameters shows significant differences between some of the adjacent stratigraphic units.

These differences are not random; rather, the means grade progressively throughout the stratigraphic column. Through four of the Wisconsinan-age tills, a retrogressive grain-size gradation may be seen, from the fine-textured Hiram Till (20 percent sand, 45 percent silt, 35 percent clay) to the coarser-textured Titusville Till (45 percent sand, 37 percent silt, 18 percent clay). This gradation does not extend across the major Sangamonian break between the Titusville Till and the Mapledale Till, nor does it include the Ashtabula Till in the lake-border moraines.

The differences in light mineralogy are even more striking. The feldspar percentages grade retrogressively from a high of 32 percent in the Ashtabula Till to a low of 6 percent in the Mapledale Till. The potassium-feldspar percentage of the total feldspar also shows a gradation throughout the entire stratigraphic column, from a low of 39 percent in the Ashtabula Till to a high of 55 percent in the Mapledale Till. In general, the younger tills are fine-grained and feldspar-rich (plagioclase dominant), whereas the older tills are coarse-grained and feldspar-poor (potassium-feldspar dominant).

Areal Gradations within Till Units

Feldspar

The relatively high standard deviations shown in Table 1 do not represent unexplained random variations; they are the result of areal gradations within each till sheet. This type of compositional gradation is seen in Figure 2, which shows the percentage of feldspar in the Titusville Till. The dots represent sampling localities, each of which contains one to eleven samples. Multiple samples from individual localities have been averaged so that the 108 localities represent 326 analyses. The 10-percent, 20-percent, and 30-percent isopleths in Figure 2 are continuous across the plateau area, with the highest feldspar contents found in the north; feldspar percentage decreases to the southeast. In very general terms, the isopleths are parallel to the glacial boundary, trending northeast-southwest, except in the area adjacent to the glacial boundary, where the isopleths curve to the southeast. A plot of the percentage of feldspar in the Kent Till, which immediately overlies the Titusville Till, is shown in Figure 3. The 103 Kent Till localities represent 141 analyses. Kent Till isopleths have the same general pattern as do Titusville Till isopleths, with the 10-percent, 20-percent, and 30-percent lines being continuous across the area. The percentage of feldspar in the Lavery Till, which immediately overlies the Kent Till, is shown in Figure 4. The 45 Lavery Till localities represent 51 analyses. Lavery Till feldspar-

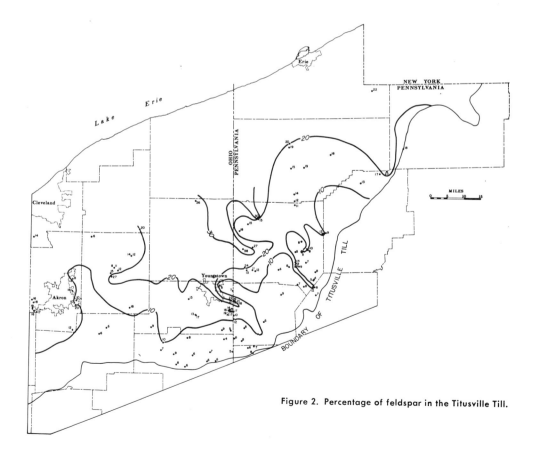

Figure 2. Percentage of feldspar in the Titusville Till.

isopleth lines show the same general pattern as do those of both the Kent and Titusville Tills, but the Lavery Till isopleths represent 20-percent, 30-percent, and 40-percent feldspar, because the younger tills have a significantly higher mean-feldspar content than do the older tills.

Of the seven till units of the Allegheny Plateau (Fig. 1), six were sampled and described in this project (Table 1). Of these six till units, isopleth maps of feldspar content were drawn for only the three units (Lavery, Kent, and Titusville Tills) for which there was available an adequate number of sampling points with a sufficiently broad geographic distribution. Data from scattered points for the other three tills (Ashtabula, Hiram, and Mapledale Tills) indicate that similar feldspar gradations probably exist in these units.

Hand-contoured maps, such as Figures 2, 3, and 4, are subject to considerable operator bias in the shaping of the isopleth lines. In an effort to confirm the general nature of these area gradations, a series of polynomial trend surfaces were calculated by a digital computer. The computer program described by Good (1964) was modified for use on the University of Illinois IBM 7094 computer.

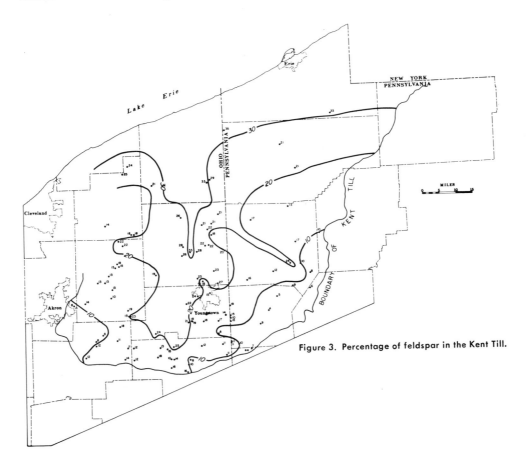

Figure 3. Percentage of feldspar in the Kent Till.

The trend-surface maps are based on an arbitrarily calculated mathematical model and are not as subject to the geologic bias of the operator as are the hand-contoured maps. The second-order trend surface of feldspar percentage in Titusville Till (Fig. 5) is of the Pennsylvania part of the Grand River Lobe, and is very similar to the hand-contoured map. Again, there is a significant decrease in feldspar percentage from north to south. The isopleths trend northeast to southwest, parallel to the southern till boundary, and there is a curving of the isopleth lines toward the southeast near the outer till boundary. This trend-surface map confirms the conclusions derived from the hand-contoured map.

Feldspar content of the Kent Till, as shown by a first-order trend surface, increases 0.38 percent per mile to the southeast (trend orientated N18°W), which is very similar to the increase of 0.41 percent per mile to the southeast (oriented N35°W) observed in the Titusville Till. The potassium-feldspar percentage of the total feldspar in the Titusville Till also increases to the southeast, increasing 0.5 percent per mile (oriented N52°W). Once again,

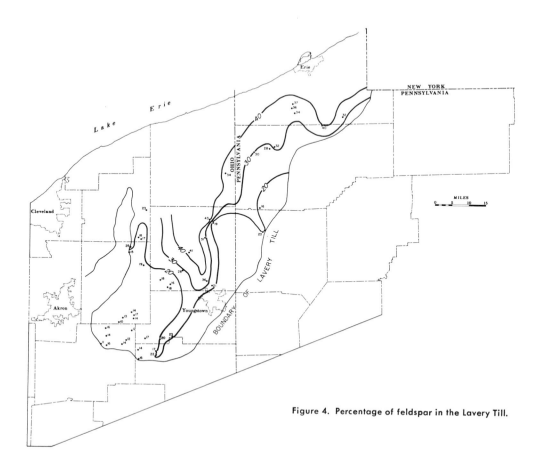

Figure 4. Percentage of feldspar in the Lavery Till.

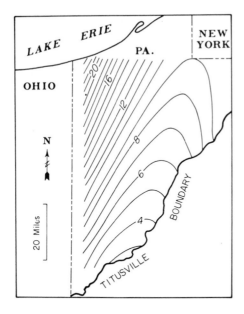

Figure 5. Second-order trend surface of the percentage of feldspar in the Titusville Till.

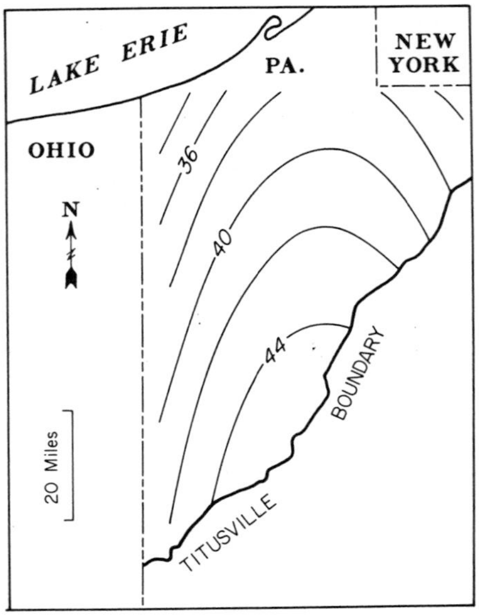

Figure 6. Second-order trend surface of the percentage of sand in the Titusville Till.

the trend of the gradation is oriented approximately normal to the southern till boundary.

Significant trends in potassium-feldspar percentage could not be proven in other till units. The standard deviations (shown in Table 1) for potassium-feldspar percentages are much larger than for any of the other measured parameters because relatively few potassium-feldspar grains were counted. The quartz-feldspar ratio is based on the total of 500 grains counted on each slide, while the plagioclase-potassium-feldspar ratio is based only on the 25 to 200 feldspar grains present on each slide. Thus, the random variation in the laboratory techniques is much larger for the potassium-feldspar percentage than for the total feldspar percentage.

Grain Size

Areal gradations have also been observed in till texture. Because the grain-size percentages are much more variable than are the feldspar percentages — that is, they have a greater random variation — the grain-size gradations can best be described through the use of trend surfaces. Figure 6 is a second-order trend surface of the percentage of sand in the Titusville Till. The gradation is one of increasing sand content toward the southeast. Once again, the isopleths trend northeast to southwest, except near the southeastern till boundary where they bend to the southeast. Titusville Till grain-size isopleth lines exhibit the same general pattern as the feldspar isopleths.

The more significant gradations in the Titusville Till are summarized in Figure 7 by means of vectors. These vectors were taken from first-order trend surfaces calculated for each of the parameters shown. All parameters change toward the southeast — the sand content increases greatly, the silt content decreases greatly, the clay content increases slightly, and the feld-spar content decreases greatly in this direction. In all cases, the trend of the gradation (the vector of change) occurs normal to the southern till boundary.

The trend-surface analyses of grain-size changes proved the presence of statistically significant gradations in all three parameters (sand, silt, and clay) only for the Titusville Till. This is logical because the most detailed information was available for the Titusville Till. Where significant grain-size gradations in just one or two of the three tested parameters were detected with trend-surface analysis on other stratigraphic units, the grada-tions were similar to those observed in Titusville Till.

Vertical Gradations within Till Units

A third type of compositional gradation is a vertical gradation within an individual till sheet. At 42 localities, two or more samples collected from a vertical section of the Titusville Till were analyzed for feldspar content. In 29 of these sections, the top of the Titusville Till sheet contained more

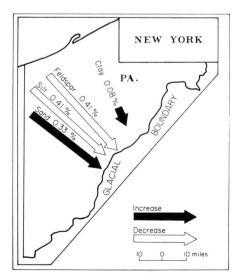

Figure 7. Summary of areal gradations in the Titusville Till. Arrows show rate of change per mile.

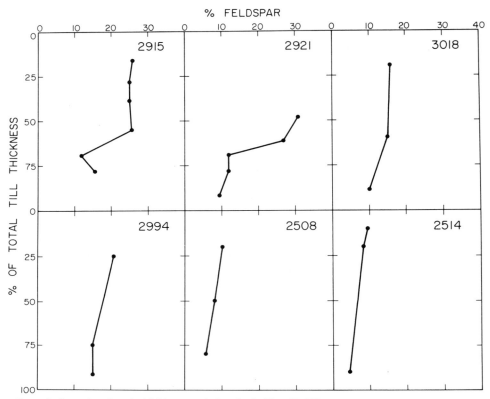

Figure 8. Examples of vertical feldspar gradations in the Titusville Till.

feldspar than did the base. Figure 8 illustrates six sections that are typical of the 29 sections demonstrating this trend. Nine sections reverse this trend, although only one of these reversals is more than the plus-or-minus-two-percent accuracy of the laboratory method, and three sections show no trends.

The simple linear decrease in feldspar content with depth, such as is shown in Figure 8, may be complicated by interruptions believed to represent thrust disruptions of bedding (see Figure 9, Moran, this volume). Even in such cases, the over-all trend of decreasing feldspar with depth within the Titusville Till is still valid. Multiple sampling of each locality was undertaken only for the Titusville Till, partly because it tends to be thicker than the other tills; therefore, such vertical gradations have not been examined in other till units. Vertical gradations in grain size were not detected.

DISCUSSION

Of the two parameters analyzed, a large number of geologic processes might affect the grain size of a till, so it is somewhat simpler to examine first the possible mechanisms that might affect the feldspar content of the 0.125-to-0.177-mm-size fraction of a till unit. Any hypothesis to explain the origin of these gradations must take into account the gradations between units, and both the areal and vertical gradations within individual till units. A list of possible hypotheses that might account for these gradations in feldspar content is given below, followed by a brief discussion of the validity of each.

1. Weathering of tills to remove feldspar.

2. Authigenic formation of feldspar.

3. Selective concentration of quartz or feldspar grains.

4. Derivation of different parent-till materials from the same source area by progressive erosion during successive glaciations.

5. Derivation of the parent-till material from different source areas.

6. Dilution of the parent-till material by various quantities of locally derived material.

The authors believe that numbers 1 through 5 can be partly, and some of them wholly, discredited, and only number 6 can be singularly responsible for the gradations.

Post-depositional weathering can remove feldspar, and the older tills of this area have the least feldspar (Table 1), but weathering can be eliminated as a possible cause for the gradations found in this study for two reasons: (1) Weathering removes feldspar from the A and B soil horizons, but more feldspar is present in the top than in the base of the Titusville Till (Fig. 8); (2) Willman, Glass, and Frye (1966, p. 40-42) reported feldspar depletion only through the B soil horizon. All samples analyzed for this study were collected below the B soil horizon.

In an extensive thin-section study of these tills, Sitler and Chapman (1955) found no authigenically formed feldspar. Therefore, authigenic formation of feldspar can be eliminated as a possible cause for the gradations.

No known glacial process could selectively sort and concentrate sand grains of relatively similar specific gravity into their mineralogical components. Feldspar grains would be more susceptible to size reduction by grinding than would quartz grains and could thereby be progressively removed from the 0.125-to-0.177-mm-size fraction. However, the larger feldspar grains should be reduced to the 0.125-to-0.177-mm-size fraction at the same time. Dreimanis and Vagners (1971) and Dreimanis (personal communication, 1969) explained that mineral grains are reduced in size through glacial comminution until they reach a terminal grade, a size characteristic for each mineral component. With additional glacial transport, mineral grains are not further reduced in size. The terminal grade for feldspar grains is 0.125 to 0.250 mm, so glacial grinding should actually tend to add feldspar to the 0.125-to-0.177-mm-size fraction. However, this study shows that, in the Allegheny Plateau area, the feldspar percentage of a till is reduced with transport, indicating that another mechanism must be working in this area, a mechanism that is masking the effects of glacial grinding of feldspar grains.

The geographical source area of the tills may have remained constant, with a progressive change in the composition of the material being eroded. The earliest till material might have been primarily derived from a quartz-rich, deeply weathered regolith. Then, as this regolith was progressively removed from the Canadian Shield, more feldspar-rich igneous and metamorphic rocks might have been exposed. Within the Woodfordian Substage, four tills were deposited in the plateau area (Fig. 1). A progressive erosion and recycling, such as postulated above, requires complete deglaciation of the Canadian Shield and the development of a weathering profile on the source-area rocks. The available time period, however, is much too short for such complete deglaciation and the development of four successive weathering profiles. No weathering profiles are present within the Woodfordian till deposits, where they could have been developed more readily than on the more resistant shield rocks that would have been exposed for a longer period of time (Frye, Willman, and Black, 1965).

Different source areas for each glaciation might explain the mean compositional differences between individual till units. Only a uniform progressive change in source area, however, could produce gradations in composition of tills from Illinoian through Wisconsinan age. Progressive shifts in source area for the deposition of each till unit would be required to explain compositional gradations within individual till units, a highly unlikely occurrence.

THE PROPOSED THEORY OF DILUTION

All of the compositional gradations described can best be explained as result of a progressive dilution of a fine-grained, feldspar-rich parent material by coarse-grained, quartz-rich debris derived from the upper Paleozoic bedrock of the Appalachian Plateau. This hypothesis of dilution of original material by enrichment with local material is best understood when the interrelation of the various types of gradations are recognized. The progressive differences, from youngest to oldest till units, are the same as the north-to-south and top-to-bottom gradations within individual till units. In all three cases, the sand content increases, the feldspar content decreases, and the relative potassium-feldspar percentage increases. The increasing sand percentage represents the addition of sand from the sandstone bedrock to the till. The sandstone is quartz-rich, thus the addition of this sand reduces the feldspar percentage of the till. Because what little feldspar is present in the bedrock is principally potassium feldspar, the addition of the sandstone increases the relative potassium-feldspar percentage of the till.

Differences among Till Units

Progressive differences in these parameters among the till sheets reflect decreased influence of local bedrock because it was progressively mantled by successive drift sheets. The glacier that deposited the Mapledale Till flowed across a bedrock surface and incorporated the greatest volume of bedrock material. Later ice sheets flowed across a surface of both till and bedrock and thus incorporated material with a composition between that of the bedrock and the parent till.

This dilution, as it may have occurred to form the Mapledale Till, is illustrated in Table 2. The till material transported south from the Lake Erie basin is assumed to have had a composition similar to that observed for the Ashtabula Till. This assumption is supported by two observations: (1) the Ashtabula Till occurs in a belt only a few miles south of Lake Erie and is probably similar to the material transported south down the Lake Erie basin; (2) all of the till of the sheets of the Allegheny Plateau increase in feldspar content to the north. All three units described in detail contain more than 30 percent feldspar near Lake Erie (Figs. 2, 3, and 4).

The glacier that deposited the Mapledale Till flowed across the predominantly bedrock surface. By adding nine parts of the composition of the local bedrock to one part of that of the parent-till material (Ashtabula Till composition), it is possible to produce almost exactly the observed composition of Mapledale Till (Table 2). The gradations in both total-feldspar content and relative potassium-feldspar content between the two tills may be produced simultaneously by the proposed dilution.

Similar calculations for each of the till units present in the drift of the Allegheny Plateau (Table 3) give similar results. In every case, the theo-

TABLE 2

How the Mapledale Till Could Be Formed by Dilution of a Till Having Ashtabula Till Composition

Material	% Feldspar	% Potassium Feldspar
Parent material (Ashtabula Till)	32	39
Local bedrock	3	66
Original material diluted by 9 volumes of local bedrock: Resultant material	5.9	52
Mapledale Till	6	55

TABLE 3

DILUTION OF ORIGINAL TILL MATERIAL

(ASHTABULA TILL COMPOSITION)

Diluting Material	Theoretical Resultant Material		Observed Till Composition		
	% Total Feldspar	% Potassium Feldspar	Till Unit	% Total Feldspar	% Potassium Feldspar
9 vol. bedrock	5.9	52	Mapledale Till	6	55
4 vol. Mapledale Till	11.2	46	Titusville Till	13	50
4 vol. Titusville Till	16.8	46	Kent Till	17	49
2 vol. Kent Till	22.0	44	Lavery Till	21	45
1 vol. Lavery Till	26.5	42	Hiram Till	27	44

retical changes may be produced by diluting material of Ashtabula Till composition, the assumed parent-till composition, by the underlying material. The first line of Table 3 shows how the Mapledale Till was formed. The next glacier, which deposited the Titusville Till, flowed across a surface of predominantly Mapledale Till material. By combining Ashtabula Till material with Mapledale Till material, a composition very close to that of the Titusville Till can be formed. This relation of parent till and underlying material can be followed throughout the stratigraphic column.

Areal Gradations within Till Units

Areal gradations in till composition reflect greater incorporation of bedrock material with increased distance of transport. The increasing sand content and decreasing feldspar content toward the southeast represent the continued incorporation of local quartz-rich sand with transport. The glacier was most actively eroding in places where it was flowing up over the plateau escarpment. In these areas, the till was diluted rather abruptly by large quantities of local bedrock, and the compositional isopleths are closely spaced. Where it flowed through the lowlands, the glacier was not eroding as actively and the till was not as quickly diluted, which would explain the wide spacing and lobate pattern of the resultant compositional isopleths.

A topographic map of the buried bedrock surface of northeastern Ohio (Cummings, 1959) is shown in Figure 9. Unfortunately, no bedrock surface map has been published for northwestern Pennsylvania. The buried bedrock surface in Ohio consists of a dissected upland in which the north-south-orientated Grand River lowland, 15 to 20 miles wide, has been eroded. Such topography is known to affect markedly the flow pattern of ice sheets, as is illustrated on a continental scale by the effects of the Great Lakes basins on the lobate pattern of the ice sheets (Hoberg and Anderson, 1956). In this area of the Allegheny Plateau, the uplands had a definite retarding effect on the flow of the ice, while the valleys, especially the Grand River lowland, acted as passageways for greater ice flow.

The main masses of the glacier flowed southwesterly down the axis of the Erie basin, but it formed sublobes to the southeast which expanded southward onto the Allegheny Plateau and down the Grand River lowland. The bedrock strike in this area (Stose and Ljungstedt, 1932) is northeast-southwest, generally normal to this local ice-flow direction. The lowlands near Lake Erie are underlain by Devonian shales, but farther south most of the bedrock consists of Carboniferous sandstones. This division between shales to the north and sandstones to the south is shown by the heavy line in Figure 9.

The 0.125-to-0.177-mm-size fraction of these Carboniferous sandstones has an average feldspar content of about 3 percent, of which 66 percent is potassium feldspar (Carswell and Burnett, 1963, p. 4, and three analyses by the authors). The low feldspar content is typical for Paleozoic bedrock and is in sharp contrast to the 6-to-32-percent means for the tills. Thus, a large part of the bedrock consists of sand-sized particles; it is quartz-rich, and what little feldspar it contains is predominantly potassium feldspar.

When the map showing the percentage of feldspar in the Titusville Till

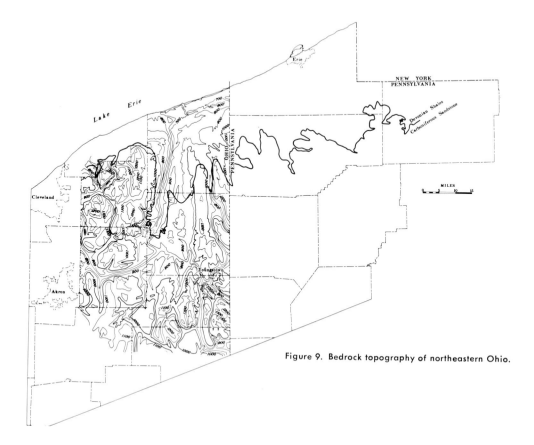

Figure 9. Bedrock topography of northeastern Ohio.

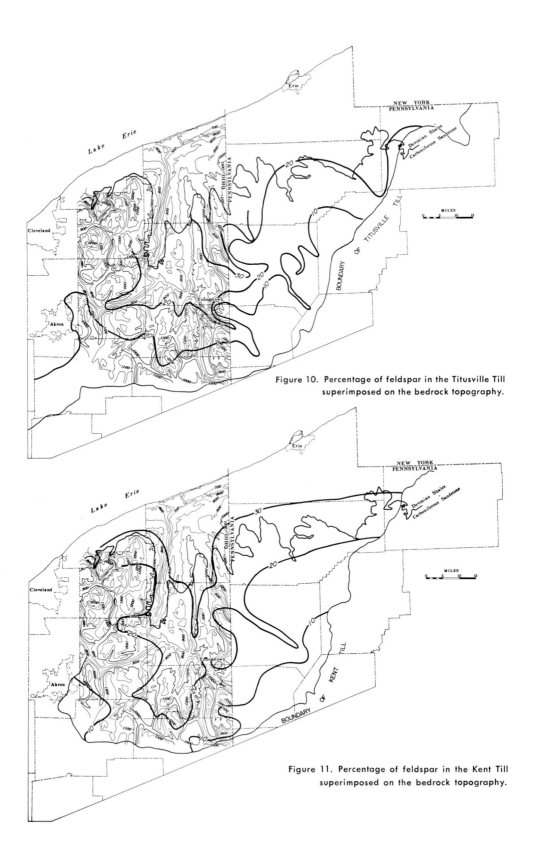

Figure 10. Percentage of feldspar in the Titusville Till
superimposed on the bedrock topography.

Figure 11. Percentage of feldspar in the Kent Till
superimposed on the bedrock topography.

(Fig. 2) is superimposed on the topographic map of the bedrock surface (Fig. 9), the Grand River lowland and even smaller lowlands are reflected by southern lobate extensions of the feldspar isopleths (Fig. 10). Feldspar-isopleth lines of both the Kent Till (Fig. 11) and the Lavery Till (Fig. 12) also reflect this variation in bedrock topography.

Vertical Gradation within Till Units

The vertical gradations in till composition reflect greater incorporation of bedrock material near the base of the till sheet. The feldspar percentage is lowest near the base of the Titusville Till sheet, because the local quartz sand was eroded and incorporated into the base of the till unit.

Applications of These Theories to Other Areas

There are three limiting prerequisites to a study of this type. (1) The glacial lithostratigraphy must be known. The identifications of the samples

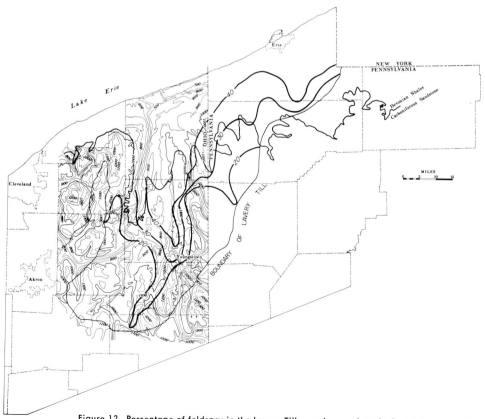

Figure 12. Percentage of feldspar in the Lavery Till superimposed on the bedrock topography.

used in this project were made on the basis of field evidence and the various parameters were calculated subsequent to the laboratory work. (2) Such a study requires a large number of samples collected throughout a broad geographic distribution. These criteria were met by the abundant exposures provided in an area of extensive strip mining and active interstate-highway construction. (3) The hypotheses involved can best be applied where the ice-flow direction is normal to bedrock strike. A cursory examination of the U.S. Geologic Map (Stose and Ljungstedt, 1932) reveals that bedrock strike is normal to ice-flow direction in a vast majority of the glaciated areas of the United States, as it is in the area of this study.

Three applications of these hypotheses are suggested. Because the compositional changes occur in the direction of transport, compositional isopleths may be considered equipotential lines for reconstruction of glacial flow lines. This is illustrated by Figure 13, in which the isopleths are a generalized representation of those in Figures 2, 3, 4, 5, and 6. The striations shown were traced from the "Glacial Map of the United States East of the Rocky Mountains" (Flint and others, 1959) and are almost universally normal to the compositional isopleths described herein. In an area of unknown ice-flow direction, the glacial flow lines could be reconstructed from compositional isopleths of some component of the till.

The second application of these hypotheses is for distinguishing between the areas where glacial erosion predominated and the areas where glacial deposition predominated. For most of the plateau, the compositional isopleths are normal to the ice-flow direction (Figs. 5 and 6). These are the areas where the glacier was actively eroding, so that the till composition was being changed with transport. In the few miles adjacent to the southeastern till boundary of each unit, the compositional isopleths are parallel

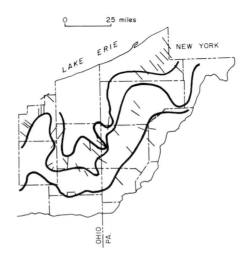

Figure 13. Relation of ice-flow indicators to generalized isopleths of feldspar content. The isopleths are a generalized representation of those of Figures 2, 3, 4, 5, and 6. The striations shown were traced from the "Glacial Map of the United States East of the Rocky Mountains" (Flint and others, 1959).

to the direction of ice flow. These are the areas where the till composition was remaining constant with transport, which in turn indicates areas where deposition was dominant over erosion.

The third application of these hypotheses is for determining the location of the source material of the tills. In very general terms, the feldspar content of the tills studied was halved with every 20 miles of transport (Fig. 5). If the diluting material had essentially no feldspar (i.e., the local bedrock), then halving the feldspar content of a till could be accomplished by dilution with an equal volume of local bedrock. Since the feldspar content of the Titusville Till is reduced by more than half in a distance of 20 miles, it is estimated that over 50 percent of the material making up the Titusville Till was derived from within 20 miles of the site of deposition.

ACKNOWLEDGMENTS

The continued advice and encouragement of Professor G. W. White are most gratefully acknowledged. Some of the samples used in the investigation were collected while the authors were working as field assistants to Professor White and Dr. S. M. Totten with the support of the National Science Foundation Grant 2675. The authors acknowledge the use of data from unpublished theses by Totten (1960), Heath (1963), and Tharin (1958).

REFERENCES

Anderson, R. C., 1957, Pebble and sand lithology of the major Wisconsin glacial lobes of the Central Lowland: Geol. Soc. America Bull., v. 68, p. 1415-50.

Carswell, L. D., and Burnett, G. D., 1963, Geology and hydrology of the Neshannock Quadrangle, Mercer and Lawrence Counties, Pennsylvania: Pa. Geol. Survey, 4th series, Bull. W15, 90 p.

Chamberlin, T. C., 1883, Preliminary paper on the terminal moraine of the second epoch: U.S. Geol. Survey, 3d Ann. Rept., p. 330-52.

Charlesworth, J. K., 1957, The Quaternary Era: London, Edward Arnold, 1700 p.

Cummings, J. W., 1959, Buried river valleys in Ohio: Ohio Div. of Water Invent. Report 10.

DeLong, R. M., and White, G. W., 1963, Geology of Stark County; Ohio Geol. Survey Bull. 61, 209 p.

DeWolf, Frank, 1929, Atlas of Pennsylvania, New Castle Quadrangle: Pa. Geol. Survey, 4th series, Bull. A5.

Dreimanis, A., and Vagners, U. J., 1971, Bimodal distribution of rock and mineral fragments in basal till: *in* this volume.

Droste, J. B., 1956a, Clay minerals in calcareous tills in northeastern Ohio: Jour. Geology, v. 67, p. 187-190.

————, 1956b, Alteration of clay minerals by weathering in Wisconsinan tills: Geol. Soc. America Bull., v. 67, p. 911-18.

Droste, J. B., and Doehler, R. W., 1957, Clay mineral composition of calcareous till in northwestern Pennsylvania: Illinois State Acad. Sci. Trans., v. 50, p. 194-98.

Droste, J. B., and Tharin, J. C., 1958, Alteration of clay minerals in Illinoian till by weathering: Geol. Soc. America Bull., v. 69, p. 61-68.

Droste, J. B., White, G. W., and Vatter, A. E., 1958, Electron micrography of till matrix: Jour. Sed. Petrology, v. 28, p. 345-50.

Flint, R. F., and others, 1959, Glacial map of United States east of the Rocky Mountains: Geol. Soc. America.

Frye, J. C., and Willman, H. B., 1960, Classification of the Wisconsinan Stage in the Lake Michigan glacial lobe: Illinois Geol. Survey Cir. 285, 16 p.

Frye, J. C., Willman, H. B., and Black, R. F., 1965, Outline of glacial geology of Illinois and Wisconsin, *in* The Quaternary of the United States, H. E. Wright and D. G. Frey (eds.), Princeton, N. J., Princeton Univ. Press, 43-61.

Gillberg, G., 1965, Till distribution and ice movements on the northern slopes of the south Swedish highlands: Geologiska Föreningens i Stockholm Förhandlingar, v. 86, p. 433-84.

————, 1967, Further discussion of the lithological homogeneity of till: Geologiska Föreningens i Stockholm Förhandlingar, v. 89, p. 29-49.

Goldthwait, R. P., White, G. W., and Forsyth, J. L., 1961, Glacial map of Ohio: U.S. Geol. Survey, Misc. Geol. Inv. Map I-316.

Goldthwait, R. P., Dreimanis, A., Forsyth, J. L., Karrow, P. F., and White, G. W., 1965, Pleistocene deposits of the Erie Lobe: *in* the Quaternary of the United States: H. E. Wright and D. G. Frey (eds.), Princeton, N. J., Princeton Univ. Press, p. 85-97.

Good, D. I., 1964, Fortran II trend-surface program for the IBM 1620: Kansas Geol. Survey Sp. Dist. Publ. 14, 54 p.

Gross, D. L., 1967, Mineralogical gradations within Titusville Till and associated tills of northwestern Pennsylvania: M. S. Thesis, Univ. Illinois.

Gross, D. L., and Moran, S. R., 1969, Textural and mineralogical gradations within tills of the Allegheny Plateau: Geol. Soc. America Abs. with Programs for 1969, Part 6, p. 18-19.

————, 1970, A technique for the rapid determination of the light minerals of detrital sands: Jour. Sed. Petrology, v. 40, p. 759-61.

Harrison, W., 1960, Original bedrock composition of Wisconsin till in central Indiana: Jour. Sed. Petrology, v. 30, p. 432-46.

Heath, C. P. M., 1963, The mineralogy of tills in the Grand River Glacial Lobe in northeastern Ohio: M. Sc. Thesis, Univ. Illinois.

Holmes, C. D., 1952, Drift dispersion in west-central New York: Geol. Soc. America Bull., v. 63, p. 993-1010.

————, 1960, Evolution of till-stone shapes, central New York: Geol. Soc. America Bull., v. 71, p. 1645-60.

Horberg, C. L., and Anderson, R. C., 1956, Bedrock topography and Pleistocene glacial lobes in central United States: Jour. Geology, v. 64, p. 101-16.

Krumbein, W. C., 1933, Textural and lithologic variations in glacial till: Jour. Geology, v. 41, p. 382-408.

Leverett, F., 1902, Glacial formations and drainage features of the Erie and Ohio basins: U.S. Geol. Survey Monograph 41, 802 p.

————, 1934, Glacial deposits outside of the terminal moraine in Pennsylvania: Pa. Geol. Survey, 4th ser., Bull. G7, 123 p.

Lewis, H. C., 1882, Report on the terminal moraine in Pennsylvania and western New York: Pa. 2d Geol. Survey Rept. 2, 299 p.

Moran, S. R., 1967a, Stratigraphy of Titusville Till in the Youngstown region, Eastern Ohio: M. Sc. Thesis, Univ. Illinois.

————, 1967b, Stratigraphic divisions of the Titusville Till near Youngstown, Ohio: Geol. Soc. America Special Paper 115, p. 393-94.

————, 1971, Glacio-tectonic structures in till: in this volume.

Preston, R. W., 1950, Glacial foreland of northwestern Pennsylvania: Guidebook, Field Conference of Pa. Geologists, 16th Ann. Meeting, Pittsburgh, Pa., 47 p.

Shepps, V. C., 1953, Correlation of the tills of northeastern Ohio by size analysis: Jour. Sed. Petrology, v. 23, p. 34-48.

————, 1958, "Size Factors," a means of analysis of data from textural studies of till: Jour. Sed. Petrology, v. 28, p. 482-85.

Shepps, V. C., White, G. W., Droste, J. B., and Sitler, R. F., 1959, The glacial geology of northwestern Pennsylvania: Pa. Geol. Survey, 4th series, Bull. G32, 64 p.

Sitler, R. F., and Chapman, C. A., 1955, Microfabrics of till from Ohio and Pennsylvania: Jour. Sed. Petrology, v. 25, p. 262-69.

Stose, G. W., and Ljungstedt, O. A., 1932, Geologic map of United States: U. S. Geol. Survey.

Tharin, J. C., 1958, Textural studies of the Wisconsin till of northwestern Pennsylvania: M. Sc. Thesis, Univ. Illinois.

Totten, S. M., 1960, Quartz-feldspar ratios of tills in northeastern Ohio: M. Sc. Thesis, Univ. Illinois.

Totten, S. M., Moran, S. R., and Gross, D. L., 1969, Greatly altered drift near Youngstown, Ohio: Ohio Jour. Sci., v. 69, p. 213-25.

White, G. W., 1949, Geology of Holmes County: Ohio Geol. Survey Bull. 47, 373 p.

————, 1951, Illinoian and Wisconsin drift of the southern part of the Grand River lobe in eastern Ohio: Geol. Soc. America Bull., v. 62, p. 967-77.

————, 1953a, Geology and water-bearing characteristics of the unconsolidated deposits of Cuyahoga County, in Winslow, J. D., White, G. W., and Webber, E. E., The water resources of Cuyahoga County, Ohio: Ohio Dept. Nat. Resources, Div. Water Bull. 26, p. 36-41.

————, 1953b, Character and distribution of the glacial and alluvial deposits, in Smith, R. C., The ground-water resources of Summit County, Ohio: Ohio Dept. Nat. Resources, Div. Water Bull. 27, p. 18-24.

————, 1953c, Sangamon soil and early Wisconsin loesses at Cleveland, Ohio: Am. Jour. Sci., v. 251, p. 362-68.

————, 1957, Wisconsin glacial deposits of northeastern Ohio (Abs.): Geol. Soc. America Bull., v. 68, p. 1902.

————, 1960, Classification of Wisconsin glacial deposits in northeastern Ohio: U.S. Geol. Survey Bull. 1121-A, 12 p.

————, 1961, Classification of glacial deposits in the Killbuck lobe, northeastern central Ohio: U.S. Geol. Survey Prof. Paper 424-C, p. 71-73.

————, 1962, Multiple tills of end moraines: U.S. Geol. Survey Prof. Paper 450-C, p. 96-98.

White, G. W., and Totten, S. M., 1965, Wisconsinan age of the Titusville Till (formerly called "Inner Illinoian"), northwestern Pennsylvania: Science, v. 148, no. 3667, p. 234-35.

White, G. W., 1967, Loess deposits in northeastern Ohio: Geol. Soc. America Spec. Paper 115, p. 400-401.

————, 1968, Age and correlation of Pleistocene deposits at Garfield Heights (Cleveland), Ohio: Geol. Soc. America Bull., v. 79, p. 749-52.

————, 1969, Glacial deposits of the Allegheny Plateau, U.S.A.: Quat. Jour. Geol. Society London, v. 124, p. 131-51.

————, 1971, Thickness of Wisconsinan tills in Grand River and Killbuck lobes: *in* this volume.

White, G. W., Totten, S. M., and Gross, D. L., 1969, Pleistocene stratigraphy of northwestern Pennsylvania: Pa. Geol. Survey, 4th series, Bull. G-55, 88 p.

Willman, H. B., Glass, H. D., and Frye, J. C., 1966, Mineralogy of glacial tills and their weathering profiles in Illinois, Part II, weathering profiles: Illinois Geol. Survey Cir. 400, 76 p.

Particle-Size and Carbonate Analysis of Glacial Till and Lacustrine Deposits in Western Ohio

Joseph R. Steiger and Nicholas Holowaychuk

ABSTRACT

Calcareous glacial till and lacustrine materials of Wisconsin age, sampled about two feet below the base of the zone of carbonate leaching, were evaluated in terms of particle-size distribution, calcium carbonate equivalent, and Ca/Mg molecular ratios. Means and standard deviations were computed for samples occurring on morainal systems of the Miami, Scioto, and Erie Lobes, and in two areas in lacustrine basins. Differences in till characteristics among adjacent areas were tested for statistical significance.

Data are compiled on the glacial map of western Ohio to show the general distribution of soil parent materials. Nine regions are differentiated — two in tills of the Miami Lobe, three in tills of the Scioto Lobe, two in tills of the Erie Lobe, and two in lacustrine deposits of northwest Ohio. The southern limits of the Wabash, Bloomer-Broadway, and Union City-Powell Moraines represent boundaries between significantly different till deposits.

Clay content is highest in the lacustrine deposits (45-68 percent) and decreases to about 18 percent in tills in the southern part of the area. Calcium carbonate equivalent is about 20 percent in the lacustrine deposits and increases to about 40 percent in tills toward southern extremities of the Miami and Scioto Lobes. The Ca/Mg ratio decreases southward from 6.5 in the Paulding Basin to 4.2 to 4.8 in the Erie Lobe and 2.1 to 3.0 in the Miami and Scioto Lobes. In these glacial deposits, Ca/Mg ratios are two to six times greater than are those in the underlying dolomitic bedrock of western Ohio.

This paper is a contribution of the Department of Agronomy, Ohio State University, OARDC journal paper No. 94-69.

INTRODUCTION

A general relationship between the distribution of a number of soil sequences and the underlying glacial materials in western Ohio has been recognized for some time. Early in this century, Coffey and Rice (1915), in their reconnaissance soil survey of Ohio, noted that there were areal differences in the clay content of soil parent materials and that these differences were reflected to some extent in the textures of the soils. Later detailed soil surveys in western Ohio have indicated that, as a rule, the upland soils in the northern part of the state (Heffner and others, 1965; Rapparlie and Urban, 1966) have a higher content of clay than do those in the southern part (Garner and Meeker, 1962; Lerch and others, 1969). These differences are related at least in part to the mechanical composition of the parent materials. There are also differences in the calcium carbonate content of these parent materials. In the southern part of the state, the calcium carbonate equivalent in the fine fraction (<2 mm) ranges from 35 to 45 percent (Garner and Meeker, 1962; Lerch and others, 1969), while in the northern area, the range is about 20 percent less (Heffner and others, 1965; Rapparlie and Urban, 1966). Goldthwait (1959) likewise has referred to areal differences in the compositions of tills of western Ohio. He suggests that such variations reflect new sources of materials or new directions of ice movement resulting from minor retreats and readvances that apparently had taken place.

The relationship of the clay content in the B horizons of Celina and Morley soils to the clay and carbonate contents in the C horizons was studied by Smeck and others (1968). Both of these soils occur extensively in western Ohio, the Celina in the southern and the Morley in the northern parts of the area. They concluded that most of the enrichment of the clay content in the B horizon was due to the removal of the calcareous diluent from the till parent material (Smeck and others, 1968). It was noted that the higher the carbonate content in the till, the more marked was the enrichment of the clay in the B horizon. They also found that only a small fraction of the total carbonate content occurred as clay-size particles. Thus, in addition to the mechanical composition inherent in the parent material, carbonate content is also an important factor influencing the characteristics of soils in this region.

The objective of this report is to summarize the data available on the grain size, the calcium carbonate equivalent, and the Ca/Mg ratios of several hundred samples of soil parent materials collected in western Ohio. Dreimanis (1962), in similar work, has used data on the calcite/dolomite ratio, calcium carbonate equivalent, and mechanical analysis to characterize glacial till in southern Ontario. As a part of soil-survey operations in western Ohio, nearly 500 sites had been sampled and analysed by the Soil Characterization Laboratory of the Ohio Agricultural Research and Development Center in Columbus. These data provide information by which the parent materials of western Ohio can both be characterized with respect to the several properties mentioned and be grouped into classes. In addition, the wide geographic scattering of these sample sites makes it possible to outline the areal distribution of these several classes of materials.

EXPERIMENTAL METHODS

The collection of the deeper soil profile samples and the analyses performed were directed primarily to the characterization of the material from which at least a part of the solum had been derived. It was shown by Wenner and others (1961) that there is appreciable alteration in the upper two feet or so of the calcareous material below the base of the soil solum. The clay content is higher, and both the calcium carbonate equivalent and the Ca/Mg ratio are lower in this altered zone. Below this zone, however, they found that the subjacent materials did not exhibit any significant changes in the several properties with depth. It is concluded, therefore, that the material below this zone of alteration is representative of at least the upper portions of the till and/or lacustrine deposits in the area investigated, and that its characteristics may be determined from the data provided by the analyses done by the Soil Characterization Laboratory.

The Study Area

The area considered in this study is confined to Wisconsin-age drift in western Ohio. This area extends northward from the Hartwell Terminal Moraine to and including the Lake Maumee Basin. The eastern boundary of the study area coincides approximately with the line of change in bedrock from calcareous limestones and dolomites to noncalcareous shales, a change reflected in the glacial tills by decrease in carbonate content to the east. The Ohio-Indiana state line is the western boundary. The dominant soil toposequences of the area are: (1) the Miami and Russell sequences across the southern part of the region, (2) the Morley sequence across the midsection, (3) the Hoytville, Paulding, and Toledo sequences in the Lake Maumee Basin, and (4) the Cardington sequence along the eastern fringe of the area, as shown by Figure 1.

Climatic differences across the area are relatively minor, so most differences in soils are attributed either to variations in parent material or to soil-moisture regimes as these are affected by topographic position. According to Goldthwait (1958), the Wisconsin deposits in Ohio range in age from 14,000 to 19,000 years from north to south. Goldthwait (1959) suggests

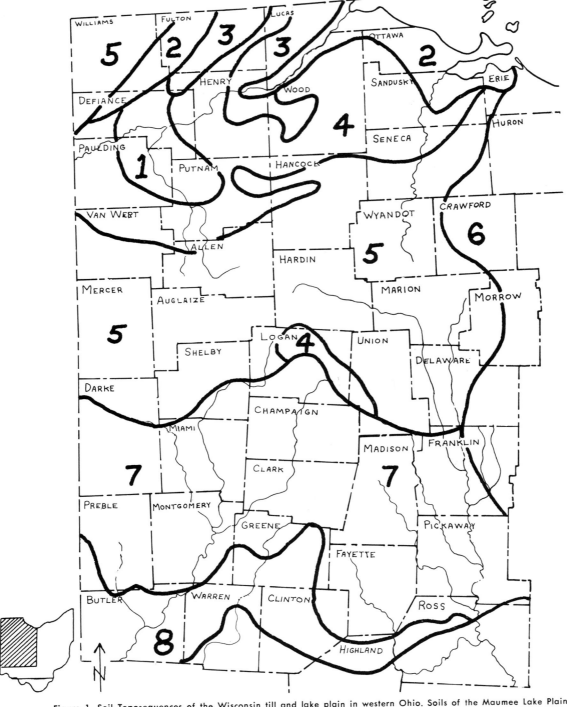

Figure 1. Soil Toposequences of the Wisconsin till and lake plain in western Ohio. Soils of the Maumee Lake Plain: (1) Paulding-Broughton Soils, (2) Toledo-Fulton Soils, and (3) mixed sandy soils. Soils of the till plains: (4) Hoytville-Nappanee-St. Clair Soils, (5) Morley-Blount-Pewamo Soils, (6) Alexandria-Cardington-Bennington-Morengo Soils, (7) Miami-Celina-Crosby-Brookston Soils, and (8) Russell-Xenia-Fincastle-Brookston Soils.

that some of the moraines crossing the area are the result of both readvances and retreats of the ice; these include the Cuba II, Farmersville, Reesville, Powell, and Wabash Moraines (Goldthwait, 1959). The bedrock underlying the glacial drift is Ordovician shales and limestones across the southwestern part of the area, Silurian dolomite in northwestern Ohio, and Devonian limestone and shale along the eastern fringe of the area and in the outlier in Logan County that divided the advancing glacier into the Scioto and Miami Lobes (Bownocker, 1947).

Site Selection

The locations of most sites were determined by soil survey field parties. Primary consideration in selecting profile samples (U.S. Dept. of Agriculture, 1951) was that sites should be representative of landscape units recognized within the county being mapped. The nearly 500 samples analyzed were screened for those that were sampled to depths of at least 20 inches below the depth of effervescence. Likewise, samples taken in profiles that showed noticeable stratification in the till were not included in this study. No attempt was made to obtain a uniform areal distribution of samples; hence, some counties lacking a recent soil survey are not represented by any samples. The total of 180 sites satisfying these criteria, and therefore included in the analysis, are plotted on Figure 2.

Laboratory Procedures

The portion of the sample passing through a 10-mesh sieve (<2 mm fraction) was used in all of the several analyses performed. Particle-size distribution was determined by a modification of the sedimentation-pipette method of Steele and Bradfield (1934), using sodium hexametaphosphate as a dispersing agent. Samples were agitated in a reciprocating shaker for 18 hours prior to sedimentation measurements. Calcite and dolomite contents were measured by the gasometric method of Dreimanis (1962), the samples (<2 mm fraction) being ground to pass 200 mesh for this measurement. The values for calcite and dolomite contents obtained by this method were combined to derive the total calcium carbonate equivalent. These values were also used for calculating the molecular Ca/Mg ratios of the carbonate fraction.

RESULTS AND DISCUSSION

Results are reported as: (1) particle-size distribution with respect to (a) sand content (2.0-.05 mm), (b) silt content (.05-.002 mm), (c) clay

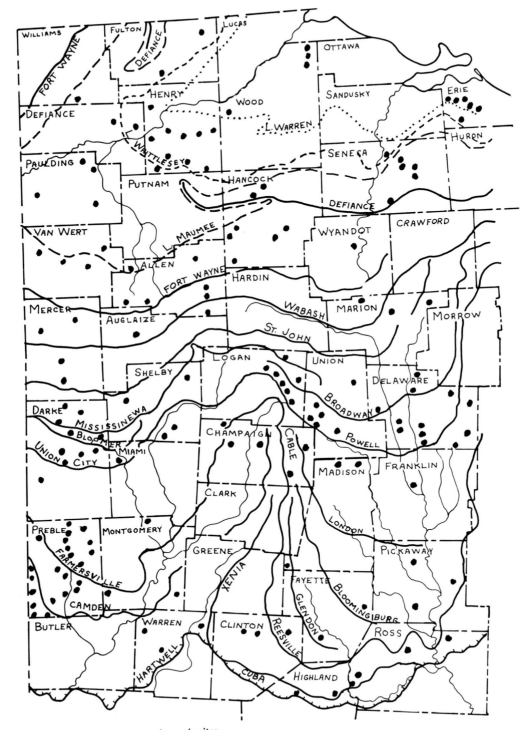

Figure 2. Moraine systems and sample sites.

content (<.002 mm); (2) total calcium carbonate equivalent ; and (3) Ca/Mg molecular ratio of the carbonate fraction. In order to evaluate the sites statistically, samples were grouped according to their occurrence on a moraine system or in a lacustrine basin, using the *Glacial Map of Ohio* (Goldthwait, White, and Forsyth, 1961) as a base. Samples from both end moraine and ground moraine of each system are assigned the name of the associated end moraine to the south. The data-sample means (x) and standard deviations (S_x) are shown in Tables 1 and 2; if less than six samples were available, the range of values is reported in parentheses. The statistical model used to test the equality of sample means (x) from adjacent morainic areas is based on the assumption of a normal distribution, where about 68 percent of observations are within one standard deviation from the mean. The model also assumes independent samples of two populations with equal variance. In this study, if three or more of the five parameters tested were significantly different between adjacent moraine systems, the materials were presumed to constitute different classes. This criterion was used as a guide in grouping glacial tills and lacustrine deposits that appear similar based on the statistical test of equality.

According to the properties evaluated and the criteria mentioned previously, the materials investigated can be grouped into nine classes or categories. To facilitate discussion, each group is defined in general terms as to the calcium carbonate equivalent, the Ca/Mg ratio of the calcareous

TABLE 1

SAMPLE MEAN (x), STANDARD DEVIATION (Sx), AND NUMBER OF SAMPLES (n) FOR PARTICLE SIZE DISTRIBUTION OF C HORIZONS

Sample Location	n	Percent		
		Sand	Silt	Clay
Lacustrine plain				
Toledo Basin	8	2.2 ± 1.6	51.8 ± 8.0	45.3 ± 4.6
Paulding Basin	6	2.3 ± 1.6	29.9 ± 10.4	67.8 ± 11.1
Erie lobe				
Defiance Moraine	19	21.7 ± 2.5	41.9 ± 2.6	36.4 ± 3.0
Fort Wayne Moraine	17	20.3 ± 3.7	43.2 ± 2.8	36.5 ± 4.1
Wabash Moraine	5	21.8°	44.8°	33.4°
		(16.6-26.2)	(40.4-48.7)	(29.1-38.2)
Miami lobe				
St. Johns-Bloomer Moraine	13	19.6 ± 5.5	44.6 ± 1.9	36.1 ± 4.6
Union City Moraine	8	23.0 ± 2.5	48.6 ± 3.3	27.5 ± 2.6
Farmersville Moraine	18	37.6 ± 8.1	41.7 ± 7.1	20.7 ± 5.8
Camden Moraine	10	42.3 ± 4.4	39.6 ± 2.2	18.1 ± 2.7
Hartwell Moraine	9	40.7 ± 8.5	41.8 ± 4.3	17.4 ± 6.6
Scioto lobe				
Broadway Moraine	8	22.5 ± 3.4	46.3 ± 5.2	31.2 ± 2.4
Powell Moraine (west)	13	16.6 ± 5.9	40.7 ± 7.3	42.7 ± 4.8
Powell Moraine (east)	11	25.1 ± 4.6	47.9 ± 3.4	27.1 ± 2.1
London Moraine	14	33.5 ± 5.9	46.0 ± 4.0	20.5 ± 3.1
Reesville Moraine	10	31.1 ± 6.4	48.1 ± 7.0	20.7 ± 2.3
Cuba Moraine	7	32.0 ± 6.7	47.7 ± 8.6	20.2 ± 3.0

°Statistical values other than sample mean not calculated for less than six samples (range of observations in parentheses).

TABLE 2

SAMPLE MEAN (x), STANDARD DEVIATION (S_x), AND NUMBER OF

SAMPLES (n) FOR PARTICLE SIZE DISTRIBUTION OF C HORIZONS

Sample Location	n	CaCO₃ Equivalent (Percent)	n	Molecular Ca/Mg Ratio
Lacustrine plain				
Toledo Basin	8	18.7 ± 6.8	8	2.5 ± 0.5
Paulding Basin	5	21.3° (17.3-25.4)	4	6.5° (6.2-7.8)
Erie lobe				
Defiance Moraine	19	20.6 ± 4.5	18	5.0 ± 0.9
Fort Wayne Moraine	16	21.1 ± 3.3	12	4.7 ± 0.6
Wabash Moraine	5	22.7° (19.3-25.1)	5	4.2° (3.8-4.7)
Miami lobe				
St. Johns-Bloomer Moraine	13	28.3 ± 5.0	10	3.0 ± 0.6
Union City Moraine	8	38.6 ± 2.5	6	2.7 ± 0.5
Farmersville Moraine	18	40.4 ± 4.4	10	2.3 ± 0.4
Camden Moraine	10	34.5 ± 7.1	7	2.3 ± 0.9
Hartwell Moraine	9	35.2 ± 8.8	9	2.3 ± 0.3
Scioto lobe				
Broadway Moraine	8	24.7 ± 5.0	6	2.9 ± 0.7
Powell Moraine (west)	13	28.8 ± 2.8	13	3.7 ± 0.7
Powell Moraine (east)	11	18.5 ± 3.8	8	2.1 ± 0.8
London Moraine	14	37.6 ± 7.1		
Reesville Moraine	10	34.5 ± 3.3	10	2.2 ± 0.3
Cuba Moraine	7	27.2 ± 6.0		

°Statistical values other than sample mean not calculated for less than six samples (range of observations in parentheses).

fraction, and the texture class (U.S. Dept. Agr., 1951) based on particle-size analysis. Each resulting group is identified by a capital letter. The geographic distribution of each group is shown in Figure 3, and the five parameters measured are discussed as they pertain to this grouping.

Area A occupies that part of the Maumee Lake Basin to the southwest of the Defiance Moraine. It has been called the Paulding Basin in a study of soils by Baker, Holowaychuk, and Schafer (1960). The sample means for particle size, based on six samples, are 2.3 percent sand, 29.9 percent silt, and 67.8 percent clay. The calcium carbonate equivalent is 21.3 percent and the ratio of Ca/Mg in the calcareous fraction is 6.5. These lacustrine materials are designated as calcareous clay with high Ca/Mg ratio. The boundary of this area corresponds to the recognized extent of the Paulding soils toposequence in the Maumee basin.

Area B, called the Toledo Basin, is composed of lacustrine sediments and lies north of the Defiance Moraine and adjacent to the shores of Lake Erie. This area has means of 2.2 percent sand, 21.8 percent silt, and 45.3 percent clay, based on eight samples. The calcium carbonate equivalent is 18.7 percent and the Ca/Mg ratio is 2.5. These lacustrine materials are defined as calcareous, silty clay with low Ca/Mg ratio. The limits of this area conform to the distribution of the Toledo soils toposequence.

The differences between these two lacustrine deposits in clay and silt

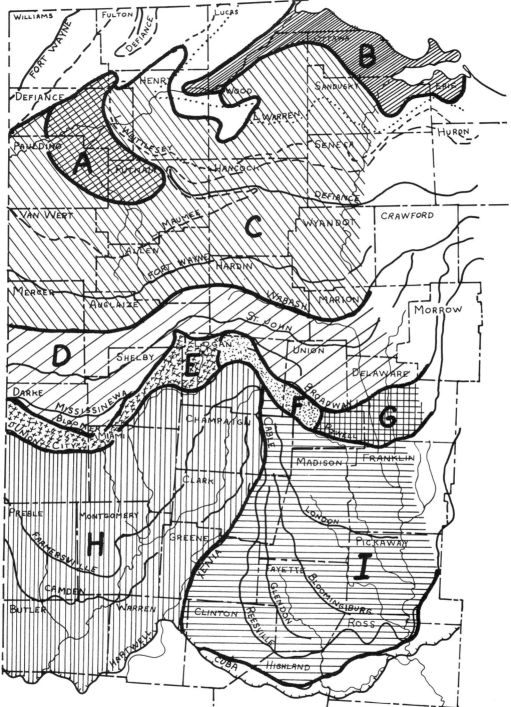

Figure 3. Distribution of glacial materials in western Ohio, based on generalization of particle-size distribution, carbonate content, and Ca/Mg ratio. Lacustrine deposits: (A) calcareous clay with high Ca/Mg ratio, and (B) calcareous, silty clay with low Ca/Mg ratio. Glacial till: (C) calcareous, heavy clay loam with high Ca/Mg ratio, (D) calcareous, heavy clay loam with intermediate Ca/Mg ratio, (E) strongly calcareous, light clay loam with intermediate Ca/Mg ratio, (F) calcareous, silty clay with high Ca/Mg ratio, (G) calcareous, light clay loam with low Ca/Mg ratio, (H) strongly calcareous loam with low Ca/Mg ratio, and (I) strongly calcareous, silt loam with low Ca/Mg ratio.

content and in Ca/Mg ratio suggests that differing environmental conditions influenced their deposition. The Paulding Basin sediments indicate a low-energy bottom environment, conducive to the separation of calcite from magnesite. Since calcite is less soluble in water than is magnesite (Hutchinson, 1957), it would most likely be precipitated in greater proportion. In contrast, the sediments in the Toledo Basin are higher in silt and sometimes somewhat stratified, suggesting either shallower water, or closer source of sediment, or a higher energy environment, with less opportunity for sorting and segregation of the carbonate fraction.

Area C, the northernmost till area considered in this study, lies adjacent to the lacustrine sediments in the Erie Lobe. The Defiance, Fort Wayne, and Wabash End Moraines are included in this area. Sand content of these moraines averages near 20 percent, silt content is about 43 percent, and clay content is about 35 percent. The latter value appears to decrease southward, but the small number of samples from the Wabash Morainal area made it impossible to test the statistical significance of this apparent decrease. Carbonate analyses show the calcium carbonate equivalent to be nearly the same as that of the lacustrine sediments, and the Ca/Mg ratio of 4 to 5 approaches that of the Paulding Basin. This till area is referred to as calcareous, heavy clay loam, with high Ca/Mg ratio. Both the Napannee and Morley toposequences of soils have been recognized in this area. The Wabash Moraine appears to be a significant boundary in terms of calcium carbonate equivalent and Ca/Mg ratio. Immediately south of this moraine, the total calcium carbonate equivalents are higher by about 8 to 10 percent, and the Ca/Mg ratios decrease from more than 4 to nearly 3. No change in particle-size distribution was observed, however, at this moraine.

Area D, south of the Wabash and north of the Bloomer-Broadway Moraines, is designated as calcareous, heavy-clay-loam till, with intermediate Ca/Mg ratios. The Morley soils toposequence are predominant on these materials.

Area E, between the Bloomer and the Union City Moraines of the Miami Lobe, is transitional in nature, compared to materials to the north and south. The calcium carbonate equivalent (38.6 percent) is similar to that of the Farmersville till to the south, whereas the Ca/Mg ratio of 2.7 is comparable to that of the materials of the St. Johns-Mississinewa-Bloomer complex to the north. The mean-particle-size distribution of 23.0 percent sand, 48.6 percent silt, and 27.5 percent clay are also intermediate in area E, resembling the materials of area G, the eastern segment of the Powell Moraine, in this respect. Soils of both the Miami and Morley toposequences have been identified on these materials. The till is designated as strongly calcareous, light clay loam, with intermediate Ca/Mg ratios.

Areas H and I lie south of the Union City and Powell Moraines. The range of calcium carbonate equivalent of the till is 35 to 40 percent, whereas the Ca/Mg ratio is about 2.2 to 2.3. These areas are coincident with the distribution of the Miami and Russell soils toposequences. A distinction can be made between the materials of area H and of area I in particle-size distribution. A higher silt content of 46 to 48 percent is found in area I, which may possibly reflect the influence of more shaly bedrock as a source of the silt. The silt content is only slightly below the 50 percent minimum for the silt-loam texture class. The till of Area I in the Scioto Lobe is designated as strongly calcareous silt loam, with low Ca/Mg ratio. Area H in the Miami Lobe is likewise strongly calcareous and has a low Ca/Mg ratio, but has a loam texture.

The segment of the Powell Moraine and till plain west of Marysville (Powell Moraine-west) has a clay content of 42.7 percent, the highest of any till studied. Area F is thus characterized as calcareous, silty clay with high Ca/Mg ratios. The eastern segment of the Powell Moraine (area G) is calcareous, light clay loam with low Ca/Mg ratios. This area has the lowest calcium carbonate equivalents in the area studied. This is attributed to the incorporation of non-calcareous shale material into the tills.

Over much of the area in this study, Ca/Mg ratios of the calcareous bedrock substrate are appreciably lower than are those of the glacial drift. Much of the bedrock which has been exposed to glacial erosion is dolomitic, most of which is rock of the Niagara and Monroe Formations (Monroe Formation, as used in this report, includes both the Silurian Bass Islands and Devonian Detroit River Dolomites [J. L. Forsyth, personal communication, 1969]). According to data reported by Stout (1941), these formations have a Ca/Mg ratio of 1.0 throughout the area shown in Fig. 4. In central Ohio, the Devonian Delaware and Columbus Limestones appear as a north-south belt five to ten miles wide, extending from Lake Erie south to Pickaway County. A remnant of Columbus Limestone is also present in the vicinity of the Bellefontaine highland. The Ca/Mg ratio of the Columbus Formation in Logan County ranges from 1.0 to 1.3. The Silurian Brassfield Limestone exposed in southwestern Ohio has a ratio of between 4.0 and 24.0. The limestones of the Ordovician system also have high Ca-Mg ratios, but are not as high in total calcium.

It is generally postulated (R. P. Goldthwait, personal communication, 1965) that a large proportion of the ice-laid materials had their origins only a few miles from their sites of deposition. If this is the case, a considerable degree of similarity in composition of the glacial till and underlying bedrock could be expected.

In the Erie Basin, presently occupied by the western part of Lake Erie, there are extensive outcrops of Columbus Limestone (Carman, 1946), with Ca/Mg ratios in the 4.0 to 6.0 range. Some of the material derived from these rocks may have been incorporated into the ice sheet as it moved southward through the basin. Since the ratios characteristic of the present till

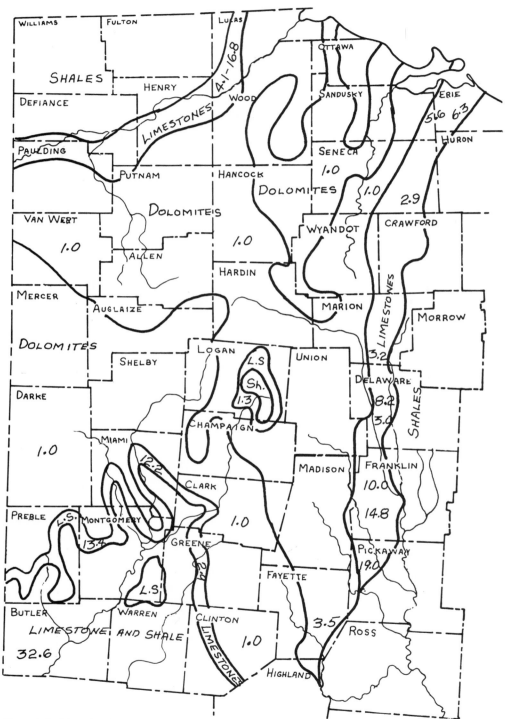

Figure 4. Ca/Mg ratios of bedrock, adapted from Bownocker (1947) and Stout (1941).

material farther south in western Ohio are more than twice those of the dolomitic rocks, some other source appreciably higher in calcite must be postulated.

The postulated source of calcite from the limestone formations does not account for the higher clay content of the northern Ohio tills. Shale deposits in the eastern Lake Erie basin (Carman, 1946) would contribute some clay to till materials, but could not supply a carbonate fraction with a high Ca/Mg ratio.

The area north of the Erie-Ohio divide is presumed to have acted as a proglacial basin whenever ice retreated north of the divide. It was the locale of lakes and lacustrine deposits prior to and during Wisconsin inter-glacial periods. The lacustrine clays and silts of this basin would have been incorporated into the till carried by subsequent readvances of the glacier and transported southward. Such a process may also have operated on a more local level and resulted in till areas such as the western segment of the Powell moraine which has such a high clay content.

The origin of the high calcium content of some lake sediments has been stated by Hutchinson (1957) as being due to the wide difference in solubility between $CaCO_3$ and $MgCO_3$. The latter, being more soluble, would be removed in waters flowing out of the lake. Whether the carbonates are precipitated as clay particles or part of the silt fraction or both is not clear. The high Ca/Mg ratios in the clay-rich materials of the Paulding Basin suggest that carbonates form part of the clay fraction, but recent work (Smeck, Wilding, and Holowaychuk, 1968) shows that, on the other hand, clays contribute only a small fraction of the carbonates in glacial till. Further investigation of the carbonate distribution in the lacustrine and till-derived parent materials is needed to resolve this question.

SUMMARY

In summary, the lacustrine materials of the Maumee Basin have higher Ca/Mg ratios than do the underlying dolomites, though these ratios are probably lower than is that of the Columbus Limestone that crops out beneath western Lake Erie. The Ca/Mg ratios of the till materials of the Erie Lobe are about four to five times greater than those of the subjacent Monroe and Niagara Dolomites. This striking difference suggests a source material higher in calcite. Furthermore, the high clay content of the till in the Erie Lobe suggests that lacustrine clays had been incorporated into this material.

The decrease in Ca/Mg ratio and increase in total calcium carbonate equivalent toward the south in both the Miami and Scioto Lobes indicates a different source of till materials. The increased influence of the dolomitic bedrock in this direction is apparent. However, the Ca/Mg ratio in these tills is about twice that in the dolomitic rocks. Lacking nearby lacustrine sources with a potentially high Ca/Mg ratio, the only other source of calcite materials appears to be the narrow bands of limestones that crop out in the western part of the Erie Basin and at a few other locations in the area studied.

If the source of the higher Ca/Mg material is to the north or northeast, the major part of the till components must have been transported for a considerable distance. The proposal that tills are composed mainly of materials carried relatively short distances by the advancing ice is not supported by the analysis of the fine fraction (<2 mm) of till materials in western Ohio.

ACKNOWLEDGMENTS

The authors wish to acknowledge the valuable assistance provided by E. Moye Ruteledge, who directed the Soil Characterization Laboratory of the Ohio Agricultural Research and Development Center. His experience and suggestions enhanced the reliability of the laboratory data and contributed to the successful completion of this project.

REFERENCES

Baker, F. J., Holowaychuk, N., and Schafer, G. M., 1960, Surficial materials and soils of Paulding County, Ohio: Ohio Jour. Sci., v. 60, p. 365-77.

Baker, F. J., Meeker, R. J., and Holowaychuk, N., 1960, Soil survey of Paulding County, Ohio: U.S. Dept. Agr. Soil Survey Series 1954, no. 12.

Bownocker, J. A., 1947, Geologic map of Ohio: Geol. Survey of Ohio.

Carman, J. E., 1946, The geologic interpretations of scenic features in Ohio: Ohio Jour. Sci., v. 46, p. 241-83.

Coffey, G. N., and Rice, T. D., 1915, Reconnaissance soil survey of Ohio: U.S. Dept. Agr. Bureau of Soils.

Dreimanis, A., 1962, Quantitative gastrometric determination of calcite and dolomite by using Chittuck apparatus: Jour. Sed. Petrology, v. 32, p. 520-29.

Forsyth, J. L., 1959, The beach ridges of northern Ohio: Geol. Survey of Ohio Inf. Cir. 25, 10 p.

———, 1961, Dating Ohio glaciers: Geol. Survey of Ohio Inf. Circ. 30, 9 p.

———, 1967, Glacial geology of the East Liberty quadrangle, Logan and Union Counties, Ohio: Geol. Survey of Ohio Rpt. of Invest. no. 66.

Garner, D., and Meeker, R., 1962, Soil survey of Clinton County, Ohio: U.S. Dept. Agr. Soil Survey Series 1958, no. 23.

Goldthwait, R. P., 1958, Wisconsin age forests in western Ohio: I. Age and glacial events: Ohio Jour. Sci., v. 58, p. 209-19.

———, 1959, Scenes in Ohio during the last ice age: Ohio Jour. Sci., v. 59, p. 193-216.

———, 1965, Great Lakes–Ohio River Valley, INQUA Guidebook for Field Conf. G., Intern. Assoc. for Quat. Research, VIIth Congress, Nebraska Acad. of Sciences, Lincoln, Neb., 109 p.

Goldthwait, R. P., White, G. W., and Forsyth, J. L., 1961, Glacial map of Ohio: U.S. Geol. Survey, Misc. Geol. Invest. Map I-316.

Heffner, R. L., Brock, A. R., Christman, R. L., and Waters, D. D., 1965, Soil survey of Allen County, Ohio: U.S. Dept. Agr. Soil Survey Series 1960, no. 24.

Hutchinson, G. E., 1957, A treatise on limnology, vol. I: New York, John Wiley and Sons, Inc.

Lerch, N. K., Davis, P. E., Tornes, L. A., Hayhurst, E. N., and McLoda, N. A., 1969, Soil Survey of Preble County, Ohio: U.S. Dept. Agr. Soil Survey Series.

Rapparlie, D. F., and Urban, D. R., 1966, Soil Survey of Wood County, Ohio: U.S. Dept. Agr. Soil Survey Series.

Smeck, N. E., Wilding, L. P., and Holowaychuk, N., 1968, Genesis of Argillic Horizons in Celina and Morley Soils of Western Ohio: Soil Sci. Soc. Amer. Proc., v. 4, p. 550-56.

Steele, J. G., and Bradfield, R., 1934, The significance of size in the clay fraction: Amer. Soil Survey Assoc. Bull. 15, p. 88-93.

Stout, Wilbur, 1941, Limestones and dolomites of western Ohio: Ohio Geol. Survey, 4th Ser., Bull. 42. 468 p.

U.S. Department of Agriculture, Soil Survey Staff, 1951, Soil Survey Manual: U.S. Dept. Agr. Handbook No. 18, p. 123-41, 327-37.

Wenner, K. A., Holowaychuk, N., and Schafer, G. M., 1961, Changes in clay content, calcium carbonate equivalent, and calcium/magnesium ratio with depth in parent materials of soils derived from calcareous till of Wisconsin age: Soil Sci. Soc. of Amer. Proc., v. 25, p. 312-16.

*Mineral and Elemental Composition of
Wisconsin-Age Till Deposits in
West-Central Ohio*

**L. P. Wilding, L. R. Drees,
N. E. Smeck, and G. F. Hall**

ABSTRACT

Clay mineralogy and elemental Ti, Zr, Ca, K, and Fe analyses are presented for representative tills at six sites sampled on nearly level interfluves in a north-south transect from the Fort Wayne ground moraine south to the Camden End Moraine. The clay-size fraction in the calcareous tills is illite (50-80 percent), vermiculite (10-20 percent), quartz (5-10 percent), kaolinite (<5 percent), chlorite (<5 percent), and montmorillonite (absent or trace). Carbonates (primarily dolomite) comprise less than 10 percent of the 2.0-0.2 μ clay fraction, and less than 1 percent of the <0.2 μ fraction. Pedogenic transformation of illite to both vermiculite and montmorillonite occurs upon weathering of these deposits in the southern sector, but proceeds primarily to vermiculite in the northern part. Elemental Ti, Fe, and K in calcareous deposits decrease by 25-40 percent in traversing from north to south; Zr remains approximately constant; and Ca increases by about 40 percent due to higher carbonate status of the tills in the south. Concentration of Ti and Fe occurs in the fine-silt (0.02-002 mm) fraction, and Zr in the coarse-silt (0.05-0.02 mm) fraction; and K is approximately equally distributed between the above fractions. Calcium occurs primarily in the form of carbonates and has a bimodal distribution, with maxima corresponding to the 2.0-0.5-mm and 0.05-0.02-mm fractions. Calcite/dolomite ratios of the above two fractions progressively decrease from approximately 2.0 in the north to 0.5 in the south. Differences in composition of these till deposits are considered to be the primary factor responsible for the more intensive weathering of the soils in the area south of the Bloomer Moraine.

This paper is a contribution of the Department of Agronomy, Ohio State University; research conducted under State Projects 371 and 106.

INTRODUCTION

For some time pedologists and geologists have been interested in the physical and chemical composition of tills in western Ohio. Pedologists have been concerned with understanding how till composition influences intensity of soil development when other soil-forming factors remain relatively constant. Geologists commonly utilize such information as supplemental aids or auxiliary tools in mapping and correlating till stratigraphy. Prior to this work, particle-size and carbonate composition of till deposits in western Ohio have been employed on a regional basis to differentiate two extensive soil associations — the Miami-Celina-Crosby-Brookston toposequence* — in the southern portion of the study area and the Morley-Blount-Pewamo in the northern reaches. Soils of these associations encompass almost all of the moderately to strongly calcareous Wisconsin-age deposits of the study area (Fig. 1).

It has been of mutual interest to both disciplines to find a satisfactory explanation for the similarity in textural development of Miami and Morley toposequence members, when in fact, the respective till deposits from which these soils have been derived vary considerably in particle-size and carbonate composition. Smeck and others (1968) have discussed the implications of till composition on the pedogenic formation of soil B horizons in this area. The purpose of this paper is to characterize the till deposits of western Ohio — with regard to carbonate particle-size distribution, clay mineralogy, and elemental analysis — and to relate these parameters to pedogenic weathering phenomena.

REVIEW OF LITERATURE

Considerable information is available regarding the clay-mineral components of till deposits in western Ohio, but only limited elemental analyses have been reported. Brown and Thorp (1942), based on total chemical analyses, suggest that the colloidal fractions of soils derived from till deposits in Indiana (similar to those in western Ohio) are predominantly hydrous mica (illite) with about 10 percent kaolinite and 10 to 25 percent montmorillonite. Bidwell and Page (1950), using thermal and X-ray analyses, have reported the effects of soil-formation processes, as influenced by differences in drainage, on the clay-mineral composition of till-derived soils of the Miami toposequence. Their investigations of the fine-clay fraction ($<.5\mu$) indicate that illite is the predominant clay mineral throughout the sola of all soils in this toposequence, and especially in the calcareous parent

---

*The Miami member of this toposequence has recently been correlated Miamian in Ohio.

till deposits. Its presence was attributed to inheritance, rather than to synthesis *in situ*. Bidwell and Page (1950) also noted slight evidence for the presence of montmorillonite in the B horizons of more poorly drained toposequence members and postulated that this clay mineral may have been synthesized by pedogenic processes subsequent to the deposition of the till.

Holowaychuk (1950) further pursued the clay mineral-drainage relationship of these soils and concluded that the more poorly drained members of this toposequence contained appreciable quantities of montmorillonite that must have been synthesized *in situ*. In addition to illite and montmorillonite, he found limited quantities of kaolinite in weathered sections of these till deposits.

Thorp and others (1959), from work in eastern Indiana, concluded that B horizons of a Miami profile (derived from till deposits of comparable age and composition to those in the study area) are developed primarily through translocation of suspended fine clay and humus from overlying surficial horizons with subsequent deposition of these colloids in B horizons. Their studies of the clay fraction indicate that illite is dominant in the parent till and in the lower B horizons, but decreases upward toward the soil surface. Montmorillonite is dominant in the horizons of maximum clay accumulation, and vermiculite is greatest in the overlying eluvial horizons. This distribution suggests that illite is being weathered to vermiculite, which in turn may be weathering to montmorillonite.

More recently, work by Andrew (1960), Gersper (1963), and Smeck and others (1968), all using X-ray-diffraction methods, have conclusively demonstrated that oxidized calcareous Wisconsin-age till deposits in the Miami and Scioto Lobes of western Ohio are dominantly illitic, and that at least a portion of the vermiculite and montmorillonite found in weathered sections of these deposits are of pedogenic origin from an illite precursor. They also have reported kaolinite in both weathered and calcareous sections, and believe it to be primarily indigenous to the parent till.

Mickelson (1942) concluded from weathering studies using resistant heavy minerals (0.02-0.05 mm) as an index, that little clay, if any, had accumulated in B horizons of a Miami profile sampled in Champaign County, Ohio, although considerable quantities had been lost from A horizons. It appeared to him that clay gains in B horizons via illuviation were nearly in balance with clay losses by weathering. Based on present knowledge, the clay losses he found in A horizons probably reflect more lithological contamination of these surfaces with loessial materials than clay eluviation (Wilding and Drees, 1968). Clay-mineral and thin-section analyses would suggest that B horizons have been clay enriched both through illuviation and pedogenic clay synthesis (Thorp and others, 1959; Smeck and others, 1968).

Green (1933), Mickelson (1942), and Smeck and others (1968) have investigated the carbonate status of these till deposits. All have reported that carbonate minerals were concentrated in sand- and silt-size fractions. Smeck and others (1968) noted that the distribution is bimodal, with greatest concentrations in the very-coarse-sand (1 to 2 mm) or coarse-sand (0.5 to 1 mm) fractions and in the coarse-silt (0.05 to 0.02 mm) fraction. No rational explanation could be offered for this phenomenon. However, since particle-size analysis of soils and geological sediments is commonly determined on samples before removal of carbonates (carbonate-basis), it is important to know the particle-size distribution of carbonates before evaluating the degree of textural development of soils derived from a specific deposit. This is true because carbonate leaching of a particular deposit will either augment or decrease the non-carbonate clay content of the altered material, depending on whether the carbonates are primarily of clay-size or coarser fractions.

The only elemental analyses reported for till deposits similar to those characterized herein are by Brown and Thorp (1942) from eastern Indiana. Analyses (as oxides) are presented on a total-soil basis; thus, they are difficult to interpret from a weathering aspect, because such results reflect both the relatively immobile sand and silt fractions and the more mobile clay fractions. Elemental analyses deviate considerably among different size fractions. Hence, as finer fractions are redistributed through pedogenic weathering processes, it is not possible , from total analyses, to determine if the data reflect weathering trends of immobile constituents, redistribution of weathering products, clay redistribution as a dilution mechanism, lithological discontinuities, or a combination of the above. In view of these limitations, elemental analyses of individual sand and silt fractions, which can subsequently be combined on a weight-fraction basis if desired, are much more meaningful in evaluating weathering transformations and pedogenesis.

Elemental Ti and Zr, because of their occurrence in relatively resistant minerals, have been of interest to pedologists as a means of evaluating parent-material homogeneity (Johnson, 1961; Alexander and others, 1962; Chapman and Horn, 1968; Calhoun, 1968). Khangarot (1969) recently reviewed the literature concerning mineralogy, stability, and utility of elemental Ti, Zr, and Ti/Zr indices of parent-material homogeneity. The advent of X-ray spectroscopy has permitted relatively rapid and accurate determinations of these elements in a native (solid) state for any specified size-fraction desired. This technique eliminates laborious chemical fractionation and does not suffer from the somewhat questionable statistical limitations of optical microscopy (Brewer, 1964, p. 48). However, X-ray spectroscopy does have the limitation that, unless correlated with mineralogy, one cannot infer the source contributing to such elements. Fortunately, Ti and Zr occur principally in only a few accessory minerals — minerals which are common to most deposits and, under the weathering conditions in the area studied, are relatively stable.

Elemental K and Ca in the size-fractions investigated occur principally in feldspars, according to the work of Mickelson (1942). He comprehensively investigated the mineralogy of several specific-gravity isolates separated from samples collected from Miami-toposequence members in western Ohio. Micas comprise only a small percentage of the light-mineral isolate separated from sand and silt fractions, but undoubtedly, as shale lithics in these separates increase, the percentage of K from this mica source would become more important. Previous work (Beavers, 1960; Johnson, 1961; Beavers and others, 1963; Jones and Beavers, 1966; Foss and Rust, 1968; and Smith and Buol, 1968) suggests that molar ratios of K_2O/ZrO_2 and CaO/ZrO_2 are useful parameters for evaluating weathering intensities. Elemental Fe, because of its occurrence in minerals readily susceptible to weathering, has also been found to be a sensitive weathering index. Commonly Ca, because of its presence in more weatherable Ca-feldspars, yields a more pronounced weathering curve with depth than does K; that is, there is a decrease in the amount of Ca from surficial to less intensively weathered subsoil or parent-material zones, while the K content remains approximately constant, or in some cases, decreases slightly with depth.

STUDY AREA AND SITE SELECTION

The study area is located in the Miami Lobe between the Glacial Lake Maumee Shoreline and the Camden End Moraine of western Ohio (Fig. 1). Soil-forming factors of this area, except for parent material, are essentially constant and similar to those previously discussed for the Morley and Blount soils (Wilding and others, 1964). The mean annual temperature and precipitation for the northern portion of the area are slightly less than precipitation for the south — 51° F and 36 inches in the north, versus 54° and 39 inches in the south. Surficial glacial deposits range in age from 14,000 to 18,000 years from north to south (Goldthwait and others, 1965). The glacial till north of the Bloomer Moraine has a clay-loam or silty-clay-loam texture, with a mean clay content of 32 percent and a mean calcium-carbonate equivalent of 24 percent (Wilding and others, 1964), whereas to the south of this moraine, the till is of loam texture, with a mean clay content of 19 percent and a mean calcium-carbonate equivalent of 40 percent. Both of these till deposits are dense (dry bulk densities of 1.8 to 1.9 g/cm^3) and compact, and range in thickness from a few to several hundred feet. Vertical fractures, coated with gray secondary carbonate segregations, extend to depths of eight to ten feet in the oxidized calcareous till. These fractures are numerous, occurring laterally at intervals of two to four inches.

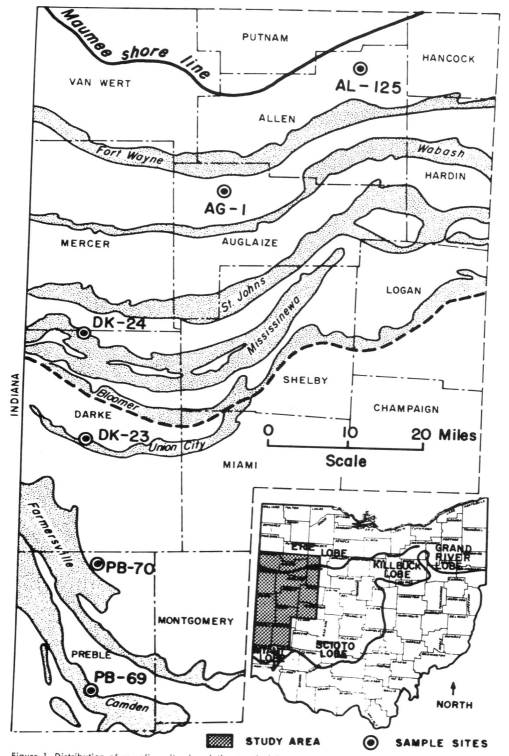

Figure 1. Distribution of sampling sites in relation to glacial end moraines in the Miami Lobe of the Wisconsin till sheet in western Ohio. The heavy dashed line paralleling the Bloomer Moraine separates the Miami-Celina-Crosby-Brookston soil association to the south from the Morley-Blount-Pewamo soils to the north.

A thin loess mantle ($<18''$ thick) is superposed and frequently admixed with the till over most of this landscape (Wilding and Drees, 1968).

Six sampling sites, three in the area of Celina soils and three in the Morley association, were selected in a north-south transect across the study area by utilizing characterization data available from previous sampling in conjuncture with the Ohio Soil Survey Program. All profiles were sampled on nearly level, relatively stable interfluves at approximately 15-mile intervals along a north-south direction across the Miami Lobe (Fig. 1).

EXPERIMENTAL METHODS

Samples collected for this work were obtained at each site from pits excavated to approximately three feet and from a three-inch core obtained with a power probe from below this depth. All samples of calcareous till were from the oxidized zone. Methods of collection, fractionation, and determination of chemical, physical, particle-size fractionation, and carbonate analyses have been reported recently by Smeck and others (1968). Clay mineralogy was determined on the total clay ($<2\mu$) fraction as follows. Sixty-gram samples of total soil (<2mm) were dispersed by adding approximately 1.8 g of sodium carbonate per 200 ml of distilled water and shaking the suspension in 500 ml bottles on a reciprocating shaker for 15 hours. Organic matter was destroyed by using 30 percent hydrogen peroxide when the organic matter content exceeded 3 percent. No pretreatment was used for removal of carbonates or iron oxides. Sand was removed by using a 300-mesh sieve, and the clay fraction was separated by an automatic fractionator, as described by Rutledge and others (1967). The clay was flocculated with sodium chloride and stored in a 20 percent methanol solution for subsequent diffraction analysis.

An aliquot of the clay fraction to be analyzed was Mg-saturated with three washings of $1N$ $MgCl_2$. Excess chloride was removed by repeated washing with distilled water and methanol until the decantate yielded a negative chloride test. The dispersed samples were then plated via vacuum on porous ceramic plates (Kinter and Diamond, 1956). Three plates were prepared for each sample. One plate was leached with a 10 percent aqueous solution of ethylene glycol, stored under room conditions for several hours, and then placed in a saturated ethylene glycol atmosphere until heat-treated. Immediately prior to X-raying, this glycolated specimen was heated to $40°$ C for several hours to remove excess ethylene glycol from the surface. The other two samples were allowed to air dry at room temperature for 48 hours. One was X-rayed at room temperature, heated to $400°$ C for two hours, allowed to cool to $100°$ C, and then X-rayed again. The other sample was heated to 550° C for two hours, cooled to room temperature, and then X-rayed.

Elemental Ti, Zr, Fe, Ca, and K were determined, using X-ray spectrographic techniques. Analyses were made for the 0.002-0.02 mm and 0.02-0.05 mm silt fractions and for the fine and very fine sand fraction (0.05-0.25 mm). For each of the above fractions, duplicate briquettes were prepared by adding boric acid (30 percent by weight) as a binding agent and pelleting the sample under a pressure of 25 tons/square inch. Standards were prepared by the addition method of Handy and Rosauer (1959), and where appropriate, National Bureau of Standard Samples were also employed as elemental references. Spectographic analyses were obtained with a Norelco Unit using a proportional counter, a vacuum path, and a Cr-tube for Ca, K, and Ti. A scintillation counter, air-path, and W-tube were used for determination of Fe and Zr. The time required for 100,000 counts was determined for all elements. More detailed machine conditions are presented elsewhere (Khangarot, 1969).

Reconstruction of elemental and clay gains and losses accompanying weathering of these till deposits were computed based on the stable-constituent method of Brewer (1964, p. 83-84). Total Zr (0.002-0.25 mm) was used as the stable constituent. The initial step in reconstruction was to determine the volume of parent till necessary to comprise a unit volume of weathered soil horizon. This constant is termed the weathering factor (W.F.) and is computed as follows:

$$\text{W.F.} = \frac{D_s R_s}{D_p R_p}$$

where

D_s = bulk density (at 1/3 atmosphere moisture tension) of the weathered soil horizon (g/cc)

R_s = % Zr (by weight) in the 0.002-0.25-mm size-fraction, multiplied by the percentage of this size-fraction in the total weathered soil (g/100g)

D_p = bulk density (at 1/3 atmosphere moisture tension) of parent till (g/cc)

R_p = % Zr (by weight) in the 0.002-0.25-mm size fraction multiplied by the percentage of this size fraction in total parent till (g/100g).

This weathering factor is the same as that computed by Westin (1953), and the reciprocal of the volume factor as defined by Brewer (1964, p. 81).

The next step was to compute, on a volume basis (g/100 cc), the weight of elemental or clay constituent present in the weathered soil horizon. The present weight in grams of constituent X (X_s) in unit volume of the soil horizon is given as

$$X_s = D_s P_x$$

where $P_x =$ the percentage by weight of constituent X in the present-day weathered horizon. The *original* weight of elemental constituent X (X_p) in the volume of parent till necessary to form a unit volume of present-day weathered soil is derived as follows:

$$X_p = W.F. \; (D_p \; P_x')$$

where P_x' is the percentage by weight of constituent X in the parent till. The reconstruction of constituent X (X_g) in g/100 cc gained or lost in the soil during weathering is given by

$$X_g = X_s - X_p .$$

Negative values indicate losses, and positive values gains.

Net changes on a volume basis (g/100 cc) can be meaningfully compared among horizons within or between profiles because such reconstructions are not confounded by differences in thickness among horizons considered. On the other hand, to determine the total gain or loss of a constituent for the entire weathered profile, or for a section thereof, it is necessary to multiply the net change for each horizon (g/100 cc) by its thickness (cm) and sum the resulting products. This may be imagined as a column of soil with horizontal dimensions of 10 cm by 10 cm (100 cm²) and vertical dimensions corresponding to the thickness of the zone in question. This yields dimensions of g/100 cm²/profile thickness. Since clay constituents occur in greater quantities than do elemental constituents, net changes of clay are reported on the basis of a vertical column of soil with a unit area of one cm².

RESULTS AND DISCUSSION

Particle-size distribution, bulk density, pH, and carbonate status for selected horizons are presented for all sites. However, carbonate particle-size distribution, clay mineralogy, elemental analyses, and reconstruction data are presented for only two of the six profiles in this transect — the representative Celina (PB-70) and Morley (AG-1) profiles. On the basis of this work and of numerous physical and chemical analyses of other samples collected in conjunction with the Ohio Soil Survey Program in the study area, it is concluded that these two sites are representative of soils derived from the respective till deposits. Detailed soil profile descriptions and physical and chemical data for all six profiles are available elsewhere (Smeck, 1966).

TABLE 1

Selected Physical and Chemical Analyses Ap, IIB2, and IIC Horizons of Sites Investigated

Horizon	Depth (Inches)	Total Sand (0.05-2mm)	Total Silt (0.002-.05mm)	Fine (<0.2)	Coarse (0.2-2)	Total (<2)	Bulk Density <2mm Basis ⅓ Atm. Moisture (g/cc)	pH 1:1 Water	% Calcite	% Dolomite	% CaCO₃ Equivalent
Celina silt loam (PB-69)											
Ap	0-8	22.6	62.5	3.8	11.1	14.9		6.5			
IIB22t	16-20	29.6	40.1	18.0	12.3	30.3	1.58	7.0			
IIC3	58-70	44.6	39.9	4.8	10.7	15.5	1.90	8.0	6.9	29.6	39.1
Celina silt loam (PB-70)											
Ap	0-7	28.9	54.4	3.6	13.1	16.7		6.0			
IIB22t	13-17	16.9	41.6	21.5	20.1	41.5	1.54	6.5			
IIC3	46-56	35.9	45.5	4.9	13.7	18.6	1.86	8.1	9.1	34.1	46.1
Celina silt loam (DK-23)											
Ap	0-7	16.3	63.3	3.4	17.0	20.4		6.6			
IIB23t	20-24	10.6	37.3	22.4	29.7	52.1	1.46	6.5			
IIC4	60-75	17.3	48.1	7.8	26.8	34.6	1.85	8.0	15.0	21.2	38.0
Morley silt loam (DK-24)											
Ap	0-6	7.4	64.9	8.6	19.1	27.7		6.5			
IIB23t	19-24	1.5	43.6	19.6	35.3	54.9	1.52	7.4			
VC4	65-80	16.6	43.6	9.8	30.0	39.8	1.78	8.0	14.3	11.5	26.8
Morley silt loam (AG-1)											
Ap	0-7	21.3	61.2	2.8	14.7	17.5		6.7			
IIB21t	12-16	13.8	41.1	18.8	26.3	45.1	1.56	5.2			
IIC4	77-89	20.8	48.8	7.4	23.1	30.5	1.80	8.1	16.2	8.1	25.1
Morley loam (AL-125)											
Ap	0-7	31.8	48.3	3.1	16.8	19.9		6.8			
IIB23	17-22	19.2	39.5	15.3	26.0	41.3	1.64	7.3			
IIC4	59-75	22.4	48.0	7.7	21.9	29.6	1.85	8.1	13.6	6.7	20.9

Selected Physical and Chemical Properties

Selected physical and chemical data for A, IIB, and IIC horizons of all six profiles are presented in Table 1 (Fig. 1). Traversing from north to south, the maximum Bt/C total-clay ratio is: 1.3, 1.5, 1.4, 1.5, 2.2, and 1.9; the fine/coarse clay ratio for the maximum clay zone is: 0.6, 0.7, 0.6, 0.8, 1.0, and 1.3; the calcium carbonate equivalent of the IIC horizon is: 20.9, 25.1, 27.7, 35.1, 46.1, and 39.1; and the calcite/dolomite ratio is 2.0, 2.0, 0.6, 0.4, 0.3, and 0.2, respectively. These data suggest a progressive increase in textural profile development in soils from north to south, associated with a parent till that becomes increasingly dolomitic and contains approximately twice the calcium carbonate equivalent in the south as in the north. In order to determine if observed differences in magnitude of the clay maxima in these soils were more apparent than real, in view of their differing carbonate status, a detailed analysis of the carbonate particle-size distribution of the parent till deposits was undertaken. If such data should reveal that most of the carbonate minerals were non-clay in size, as suggested previously by Green's (1933) and Mickelson's (1942) work, then differences in total clay distribution between Celina and Morley soils could be largely accounted for by carbonate dilution. Differences which persist after removal of this material could then be attributed to other pedogenic processes.

Carbonate Particle-Size Distribution

The carbonate particle-size distribution (Table 2) for IIC horizons of Celina and Morley profiles indicates that the very-coarse-sand and coarse-silt fractions have the highest carbonate equivalents, whereas coarse clay contains less than 10 percent and fine clay less than one percent carbonate minerals. This bimodal size distribution was consistent for all six profiles and is not clearly understood. If this reflects degree of glacial attrition, then one might expect a more random distribution at specific sites. However, Dreimanis and Vagners (1971; this volume, p. 237) reported that carbonates found in glacial-till deposits commonly fall into two modes, one in the clast-mode which is a function of proximity to the glacier (varies from the order of meters close to the source to several mm with distance) and the other the matrix-mode or "terminal grades" (.031-.062 mm for carbonates), which depend upon original sizes of mineral grains in rocks and upon the resistance of each mineral to comminution during glacial transport. Perhaps the carbonates in the coarse-sand fractions, as observed herein, represent a part of the clast-mode in these till deposits, and the coarse-silt fraction the terminal mode in the till matrix.

On the other hand, one could argue that carbonate abundance should decrease with decreasing particle-size, because of greater solubility with increase in surface area. This phenomenon may explain the meager quantities of carbonates in clay fractions, but would imply solution of such carbonates during or since till deposition. The higher calcite/dolomite ratios of the coarse-clay fraction suggest that, in this size, it may be largely of secondary origin. Thin-section analysis of gray surface coatings oriented parallel to vertical fracture planes in the zone of oxidized calcareous till indicates that these coatings are primarily composed of secondary calcite (calcans) precipitated in fine-silt- and clay-size aggregates. It is probable that these calcans explain the higher calcite composition in the coarse-clay fraction.

Two important concepts evolve from the knowledge of carbonate particle-size distribution in these deposits. First, the clay percentage reported in a particle-size analysis of these tills is primarily of noncarbonate mineralogy, and thus leaching of carbonates at the initiation of weathering in these deposits will enrich the clay percentage (on a weight basis) in proportion to the carbonate contents. For example, if a till contains 20 percent clay and has a carbonate equivalent of 50 percent, total leaching of carbonates will result in a noncarbonate residue containing 40 percent clay (by weight) prior to any subsequent pedogenic clay redistribution or synthesis. Secondly, because different particle-size fractions have variable carbonate contents, one must be cautious in selecting a size fraction that is representative of the total soil when attempting to reconstruct gains and losses of constituents accompanying weathering. If one chooses a fraction, such as the coarse-silt or very-coarse-sand fraction, which is considerably higher in carbonates than is the total-sand or total-silt fraction, actual gains or losses will be underestimated. On the other hand, if a fraction lower in carbonates than representative is selected, gains or losses will be overestimated. Likewise, in attempting to document differences in lithology between deposits, one must be careful that observed differences or similarities are not simply a confounding, due to difference in carbonate content of the fractions investigated.

Elemental Analyses

Elemental Ti, Zr, Fe, Ca, and K for the fine- and very-fine-sand (0.05-0.25 mm), coarse-silt (0.02-0.05 mm), and fine-silt (0.002-0.02 mm) fractions are presented in Table 3 for selected horizons of profiles PB-70 and AG-1. Relative percentages of these elements (except Ca), in the above size fractions, as averaged over all calcareous till deposits collected from the six sites, are illustrated in Figure 2. Elemental assays on a total basis (0.002-0.25 mm) were computed by summing the elemental percentages of the above separates in proportion to the weight fraction and are presented as functions of depth in Figure 3. Trends in total elemental analyses, on both natural and carbonate-free bases, for calcareous till deposits at sites in the north-south transect are depicted in Figure 4.

Distribution of elements in the total fraction (0.002-0.25 mm) indicates that Ti and Fe are concentrated in the fine silts, Zr in the coarse silts, and that K is approximately equally distributed among these fractions (Fig. 3). Elemental assay of the various size fractions (Table 3) generally follow this same trend.

The occurrence of Ca is primarily in the form of carbonates, with a bimodal distribution corresponding to coarse- and very-coarse-sand (2-0.5 mm) and coarse-silt (0.05-0.02 mm) fractions (Table 2). However, on a total-soil basis, the two silt fractions contribute most of the calcium-carbonate

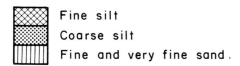

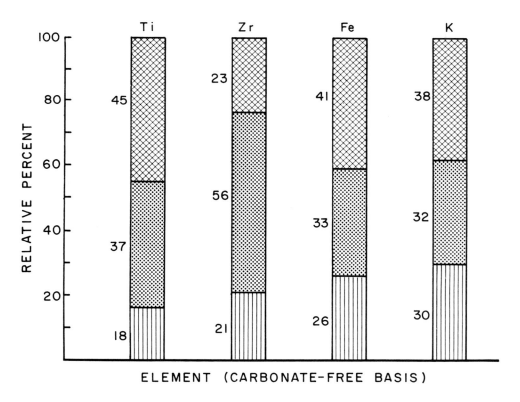

Figure 2. Relative distribution of elements by size fraction (expressed on a carbonate-free basis) in the total fraction (0.002-0.25 mm) of calcareous till samples.

equivalent of the fine-earth fraction (<2mm). Noncarbonate Ca is highest in the sand fractions and becomes less abundant with decrease in silt size (Table 3).

Similar depth-distribution trends of elements are noted, both when considered by individual size-fractions (Table 3) or when expressed on a total-fraction basis (Fig. 3). The sharp decrease in all elements with depth below the IIB2 horizons is due to carbonate dilution in the calcareous IIB3 and IIC horizons. Characteristically, Ti remains relatively constant in the weathered section or exhibits a very slight weathering trend with depth.

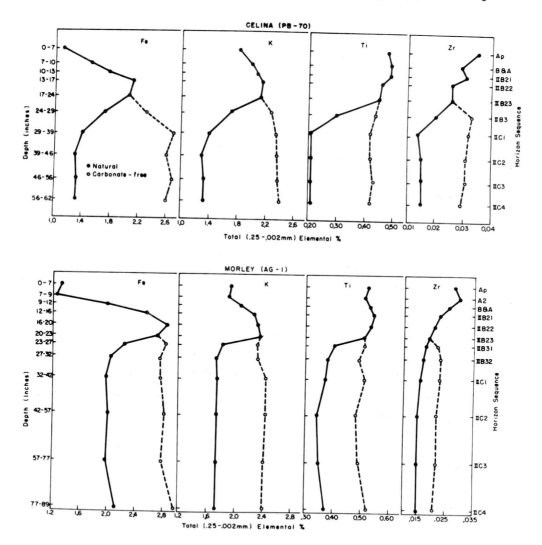

Figure 3. Elemental depth distributions of the total fraction (0.002-0.02 mm), expressed on both a natural and a carbonate-free basis for Celina and Morley profiles.

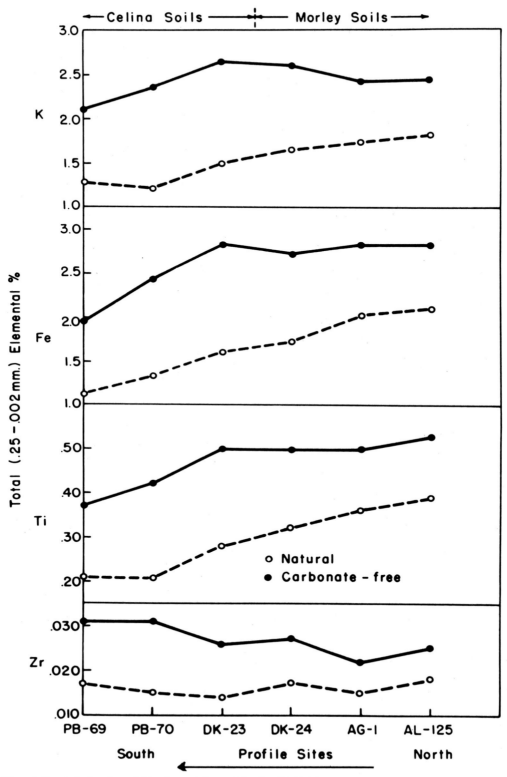

Figure 4. Trends in the elemental analyses of the total fraction (0.002-0.25 mm) expressed on both a natural and a carbonate-free basis for sites located in the north-south transect.

TABLE 2

CARBONATE DISTRIBUTION BY PARTICLE-SIZE FRACTIONS IN C HORIZONS

OF CELINA AND MORLEY PROFILES

Particle-Size[*] Fraction	% Calcite Of Particle-Size Fraction	% Dolomite Of Particle-Size Fraction	Calcite/ Dolomite Ratio	% CaCO₃ equivalent	
				Of Particle-Size Fraction	Of Total[†] Soil (<2mm)
Celina silt loam PB-70 (II-C3)					
VC sand	41.6	33.4	1.2	77.8	3.3
C sand	26.8	32.9	0.8	62.6	4.1
M sand	23.6	18.6	1.3	43.7	2.3
F sand	15.1	20.3	0.7	37.1	4.3
VF sand	20.2	25.9	0.8	48.3	4.1
Total sand	22.7	25.1	0.9	50.0	18.1
C silt	15.3	43.2	0.3	62.2	10.4
F silt	17.0	26.2	0.6	45.4	13.0
Total silt	16.3	32.4	0.5	51.6	23.4
C clay	5.4	2.9	1.9	8.5	1.2
F clay	0.2	0.4	0.5	0.6	tr‡
Total clay	4.0	2.1	1.9	6.4	1.2
Morley silt loam AG-1 (II-C4)					
VC sand	23.9	10.9	2.2	35.8	0.7
C sand	16.6	11.9	1.4	29.5	0.9
M sand	21.5	9.2	2.3	31.6	0.9
F sand	15.6	7.8	2.0	24.1	1.7
VF sand	15.2	9.5	1.6	25.5	1.4
Total sand	17.2	9.4	1.8	27.5	5.6
C silt	18.1	17.5	1.0	37.1	4.4
F silt	16.4	9.9	1.6	27.2	10.0
Total silt	16.7	11.7	1.4	29.6	14.4
C clay	5.4	1.7	3.2	7.3	1.7
F clay	0.2	0.3	0.7	0.6	tr‡
Total clay	4.1	1.3	3.2	5.5	1.7

*VC sand (1-2mm)
  C sand (0.5-1mm)
  M sand (0.25-0.5mm)
  F sand (0.1-0.25mm)
  VF sand (0.05-0.1mm)

C silt (0.02-0.05mm)
F silt (.002-0.02mm)
C clay (0.2-2μ)
F clay (<0.2μ)

†Contribution of each of the respective size fractions to calcium carbonate equivalent of total soil.
‡trace

TABLE 3

DISTRIBUTION OF ELEMENTS IN SAND AND SILT FRACTIONS OF CELINA (PB-70)

AND MORLEY (AG-1) PROFILES

Horizon	Depth (Inches)	Fine and Very Fine Sands (0.05-0.25 mm) %				
		Ti	Zr	Fe	Ca	K
Celina silt loam (PB-70):						
Ap	0-7	0.18	0.021	1.40	0.73	1.79
B&A	7-10	0.22	0.015	1.98	0.67	1.95
IIB21t	10-13	0.24	0.018	2.40	0.66	1.96
IIB22t	13-17	0.23	0.017	2.36	0.75	2.03
IIB23t	17-24	0.23	0.019	2.14	0.85	1.88
IIB3	24-29	0.15	0.013	1.41	Ca*	1.45
IIC1	29-39	0.13	0.012	1.12	Ca	1.21
IIC3	46-56	0.13	0.014	1.08	Ca	1.26

TABLE 3 — *Continued*

Horizon	Depth (Inches)	Fine and Very Fine Sands (0.05-0-25 mm) %				
		Ti	Zr	Fe	Ca	K
Morley silt loam (AG-1):						
Ap	0-7	0.22	0.023	1.76	0.77	2.17
A2	7-9	0.23	0.022	1.75	0.80	2.18
B&A	9-12	0.24	0.019	2.15	0.57	2.24
IIB21t	12-16	0.28	0.021	2.79	0.57	2.30
IIB22t	16-20	0.30	0.017	3.15	0.68	2.36
IIB23t	20-23	0.27	0.017	2.48	Ca*	2.24
IIB31t	23-27	0.21	0.013	1.98	Ca	1.68
IIB32	27-32	0.19	0.013	1.79	Ca	1.68
IIC1	32-42	0.18	0.014	1.79	Ca	1.63
IIC4	77-89	0.20	0.014	1.77	Ca	—

		Coarse Silt (.05-.02 mm) %				
		Ti	Zr	Fe	Ca	K
Celina silt loam (PB-70):						
Ap	0-7	0.49	0.063	0.86	0.37	1.73
B&A	7-10	0.41	0.057	1.07	0.36	1.85
IIB21t	10-13	0.42	0.058	1.32	0.39	1.95
IIB22t	13-17	0.38	0.042	1.54	0.44	1.97
IIB23t	17-24	0.40	0.049	1.57	0.65	1.99
IIB3	24-29	0.25	0.035	1.06	Ca	1.11
IIC1	29-39	0.19	0.025	0.89	Ca	0.89
IIC3	46-56	0.19	0.026	0.86	Ca	0.83
Morley silt loam (AG-1):						
Ap	0-7	0.42	0.047	0.91	0.37	1.79
A2	7-9	0.39	0.051	0.86	0.35	1.81
B&A	9-12	0.38	0.049	1.19	0.34	1.94
IIB21t	12-16	0.36	0.038	1.93	0.41	2.12
IIB22t	16-20	0.40	0.038	2.24	0.50	2.14
IIB23t	20-23	0.41	0.036	2.04	Ca	2.14
IIB31t	23-27	0.30	0.030	1.63	Ca	1.61
IIB32	27-32	0.31	0.027	1.59	Ca	1.52
IIC1	32-42	0.31	0.028	1.51	Ca	1.42
IIC4	77-89	0.30	0.024	1.70	Ca	1.47

		Fine Silt (.02-.002 mm) %				
		Ti	Zr	Fe	Ca	K
Celina silt loam (PB-70):						
Ap	0-7	0.63	0.027	1.11	0.33	1.86
B&A	7-10	0.61	0.023	1.52	0.27	2.04
IIB21t	10-13	0.59	0.025	1.75	0.23	2.15
IIB22t	13-17	0.59	0.023	2.28	0.32	2.25
IIB23t	17-24	0.57	0.020	2.23	0.39	2.27
IIB3	24-29	0.44	0.014	2.03	Ca	1.83
IIC1	29-39	0.28	0.010	1.52	Ca	1.36
IIC3	46-56	0.28	0.011	1.62	Ca	1.43
Morley silt loam (AG-1):						
Ap	0-7	0.63	0.024	1.33	0.28	1.92
A2	7-9	0.62	0.027	1.25	0.22	1.84
B&A	9-12	0.66	0.022	2.14	0.22	2.07
IIB21t	12-16	0.66	0.019	2.64	0.23	2.31
IIB22t	16-20	0.64	0.018	2.90	0.33	2.36
IIB23t	20-23	0.62	0.016	2.93	Ca	2.44
IIB31t	23-27	0.50	0.015	2.52	Ca	1.94
IIB32	27-32	0.45	0.014	2.33	Ca	1.85
IIC1	32-42	0.44	0.013	2.34	Ca	1.88
IIC4	77-89	0.45	0.012	2.37	Ca	1.94

*Horizons calcareous

Zirconium, on the other hand, yields a concentration curve that decreases with depth. In the more highly weathered zones, Zr concentrates as other more susceptible minerals are removed, and then as weathering intensities decrease, Zr contents decrease slightly to a point of constancy in the parent till. Elemental Fe and Ca yield distinct weathering trends with depth, while K suggests only a slightly tendency for such a pattern. Higher Ca values in surficial horizons are attributed to contamination from agricultural limestone amendments. Because of the likelihood of loessial contamination in the upper 10 to 12 inches of these till-derived profiles, elemental depth distributions are difficult to interpret. Elemental Ti and Zr or ratios of these elements do not conclusively either support or negate the admixture of loess with till deposits in surficial horizons; however, higher silt contents in surficial horizons (Table 1) and sponge-spicule data (Wilding and Drees, 1968) do lend credence to this concept. In spite of this probable loess-till mixing in the upper sola, elemental trends previously discussed generally hold when only the till-derived portion of the profile is considered (IIB and IIC horizons).

Although not readily apparent from Table 3, Fe and K undergo considerably more weathering in fine silts than in coarser fractions. This trend becomes evident when elemental constituents are reconstructed by specific size fractions (data not shown). In fact, losses of Fe in the fine-silt fraction are four to eight times greater than in the coarse-silt fraction. Iron constituents are being gained in the sand-size fraction, probably through reconstitution of weathering products of finer fractions. All K losses are in the fine-silt fraction, while gains occur in coarser fractions. This implies that K-feldspars and micas in coarse-silt and sand-size fractions are sufficiently stable to concentrate, while more weatherable constitutents are being removed.

Elemental changes with depth, expressed on a total-fraction basis (0.002-0.25 mm), are presented to illustrate two points. First, Ti and Zr tend to yield concentration curves indicative of their occurrence in resistant minerals, while both K and particularly Fe yield characteristic weathering curves. Expression on a carbonate-free basis in calcareous horizons emphasizes the degree of dilution due to carbonates. The second aspect is the apparent elemental uniformity of these constituents in calcareous horizons of these deposits. These data suggest greater elemental homogeneity for till deposits than was anticipated by the authors. In profile PB-70, the Zr curve suggests a possible lithological discontinuity between the IIB2 and underlying calcareous IIB3 and IIC horizons. A slight break at the same point is also noted in profile AG-1. However, this disparity may only reflect a very slight error in correction for carbonate dilution and matrix absorption effects. Titanium curves do not support a lithological break at this position. Likewise, Ti/Zr ratios of these elements on a natural basis do not support a discontinuity. On other other hand, Ti/Zr ratios in surficial horizons, as well as absolute distributions, suggest the probability of a discontinuity at the top of the IIB horizons. It should be mentioned, however, that extreme caution must be used in interpreting lithological discontinuities from Ti/Zr ratios alone. Based on present laboratory evidence, it is possible to obtain similar Ti/Zr ratios across evident textural breaks because both resistant elements decrease or increase proportionally across the discontinuity. Thus, the ratio of the elements remains essentially constant. Data from profile DK-24 (not shown) provides an excellent example of this phenomenon because distinct textural stratification was observed and described in the calcareous drift deposits at time of sampling. Our experience also suggests that distribution of Ti, Zr, or Ti/Zr ratios may support a lithological discontinuity for some particle-size fractions, but not for others. If indeed such a break does occur, one would expect elemental analyses of all fractions to support it. It would seem that the explanation for this apparent anomaly must lie in the density, size distribution, and provenance of the host mineral, as integrated with the energy of transportation and sedimentation mechanisms. At the present time, the role of each of these factors with regard to the above elements is not clearly understood. Therefore, one must examine closely not only ratios of elements, but also absolute distributions, in addition to other physical or chemical evidence supporting a disconformity. Elemental analysis simply provides another tool for this purpose.

The trend in elemental Ti, Zr, K, and Fe from north to south across the study area (Fig. 4) suggests that, on a natural basis, all of the above elements except Zr progressively decrease by 25 to 40 percent to the south. On this evidence, there is no clear geographical boundary differentiating the till deposits from north to south in western Ohio. However, when elemental distributions are computed on a carbonate-free basis (Fig 4), a distinct break in K, Fe, and Ti occurs at the Union City Moraine. Elemental Zr also supports this boundary. The confounding influence of carbonate dilution on trends is apparent. Higher Fe, K, and Ti contents of the northern deposits may in part reflect higher shale contents in the till. Scanning thin-sections of calcareous till from each site indicates that the northern three sites contain perceptibly more shale fragments, particularly in the sand-size range.

The compositions of shales vary, but the chief mineral constituents are hydrated aluminum silicates (frequently micas), quartz, rutile, apatite, calcite, dolomite, and iron oxides (Lamborn and others, 1938, p. 13). The chemical composition of the Devonian shales that would have contributed to the till deposits in western Ohio are all high in K (3-5% K_2O), Fe (3-7% Fe_2O_3 and 1-3% FeO), and Ti (0.9-1.3% TiO_2) (Lamborn and others, 1938, p. 256, 258). Because the above elements are minor constituents of limestones and dolomites in the study area (<0.1% TiO_2, <0.3% Fe_2O_3, <0.3%

FeO, and 0.03% K_2O, in Stout, 1941, p. 424), and because there is no evidence that such elements covary systematically as a function of carbonate mineralogy, it seems highly improbable that variations in the carbonate status of the respective tills could account for the observed differences in Ti, Fe, and K. One could also speculate that till deposits in the northern sector of the study area are richer in crystalline relicts (such as granites, diorites, and schists), that would be important contributors of these elements. Such deposits are in closer proximity to the Canadian Shield provenance in Ontario. However, without pebble-lithology data, and considering the fact that Zr contents do not reflect higher crystalline components in the northern area, such speculation is not justified.

TABLE 4

MINERALOGY OF THE TOTAL-CLAY FRACTION ($<2\mu$) FOR SELECTED HORIZONS OF CELINA AND MORLEY SOILS

Horizon	Depth Inches	Clay Minerals° (%)						
		I	V	E	Q	K	C	INT.
Celina silt loam (PB-70)								
Ap	0-7	25	30	10	20	5	5	T†
IIB22t	13-17	25	30	20	20	<5	T	ND‡
IIB3	24-29	50	20	<5	15	<5	<5	T
IIC1	29-39	50	10	ND	10	<5	<5	T
IIC3	46-56	65	10	ND	5	<5	<5	5
Morley silt loam (AG-1)								
Ap	0-7	25	40	ND	15	<5	5	10
IIB21t	12-16	40	30	5	15	<5	<5	T
IIB31t	23-27	65	20	T	10	<5	<5	T
IIC1	32-42	70	20	ND	5	<5	<5	T
IIC4	77-89	80	10	ND	<5	<5	<5	T

° I = Illite (mica)
   V = Vermiculite
   E = Minerals which have variable expansion beyond 14A upon glycolation (mostly montmorillonite)
   Q = Quartz
   K = Kaolinite
   C = Chlorite
INT = Interstratified 10-14A clay minerals
† T = Trace
‡ND = Not detectable

Clay Mineralogy

Clay-mineral composition of selected horizons from profiles PB-70 and AG-1 are presented in Table 4. These data support work of earlier researchers (discussed previously) and illustrate the predominance of illite in calcareous till deposits, with secondary amounts of vermiculite, kaolinite, chlorite, interstratified clay minerals, and quartz. The reciprocal relationships between decreasing illite contents and increasing contents of vermiculite and expandable 14A minerals (mostly montmorillonite) upward from IIC to IIB horizons indicate that illite is being pedogenically transformed into the latter minerals. Transformations of this order are in agreement with earlier work (Jackson, 1968; Droste, 1956; Thorp and others, 1959; Ross and Mortland, 1966; Gersper, 1963; and Smeck and others, 1968). According to Jackson (1968), who has classified various clay-size minerals in soils and sedimentary deposits into 13 weathering indices from most weatherable (index 1) to least weatherable (index 13), illite (index 7) is more easily weathered than is vermiculite (index 8), montmorillonite (index 9), Al-chlorite (index 9) or kaolinite (index 10). Quartz is listed as index 6, but still tends to concentrate in more highly weathered zones of these profiles. The low, nearly constant amounts of kaolinite throughout the profile suggest that it is inherited from parent till deposits. The small quantities of chlorite reported in surficial horizons is probably of a pedogenic variety derived from thermally stable hydroxy-aluminum polymers sandwiched as interlayers between vermiculite and montmorillonite components. In these same horizons, there is evidence of less thermally stable interlayer hydroxy-aluminum components, identified by the difficulty in completely collapsing the 14A peaks with successively higher heat treatments. Such interlayer components are pedogenic weathering products and decrease progressively with depth. Likewise, interstratified 10-14A components follow this same trend and appear to be related to the more strongly weathered zones.

Pedogenic transformation of illite to vermiculite, and subsequently to montmorillonite, occurs upon the weathering of calcareous till deposits in the southern sector, but in the northern part the transformation proceeds primarily only to vermiculite (Table 4). Smeck and others (1968) analyzed the fine-clay fraction ($<0.2\mu$) of these two sites and found distinct montmorilonite (17.7A) peaks in the IIBt horizon of PB-70, but evidence of such discrete expandable components was not evident in samples from the more northerly AG-1 site. Montmorillonite in sola of Celina soils is considered pedogenic, because it is not detectable in calcareous IIC horizons and because there is a progressive increase in the size of the montmorillonite peak, beginning in the partially leached IIB3 horizons and reaching a maximum in IIB2t horizons. The presence of this expandable mineral in southern till deposits is indicative of a more advanced stage of weathering, according to Jackson's (1968) sequence. It is believed that the formation of montmorillonite in soils of the southern area is related to one or all of the following: cycling of greater quantities of Ca and Mg as a result of higher carbonate status and more dolomitic-rich parent till, lower content of K-bearing silt and sand-size minerals, greater chronological age, or an interaction of the above conditions.

Montmorillonite is known to have been synthesized both artificially and under natural environments where the ionic regime is rich in alkalies and alkaline earths (Grim, 1968, p. 479-574). The abundance of Mg, in particular, seems to favor the formation of this mineral. Under the more acid-weathering conditions of soils in this region, it is proposed that the first step in pedogenic transformations of illite to vermiculite is complete or partial removal of interlayer K^+ by H^+. This is accompanied by partial expansion from 10 to 14A, hydration, and equilibrium exchange of H^+ and K^+ sorbed on the clay surface with divalent cations, such as Ca^{++} and Mg^{++}, in soil solution. Removal of interlayer K^+ is commonly associated with a layer-charge reduction, resulting from oxidation of octahedral ferrous to ferric iron, proton incorporation in the silicate sheet, and/or imperfections developed in the silicate sheet resulting from attack and distintegration of tetrahedrons with partial Al substitution for Si (Conyers and others, 1969).

The sorption of larger, more highly hydrated divalent cations at the expense of smaller less hydrated K^+ ions would favor maintaining the partially expanded structure at 14A. Subsequent acid attack of such structures, associated with additional layer-charge reduction and sorption of divalent cations (particularly Mg because of its high hydration energy), would favor complete expansion of such structures, conforming to those of the montmorillonite group (Conyers and others, 1969). It is argued that soil-weathering conditions in the northern part of the study area do not foster this latter step, because of lower levels of Ca^{++} and Mg^{++} and greater cycling of K ions, which would collapse and regrade degraded illite and vermiculite components. If these mechanisms do indeed explain the differences in clay mineralogy across the study area, it emphasizes the importance of till composition on pedogenic weathering processes.

Clay and Elemental Reconstruction

With the knowledge of particle-size distribution, carbonate-size distribution, and noncarbonate residues in silt- and sand-size carbonates (data not presented), the clay distribution with depth in a representative Celina and a representative Morley profile is presented on both a natural and a carbonate-free basis (Fig. 5). The sharp break in these curves at a depth of 10 to 12 inches is due primarily to a lithological discontinuity — a shallow loess mantle, or loess-till admixture — in the upper sola of both soils. Clay-distribution curves (on a weight basis) illustrate only a slight total-clay maxima on a carbonate-free basis, indicating that most of the difference in textural development and Bt/C clay ratios noted earlier in these soils is in fact due to differences in carbonate dilution in IIC horizons. The southern representative, Celina, does have a slightly greater total-clay maximum. A slight total-clay increase in both soils corresponds to a prominent fine-clay maximum. Coarse-clay content in both soils is less in IIBt than in IIC horizons, which suggests that coarse clay is weathering to fine clay during profile development. This particle-size alteration is very likely associated with the

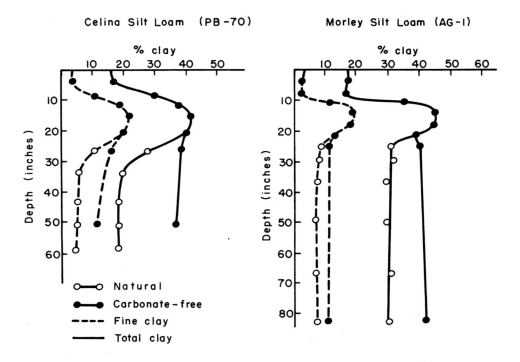

Figure 5. Fine-clay and total-clay depth distributions on both a natural and a carbonate-free basis for Celina and Morley profiles.

pedogenic transformation of illite to vermiculite and montmorillonite, noted in the clay-mineralogy section above.

When making interpretations concerning soil genesis, based on weight relationships, volume changes accompanying soil formation are not taken into consideration. Calhoun (1968) and Khangarot (1969) present excellent reviews of various reconstruction attempts and stable constituents considered for this purpose. According to Brewer (1964, p. 63-87), two basic conditions must be satisfied before reconstruction of grains and losses is possible; first, the parent material must be uniform throughout the section being reconstructed, and second, the constituent used as the basis of calculations must be stable and immobile under environmental conditions of the area. In this investigation, Zr in the total fraction (0.002-0.25 mm) was used as the stable constituent to record volume changes accompanying pedogenesis. The total fraction was chosen to minimize any errors due to differential carbonate distribution among various particle-size separates and because Zr constituents of this size are relatively stable as far as differential movement is concerned. Differential carbonate distribution could result in substantial errors in reconstruction unless one is certain that the fraction under investigation is representative of the total soil.

In addition to sponge-spicule (Wilding and Drees, 1968) and particle-size data, absolute distributions of total Ti and Zr, as well as Ti/Zr ratios, indicate that the IIBt horizons of profiles PB-70 and AG-1 have developed from till deposits similar to underlying IIC horizons. There is strong evidence that the A, and in places the upper B, horizons of these soils have loessial influence. Consequently, most of the emphasis in discussions of reconstruction will be restricted to the IIBt section of the profile. Table 5 presents weathering factors (Zr concentration) and reconstruction of elemental Ti, Fe, and K, togther with various clay-size fractions, by individual horizons (g/100 cc) and for specified profile sections (g/unit area/profile section). To illustrate differences in interpretations from such data, depending on whether or not a lithological discontinuity is recognized within the profile, data for the entire profile, as well as for only the IIB2t and IIB3 horizons, have been summarized.

TABLE 5

RECONSTRUCTION OF GAINS AND LOSSES OF ELEMENTAL AND CLAY COMPOSITION FOR CELINA AND MORLEY SOILS BASED ON TOTAL-ZIRCONIUM (0.25 TO .002 MM) FRACTION AS STABLE CONSTITUENT

Horizon	Depth (inches)	Weathering Factor[*]	Fe	Ti	K	Fine Clay	Coarse Clay	Total Clay
					gm/100 cc			
			Celina silt loam (PB-70)					
Ap	0-7	1.91	−1.727	−0.013	−1.017	−13.2	−30.6	−43.9
B&A	7-10	1.49	−0.825	0.068	−0.371	0.9	−9.5	−8.6
IIB21t	10-13	1.52	−0.682	0.047	−0.397	14.8	−9.1	5.6
IIB22t	13-17	1.18	−0.025	0.078	0.004	21.9	1.0	23.0
IIB23t	17-24	1.20	−0.095	0.064	−0.040	18.8	0.4	19.3
IIB3	24-29	1.17	−0.034	0.025	−0.078	8.8	−1.7	7.0
Summation	Ap-B3†		−45.54	2.93	−26.15	5.2	−7.0	−1.4
	IIB21-IIB3		−7.86	2.63	−4.87	7.9	−0.7	7.1
			Morley silt loam (AG-1)					
Ap	0-7	1.69	−2.437	−0.144	−1.323	−18.2	−50.7	−68.9
A2	7-9	1.95	−3.010	−0.226	−1.762	−21.9	−62.2	−84.1
B&A	9-12	1.35	−1.328	−0.063	−0.845	−0.3	−22.1	−22.4
IIB21t	12-16	1.06	−0.389	0.003	−0.285	15.5	−3.1	12.3
IIB22t	16-20	1.00	−0.003	0.022	−0.116	15.6	0.1	15.8
IIB23t	20-23	1.01	0.089	0.048	0.087	7.8	−0.4	7.4
IIB31t	23-27	1.15	−0.165	−0.018	−0.269	0.7	−9.1	−8.3
Summation	Ap-IIB31†		−74.41	−3.76	−45.37	−0.5	−15.2	−15.8
	IIB21-IIB31t		−4.86	0.46	−6.00	3.8	−1.2	2.5

[*] All values based on average of C horizons sampled for the total Zr fraction (0.25 to 0.002 mm). These values represent the volume of parent material required to form a unit volume of respective horizon, assuming lithological uniformity of the profile. Calculations based on ⅓-atmosphere bulk densities.

†These values represent the total gain or loss in elemental and clay composition for a unit cross-sectional area the thickness of entire profile (Ap-B3) or for the B horizon only (IIB21-IIB31). Units for elemental analyses are g/100 cm²/profile thickness and for clay composition g/cm²/profile thickness.

First, as would be expected from carbonate data, it is apparent from such computations that profiles derived from till deposits of the southern sector have undergone larger volume changes during pedogenesis than have those of northern sectors (Table 5). For example, weathering factors of the IIB2t horizons of the Celina profile suggest that approximately 1.2 to 1.5 volumes of calcareous till deposit are required to form a unit volume of present-day soil. Similar data for the Morley profile indicate that only slightly more than a unit volume of parent till are required to form a comparable volume of IIB2t horizon. Weathering factors supported by gains or losses of elemental and clay constituents indicate that the upper IIB2t horizon of the Celina profile is more intensely weathered than are the lower IIB2t horizons of this profile, or comparable horizons of the Morley profile. For example, the upper IIB2t horizon has lost more Fe and K, gained less fine clay, lost more coarse clay, and gained less total clay than have either underlying IIB2t horizons or similar horizons of its northern counterpart. Field observations have indicated stronger degradational phenomenon in the upper IIBt horizons of these soils in the southern reaches of the study area, as compared with those in the northern. These are manifest in gray or grayish-brown silty surface coatings (skeletans) along structural ped faces, suggesting a depletion of Fe and clay in these zones. Such evidence has been corroborated by thin-section analysis (Smeck and others, 1968) and helps to explain the results of clay reconstruction. Lower gains in fine clay and total clay, accompanied by greater losses in coarse clay, represent a conversion of coarse clay to fine clay, which has subsequently been translocated to underlying IIB2t zones or lost through weathering by dissolution. Note in Table 3 that, in profile PB-70, the IIB22t horizon has a sharp gain in fine and total clay over that in the IIB21t, while in profile AG-1, the gain in both upper IIB2t horizons is essentially the same. These results support the clay-mineral evidence for transformation of the coarser grained illite or micas to finer grained vermiculite and montmorillonite clays, and the more intensive weathering processes of the southern region.

Summaries of total gains or losses of elemental and clay constituents per unit area per profile thickness from the IIB2t to IIB3 likewise indicate that the southern profile has lost almost twice as much Fe, gained more Ti (function of weathering intensity), gained about twice as much fine clay, and gained almost three times as much total clay as the Morley profile to the north. The coarse-clay balance is dependent on transformation of this component to fine clay, which is countered by contributions to this size fraction from weathering of coarser fractions. Elemental K is being lost in both profiles at approximately the same rate, although absolute differences are slightly higher for the northern site.

SUMMARY

Results of this investigation indicate that carbonate and elemental composition of till deposits have a substantial impact on pedogenic weathering processes. Because carbonates in Wisconsin till deposits of this region are concentrated in the sand- and silt-size fractions, clay contents as commonly reported on the fine-earth fraction (<2mm) reflect primarily noncarbonate clays. Although carbonate dilution accounts for most of the difference in degree of textural development between Morley and Celina soils, the composition of the parent till has influenced the mode and magnitude of weathering processes responsible for development of the Bt horizons.

Illite is the predominant clay mineral in the calcareous till deposits of both areas. However, the more dolomitic, higher carbonate contents of the parent till in the southern sector favor pedogenic transformation of illite to vermiculite and to montmorillonite upon weathering, whereas in tills of the northern sector, this transformation proceeds primarily just to vermiculite. In addition to carbonate composition, differences in pedogenic transformations of illite are attributed to higher K-bearing minerals and to younger, less intensively weathered till deposits of the northern area, which would foster collapse of pedogenically formed vermiculite to illite and prevent its subsequent transformation to montmorillonite.

Differences in intensity of weathering of till deposits from south to north are also evident from reconstruction data. In the southern portion, IIBt horizons of soil have lost more Fe and K, but gained about twice as much fine clay and three times as much total clay. Further, degradation and stripping of Fe and clay from structural surfaces is more pronounced in the upper portion of Bt horizons in the southern area, as compared with the northern. Reconstruction results indicating a shift in particle-size to smaller fractions, particularly in the clay separate, complement observed alterations of clay minerals.

Work discussed herein illustrates how particle-size distribution, carbonate chemistry, clay mineralogy, elemental analyses, and reconstruction may be used as tools to investigate the impact of composition on weathering mechanisms in till deposits. It also emphasizes the need to closely examine the evidence for lithological discontinuties within the profile prior to interpretation of weathering mechanisms. Alternate interpretations of reconstruction are indicated, depending on whether or not a discontinuity is recognized.

REFERENCES

Alexander, J. D., Beavers, A. H., and Johnson, P. R., 1962, Zirconium content of coarse silt in loess and till of Wisconsin age in northern Illinois: Soil Sci. Soc. Amer. Proc., v. 26, p. 189-91.

Andrew, R. W., 1960, The relationship of natural drainage and the clay mineralogy of Miami and Brookston soils in central Ohio: M. Sc. thesis, Ohio State University, 61 p.

Beavers, A. H., Fehrenbacher, J. B., Johnson, P. R., and Jones, R. L., 1963, CaO-ZrO_2 molar ratios as an index of weathering: Soil Sci. Soc. Amer. Proc., v 27, p. 408-12.

Beavers, A. H., 1960, Use of x-ray spectrographic analysis for the study of soil genesis: Trans. Seventh Intl. Cong. Soil Sci., Madison, v. 2, p. 1-9.

Bidwell, O. W., and Page, J. B., 1950, The effect of weathering on the clay minerals of soils of the Miami catena: Soil Sci. Soc. Amer. Proc., v. 15, p. 314-18.

Brewer, R., 1964, Fabric and mineral analysis of soils: New York, John Wiley and Sons, Inc., 470 p.

Brown, I. C., and Thorp, J., 1942, Morphology and composition of some soils of the Miami family and the Miami catena: U.S. Dept. of Agr. Tech. Bull. 834, 55 p.

Calhoun, F. G., 1968, A weathering characterization of two northeastern Ohio **Aquic Fragiudalfs**: M.Sc. thesis, Ohio State University, 222 p.

Chapman, S. L., and Horn, M. E., 1968, Parent material uniformity and origin of silty soils in northwest Arkansas based on Zr, Ti contents: Soil Sci. Soc. Amer. Proc., v. 32, p. 265-71.

Conyers, E. S., Wilding, L. P., and McLean, E. O., 1969, Influence of chemical weathering on basal spacings of clay minerals: Soil Sci. Soc. Amer., v. 33, p. 518-23.

Dreimanis, A., and Vagners, U. J., 1971, Bimodal distribution of rock and mineral fragments in basal till: *in* this volume.

Droste, J. B., 1956, Alteration of clay minerals by weathering in Wisconsin tills: Geol. Soc. America Bull., v. 67, p. 911-18.

Foss, J. E., and Rust, R. H., 1968, Soil genesis study of a lithologic discontinuity in glacial drift in western Wisconsin: Soil Sci. Soc. Amer. Proc., v. 32, p. 393-98.

Gersper, P. L., 1963, A mineralogical investigation of soil clay subfractions: M.Sc. thesis, Ohio State University, 154 p.

Goldthwait, R. P., Dreimanis, A., Forsyth, J. L., Karrow, P., and White, G. W., 1965, Pleistocene deposits of the Erie lobe, *in* Wright and Frey (eds.), The Quaternary of the United States: Princeton, N. J., Princeton University Press, p. 85-97.

Green, T. C., 1933, Some characteristics of calcareous parent materials: Amer. Soil Survey Assoc. Bull., v. 14, p. 69-73.

Grim, R. E., 1968, Clay mineralogy: New York, McGraw-Hill (2d ed.), 596 p.

Handy, R. L., and Rosauer, E. A., 1959, X-ray fluorescence analysis of total iron and manganese in soils: Iowa Acad. of Sci. Proc., v. 66, p. 237-45.

Holowaychuk, N., 1950, Clay mineral studies of Miami, Hermon, and Worthington catenas: Ph.D. dissertation, Ohio State University, 63 p.

Jackson, M. L., 1968, Weathering of primary and secondary minerals in soils: Trans. Ninth Intl. Cong. Soil Sci., Adelaide, v. 4, p. 281-92.

Johnson, P. L., 1961, X-ray spectrographic analysis of loess deposits in Illinois: Ph.D. dissertation, University of Illinois, 130 p.

Jones, R. L., and Beavers, A. H., 1966, Weathering in surface horizons of Illinois soils: Soil Sci. Soc. Amer. Proc., v. 30, p. 621-24.

Khangarot, A. S., 1969, Relative intensity of soil weathering of Wisconsin and Illinoian-age Terraces near Newark, Ohio: Ph.D. dissertation, Ohio State University, 214 p.

Kinter, E. B., and Diamond, S., 1956, A new method for preparation and treatment of oriented aggregate specimens of soil clays for x-ray diffraction analysis: Soil Sci., v. 8, p. 111-20.

Lamborn, R. E., Austin, C. R., and Schaaf, D., 1938, Shales and Surface Clays of Ohio: Geol. Survey of Ohio, Fourth Series, Bull. 39, 281 p.

Mickelson, G. A., 1942, Mineral composition of three soil types in Ohio with special reference to changes due to weathering as indicated by resistant heavy minerals: Ph.D. dissertation, Ohio State University, 95 p.

Ross, G. J., and Mortland, M. M., 1966, A soil beidellite: Soil Sci. Soc. Amer. Proc., v. 30, p. 337-43.

Rutledge, E. M., Wilding, L. P., and Elfield, M., 1967, Automated particle-size separation by sedimentation: Soil Sci. Soc. Amer. Proc., v. 31, p. 287-88.

Smeck, N. E., 1966, Genesis of argillic horizons in Celina and Morley soils of western Ohio: M.Sc. thesis, Ohio State University, 180 p.

Smeck, N. E., Wilding, L. P., and Holowaychuk, N., 1968, Genesis of argillic horizons in Celina and Morley soils of western Ohio: Soil Sci. Soc. Amer. Proc., v. 32, p. 550-56.

Smith, B. R., and Buol, S. W., 1968, Genesis and relative weathering intensity studies in three-semi-arid soils: Soil Sci. Soc. Amer. Proc., v. 32, p. 261-65.

Stout, W., 1941, Dolomites and Limestones of western Ohio: Geol. Survey of Ohio, Fourth Series, Bull. 42, 468 p.

Thorp, James, Cady, John G., and Gamble, Erling E., 1959, Genesis of Miami silt loam: Soil Sci. Soc. Amer. Proc., v. 23, p. 156-61.

Wilding, L. P., and Drees, L. R., 1968, Distribution of sponge spicules in Ohio soils: Ohio Jour. Sci., v. 68, p. 92-99.

Wilding, L. P., Schafer, G. M., and Jones, R. B., 1964, Morley and Blount soils: a statistical summary of certain physical and chemical properties of some selected profiles from Ohio: Soil Sci. Soc. Amer. Proc., v. 28, p. 674-79.

Westin, F. C., 1953, Solonetz soils of eastern South Dakota: their properties and genesis: Soil Sci. Soc. Amer. Proc., v. 17, p. 287-93.

6

Fabric

Till/a Symposium

Methods in the Analysis of Till Fabrics

J. T. Andrews

ABSTRACT

In the last three decades, analyses of till fabrics have increased enormously. Few comparative studies have emerged from this large body of data, largely because of disparate methods of data collection and reporting. Furthermore, there is no concensus as to how, why, and when a preferred orientation and dip is achieved by till-stones within the matrix. Meaningful statistical analysis of till fabrics is hindered by the use of the horizontal plane to determine the sense of orientation. In contrast, three-dimensional vector techniques have been found to be a satisfactory statistical method.

A till fabric is a sample of a "population" of till-stones at the exposure; as such, it is essential that, before a regional survey is undertaken, an analysis is made of within- and between-site variability at selected exposures. Detailed studies of this problem in the North Yorkshire (England) cliffs indicates the presence of considerable fluctuations within a single till unit that are most apparent in the vertical plane. Finally, an investigation into the importance of stone geometry on orientation reconfirmed Holmes' (1941) findings. Tabular stones and rods lay 90° apart.

In conclusion, the most immediate problems in till-fabric analysis are 1) standardization of measurements, 2) proper designs of experiment, 3) equivalent methods of analyzing the data (statistical and graphical), and 4) consideration of the effect of stone geometry. These steps have to be taken prior to any discussion of the genesis of a till fabric.

INTRODUCTION

Progress in the understanding of geological processes is largely dependent upon the sophistication of the techniques and the ability to ask the right questions. However, the true value of a technique, especially a field technique, can only be realized once a standard method of data collection, portrayal, and analysis has been agreed upon. Such reproducibility of methods is standard in laboratory procedures, but it appears to be much more difficult to come to a consensus on appropriate standards and procedures for field techniques. It is the purpose of this paper to urge that the study of till fabric be put onto a more reproducible and rigorous basis. This paper is, therefore, a plea for standardization at one or more levels of inquiry. These levels may be identified as: 1) regional field determination and differentiation of ice movement, and 2) problems connected with the genesis of a till fabric.

To outline some of the problems, a survey of till-fabric variability in lithologically homogeneous till units is presented, based on work from England. The writer has lately been especially concerned with the within-sample variability of till fabrics (Andrews and Shimizu, 1966; Andrews and Smith, 1966; Andrews and King, 1968; Andrews and Smith, 1970). This present paper summarizes some of the conclusions that have emerged from the above studies with the hope that some of the possible dangers in the interpretation of these directional data may be seen. Kauranne (1960) has reported similar problems in a study on Scandinavian drift. An overriding ice sheet (McClintock and Dreimanis, 1964) may produce changes in fabric orientation within a single till unit; such a process would produce a vertical shift in the pattern in contrast to horizontal variability, which is more readily related to local changes in stone geometries.

PROBLEMS OF MEASUREMENT AND PORTRAYAL

Collection of till fabrics has been going on for about thirty years, so that there should be, at this moment, a large body of basic factual information that could be used to proceed from the collection phase into one of comparisons, contrasts, and general models. Is it possible to realistically and honestly proceed to this second phase? My own personal belief is that it is not possible. Why? The reasons for this disturbing (to me, at least) conclusion are:

1. Lack of standardization in field measurements. These can be subdivided into:

   a. Some people measure elongation of the long, or *a*, axis, but not its imbrication.

   b. Some workers limit the measurement of dips to $\pm 45°$ or less from the horizontal.

   c. Different ratios of the *a:b* axes are used to delimit the lower limit of permissible elongated till-stones.

d. Some fabrics are based on a restricted stone geometry.

e. Some samples are taken from horizontal faces, and others from vertical faces, which could result in bias, and could account for differences between fabrics.

f. Some studies take into account the influence of size and stone geometry on orientation and dip, whereas others do not.

2. A lack of adequate sampling control, specifically on the within-site variability of fabrics, a point to be illustrated later, but internal variability has to be examined prior to any account of regional changes in fabrics. Before interpretation of differences between till units can be carried out, it is mandatory that each unit be effectively sampled. Such a procedure necessitates the design of an experiment using analysis-of-variance. Single sample per exposure of a till unit is a statistically unsound basis for any general statement on fabric characteristics.

3. A variety of different graphical methods. For the portrayal of the fabrics, a graphic display is a necessary part of the study, although certain facets are best handled by statistical analysis. Several common techniques result in a loss of information as they group a fabric into class-intervals, as in the conventional rose-diagram, or a contour diagram. To effect a comparison between a rose-diagram and a polar-equidistant plot, the only method is to reduce the latter into class frequencies, and the effectiveness of the comparison has to suffer. Accordingly, I believe it is preferable to illustrate a till fabric such that the identity of individual observations is maintained. In a recent paper, Andrews and Smith (1970) illustrated the different visual impressions caused by presenting the same five till fabrics on five different types of diagram.

4. The lack of quantification, in the sense of statistical analysis of a till fabric, has been marked, and despite the early call for such an approach by Krumbein (1939) and Harrison (1957), the majority of papers still rely on a visual appraisal of the data. This approach is satisfactory for limited numbers of samples, but it is clearly impossible to take the same approach if the sample number exceeds about 20. Despite the overall absence of statistical analyses, there are a number of rival methods that have been proposed. This is understandable, but it must be borne in mind that the theoretical underpinnings of the methods may be completely different. Such is the case when comparing the sampled fabric with a theoretical uniform or randomly distributed one. Some methods only examine

the directional data, whereas others provide analyses of all three dimensions. Two- and three-dimensional vector means may be quite different for the same till fabric (Andrews and Smith, 1970).

5. The final problem is connected with the problems caused by adopting the conventions that the imbrication of till-stones is measured from the horizontal and that the fabric is plotted on a polar projection. Often the result is a symmetric bi-model distribution that is very difficult to handle by vector-summation methods. In sedimentological studies, the dip of pebbles in beds is subtracted from the regional trend, but this is difficult to envisage in glacial drift, as there are few signs of accumulation planes. Thus the details of an exposure of apparently homogenous till could resemble a series of units with accumulation planes of varying strike and dips. Such a possibility could be demonstrated by till-fabric analysis and may be solved by studies of till fabric, but only if the determination of central tendency, etc., are independent of the co-ordinate system used.

STATISTICAL METHODS

The variety of statistical methods that are available will not be reviewed; rather some of the advantages of three-dimensional vector methods will be outlined. Unless the prime interest is focused on the degree of dispersion of the fabric, in which case x^2 can be used, a method is required that will: 1) provide an estimate of mean direction, 2) provide an estimate of dispersion about this direction, and 3) allow the distribution to be tested against a suitable alternative, such as a random distribution. The three-dimensional vector method has been used extensively in palaeomagnetism; its advantages are that it not only provides suitable answers to the above, but it also can be used in an analysis-of-variance design and can thus provide insight into the number of samples and number of pebbles per sample for any given level of confidence, as well as determining the significance of within- and between-site variations. One stumbling block to the method is that, to provide a least-squares solution, data have to be rotated about a number of poles, because of the zero-dip convention. However, most computers now allow dips to be scored + or −, thus circumventing this tedious procedure.

WITHIN- AND BETWEEN-SITE VARIABILITY

The degree of within- and between-site variability may be illustrated by examining till fabrics taken from the coastal cliffs of North Yorkshire, England, specifically those exposed at Robin Hood's Bay near Whitby (Andrews and Smith, 1970). Samples were taken from the face of the cliff in such a manner that lateral and vertical variations in fabric means could be evaluated. The samples were taken from what appeared to be a single till sheet, that is, the color, lithology, grain size, and stone frequency were similar over the entire length and breadth of the unit. Nine samples were collected in a 3 x 3 array from a 3 x 15-m face; six of the fabrics were collected by different operators. Three-dimensional vector methods and analysis-of-variance indicated a low within-site precision of 5.0 (Watson and Irving, 1957; Steinmetz, 1961) and a moderately high between-site precision of 50; vector means were significantly different. Basal fabrics had an overall mean of 09° toward 079°, middle-layer fabrics lay at 10° toward 099° and 12° toward 094°. Thus, within this visually homogenous unit, significant and consistent changes in till-stone orientation occur.

Sampling in other cliffs in North Yorkshire led to the following conclusions: (1) Till fabrics are usually quite dispersed and within-site precision estimates are low and similar to estimates on cross-beds. (2) Between-site precision estimates ranged from low to very high, indicating that the relation of a fabric sample to neighboring samples can vary from identical to very different. These conclusions emphasize the importance of adequate sampling and the need to obtain fabrics that cover the entire vertical and horizontal dimensions of the exposure. (3) Estimates of precision can be used to compute that number of samples and that number of till-stones per sample necessary to obtain a specified cone of confidence. Assume that within-site precision equals 5.0, between-site precision equals 50, and an overall sample mean ± 5° from the true mean at the 95 percent confidence level is required. Sample size will be specified as 10 or 100; how many samples should be taken? With sample size 100, the number of fabrics needed is 100 x 17 or 1,700 individual measurements; if a sample size of 10 is used, the figure is 10 x 31 = 310, or 310, or a difference between the two estimates of 1390 stones (details of calculations in Andrews and Smith, 1970). The point is that it is desirable to take a larger number of samples with less observations per sample, rather than to increase the number of till-stones per sample.

IMPORTANCE OF STONE GEOMETRY

Holmes, in his monumental paper (1941), shows the effect of orientation of different stone geometrics. His conclusions have not led to the attention they deserve because a great number of workers (including myself) have sought to ignore this complication in an already difficult field. However, in conjunction with C. A. M. King at the University of Nottingham, a study of the effects of stone size and shape on the orientation and dip characteristics of till fabrics in ground moraine and in drumlins was undertaken (Andrews and King, 1968). The first point of note is that, in the Gipping Till, stone shape varied significantly between samples collected only one meter apart! This variability must affect the character of fabrics taken from these samples. The second point is that, though the direction of ice movement in the area was known, the till fabrics were 90° out of phase. Only when the till-stones were separated into shape categories was it found that tabular stones lay parallel to the inferred ice movement, whereas rods, ovoids, and wedges lay transverse to this direction.

In this study of drumlin till-fabric and basal drift, the effects of changes in length of the long axis and of changes in the $a:b$ axial ratios were analyzed. Significant changes in fabric orientation accompanied changes in till-stone length and morphology. These conclusions are not new — they merely repeat the earlier findings of Holmes (1941), but they serve to emphasize how essential a careful consideration of these points is to the meaningful study of till fabrics, although this type of analysis is not as common as it should be.

In conclusion, it is my opinion that the use of till fabrics to ascertain regional directions of ice movement are likely to be invalid, at least methodologically, if the following are not included:

1. Measurement of several samples taken from the same level in the same till unit.

2. Measurement of axes of till-stones.

3. Examination of individual samples and grouping into various classes to ascertain the effect of stone morphometry on orientation and dip.

Only at this point can regional interpretations of ice movement be undertaken.

I believe that we have an uncertain understanding of the origin of till fabric, which, in my view, is caused in basal tills by movements in the till mass itself, rather than by the letting-down of englacial debris or the ploughing of ice along its bed. Movement within the till is probably in response to stress differences caused by the profile of the overlying ice, and by the profile of the underlying rock surface.

ACKNOWLEDGMENTS

Funds for the computer analyses presented here were provided by the Graduate School of the University of Colorado, and are gratefully acknowledged.

REFERENCES

Andrews, J. T., and Shimizu, K., 1966, Three-dimensional vector analysis for till fabrics; discussion and FORTRAN program: Geog. Bull., v. 8, p. 151-65.

Andrews, J. T., and Smith, D. I., 1966, The variability of till fabric: British Geomorph. Res. Group, Occas. Papers, no. 3, p. 33-37.

Andrews, J. T., and King, C. A. M., 1968, Comparative till fabrics and till fabric variability in a till sheet and a drumlin; a small-scale study: Proc. Yorkshire Geol. Soc., v. 36, p. 435-61.

Andrews, J. T., and Smith, D. I., 1970, Till fabric analysis; methodology and local and regional variability (with particular reference to the North Yorkshire cliffs): Quart. Jour. Geol. Soc. London, v. 125, p. 503-42.

Harrison, P. W., 1957, A clay till fabric; its character and origin: Jour. Geology, v. 65, p. 57-71.

Holmes, C. D., 1941, Till fabric: Geol. Soc. America Bull., v. 51, p. 1299-1354.

Kauranne, L. K., 1960, A statistical study of stone orientation data in glacial till: Bull. de la Comm. Geol. de Finlande, no. 188, p. 87-97.

Krumbein, W. C., 1939, Preferred orientation of pebbles in sedimentary deposits: Jour. Geology, v. 47, p. 673-706.

McClintock, P., and Dreimanis, A., 1964, Reorientation of till fabric by overriding glacier in the St. Lawrence Valley: Amer. Jour. Sci., v. 262, p. 133-42.

Steinmetz, R., 1962, Analysis of vectorial data: Jour. Sed. Petrology, v. 32, p. 801-12.

Watson, G. S., and Irving, E., 1957, Statistical methods in rock magnetism: Monthly Notices Royal Astron. Soc., Geophys. Suppl., v. 7, p. 289-300.

Pebble Orientation and Ice Movement
in South-Central Illinois

Jerry A. Lineback

ABSTRACT

The direction of dip of the long axes of elongate pebbles provides valuable petrofabric data for Kansan and Illinoian tills in south-central Illinois. Elongate pebbles are imbricated with their long axes dipping in the direction from which the glaciers are presumed to have advanced.

An unnamed till of Kansan age is sandy and contains abundant illite and calcite. Pebble-orientation data indicate that this till was deposited by a glacier advancing to the south-southwest. The Vandalia Till* (Illinoian) is sandy and contains abundant illite and dolomite. Pebble orientations indicate deposition of the Vandalia Till also by a glacier advancing to the south and southwest. The Smithboro Till (Illinoian), which lies between the Kansan and Vandalia Tills, is silty, contains more expandable clay minerals than the other tills, and is low in total carbonate. Clay mineralogy of this till most closely resembles those of Kansan and some Illinoian tills in western Illinois that contain high concentrations of expandable clay minerals. The orientation of pebbles in the Smithboro Till in south-central Illinois indicates that it was deposited by a glacier advancing to the southeast.

TILL STRATIGRAPHY

Recent stratigraphic studies of Pleistocene deposits in south-central Illinois by Jacobs and Lineback (1969) demonstrate the presence of three major till units of differing composition — an unnamed Kansan till and the Vandalia and Smithboro (Illinoian) Tills. The Vandalia and Smithboro Tills, first defined by Jacobs and Lineback (1969), have recently been

*Terminology follows that of Jacobs and Lineback (1969).

established as members of the Glasford Formation by Willman and Frye (1970).

The unnamed Kansan till underlies the Yarmouth Soil, and is moderately sandy and high in illite, and produces more calcite X-ray counts per second than dolomite X-ray counts (Fig. 1). The Smithboro Till is silty, contains more expandable clay minerals and less illite than the other tills, and is dolomitic. The Vandalia Till is sandy, higher in illite than is the Smithboro Till, and dolomitic. The Mulberry Grove Silt separates the Smithboro and Vandalia Tills in places. The Hagarstown Member of the Glasford Formation overlies the Vandalia Till and includes sand and gravel in elongate ridges, and poorly sorted gravel and gravelly till on level drift plains. These deposits are believed to represent drift accumulated in ice-walled channels and on the surface of the stagnant and ablating glacier that deposited the Vandalia Till. The Sangamon Soil was formed on the Hagarstown Member or, where it is absent, on the Vandalia Till. Wisconsinan loess blankets the Illinoian drift plain in south-central Illinois.

Grain size was determined by a combination of dry sieving and hydrometer technique. Relative abundance of clay minerals and carbonates in the less-than-2-micron fraction of the tills is interpreted from X-ray data.

	Sand %	Silt %	Clay %	Expandable clay minerals %	Illite %	Kaolinite Chlorite %	Calcite (X-ray counts per sec.)	Dolomite (X-ray counts per sec.)
Wisconsinan loess								
Sangamon Soil								
Hagarstown beds								
Vandalia till	40 (s = 8.0) (n = 161)	40	20 (s = 3.8) (n = 161)	16 (s = 6.3) (n = 95)	66 (s = 6.1) (n = 95)	18 (s = 4.6) (n = 95)	21 (s = 11.0) (n = 95)	25 (s = 9.1) (n = 95)
Mulberry Grove silt								
Smithboro till	23 (s = 5.9) (n = 47)	50	27 (s = 4.6) (n = 47)	34 (s = 10.7) (n = 28)	45 (s = 10.5) (n = 28)	21 (s = 3.7) (n = 28)	12 (s = 5.0) (n = 28)	15 (s = 5.4) (n = 28)
Yarmouth Soil	31 (s = 3.6) (n = 11)	44	25 (s = 3.3) (n = 11)	9 (s = 2.4) (n = 8)	65 (s = 6.6) (n = 8)	26 (s = 3.4) (n = 8)	23 (s = 8.2) (n = 8)	13 (s = 5.3) (n = 8)
Unnamed till								

Figure 1. Generalized stratigraphic column and compositional data for the Pleistocene units in south-central Illinois for which fabric analyses were made. (s = standard deviation; n = number of samples.)

SIGNIFICANCE OF TILL COMPOSITION

The Kansan and Vandalia Tills are both high in illite, which is consistent with the illite-rich Paleozoic bedrock over which the glaciers moved as they advanced from the northeast as the Lake Michigan or Erie Lobes. The Smithboro Till, in contrast, contains more expandable clay minerals than tills having an eastern source. It increases in expandable-clay-mineral content westward, but does not contain as much expandable clay minerals on the average as Kansan tills in western Illinois (Willman, Glass, and Frye, 1963). The western-source Kansan till also contains more epidote than garnet, but the Smithboro Till contains either equal amounts of these minerals or more garnet, indicating different source areas (Willman, Glass, and Frye, 1963).

Kansan tills in western Illinois were deposited by a glacier moving southeastward across Iowa into western Illinois. Tills from this western source are rich in expandable clay minerals and poor in illite, reflecting the montmorillonite-rich Cretaceous bedrock over which the glacier passed. The Smithboro Till (Illinoian) could not have been deposited by a glacier advancing from the same direction as the one that deposited the montmorillonite-rich Kansan till, because the Kansan till in Iowa and Missouri is not overlain by recognizable Illinoian till in the region across which a glacier with a northwestern source would have had to advance. Early Illinoian loess deposits (Petersburg Silt) along the Illinois and Mississippi Rivers were derived in part from outwash also from a western glacial lobe and therefore are rich in montmorillonite.

Later glaciers advancing from the east deposited illite-rich tills in eastern Illinois, but in places assimilated tills and loess enriched in expandable clay minerals as they pushed westward across the area underlain by Petersburg Silt and the Kansan tills of western Illinois. The Smithboro Till, therefore, was probably deposited by a glacier from the north or northeast that, before reaching south-central Illinois, passed over an area along the Illinois Valley that was underlain by Kansan deposits or early Illinoian loess, rich in expandable clay minerals. The explanation is consistent with the clay mineralogy of the tills, but requires glacier movements not previously recognized in Illinois.

Striations are not commonly exposed. Most previously discovered striae in south-central Illinois are associated with the southwestward movements of the glaciers that deposited the more pebbly Kansan and Vandalia tills. The Smithboro Till contains less than one percent by weight of material coarser than 2 mm, suggesting considerable incorporation of older silts, such as the Petersburg Silt. The small percentage of coarse material in the Smithboro Till may account for the rare occurrence of striations below the till.

TILL-FABRIC ANALYSIS

The direction of glacier movements was investigated by till-fabric analysis. Studies of the orientation and dip angles of the long and short axes of pebbles showed that pebbles in the tills studied were well oriented (Fig. 2).

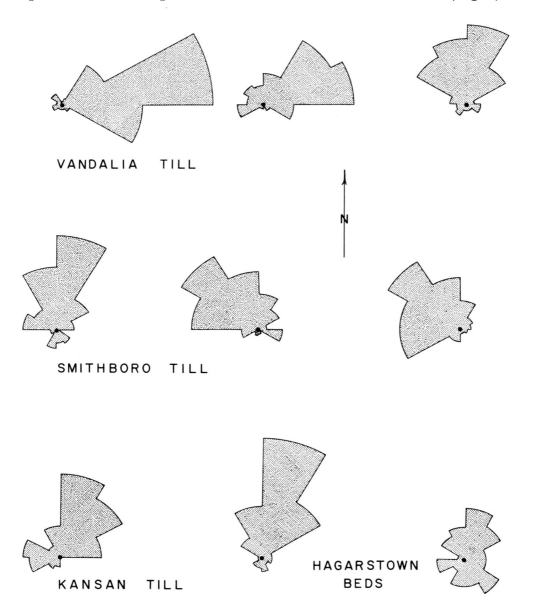

Figure 2. Rose diagrams of long-axis dip directions at representative localities for each of the units studied. Direction of the glacier movement lies 180 degrees from the means of the dip directions of long axes.

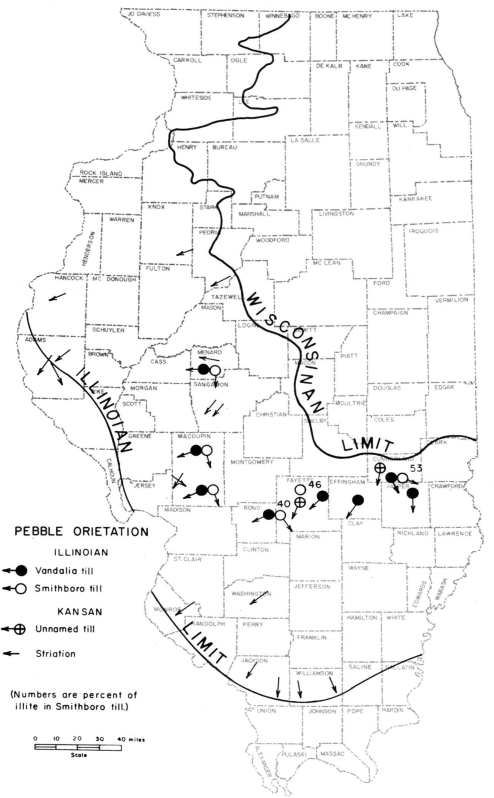

Figure 3. Direction of glacier movement in south-central Illinois determined from pebble orientations from each of the tills studied. Also shown are the orientation of striations, in part taken from the literature, and the percent of illite in the Smithboro Till from three multi-sample localities.

Most elongate pebbles have their long axes oriented parallel to the presumed glacial movement and have their axes dipping 20 to 30 degrees down from the horizontal in an up-glacier direction. Consequently, the dip directions of long axes of pebbles were taken at several localities (Fig. 3).

The method for determining the orientations of the pebbles was as follows. A convenient spot was first leveled on the outcrop. The orientation of a pebble was then determined by excavating around it until all axes were visible. An aluminum knitting needle was placed parallel to the long axis to aid in aligning the Brunton compass. Azimuths were measured in the direction of dip of the long axes. The direction of glacier movement was interpreted to be 180° from the mean of the pebble orientations taken by this method. Most pebbles measured were between 5 and 25 mm long. A minimum of 100 pebbles were measured at each site, a process taking two to three hours.

The direction of glacier movement determined from pebble orientation closely matched the orientation of glacial striae where present. Orientation studies of the particles less than 2 mm in size in oriented slabs of till also yielded data closely comparable, in most cases, to that of the larger pebbles.

RESULTS OF STUDY

The dip direction of the long axes of elongate pebbles in the Kansan till confirms that the Kansan ice in south-central Illinois moved to the southwest (Fig. 3). The glacier that deposited the Smithboro Till also apparently moved, initially, to the south or southwest across north-central Illinois, and extended into the Illinois Valley region, where it picked up material rich in expandable clay minerals (Fig. 3). Pebble-orientation data confirm that the glacier then veered south and southeastward across south-central Illinois. The Smithboro Till became diluted with illite from the bedrock and older tills high in illite as the glacier moved southeastward. The average content of illite in the Smithboro Till is shown for three multi-sample localities, illustrating the eastward increase in illite (Fig. 3).

The southeastward movement of the ice that deposited the Smithboro Till may be related to interference by different Illinoian lobes. It is possible that a slightly earlier Erie Lobe glacier extending into northeastern Illinois deflected the ice of the Lake Michigan Lobe westward into the Illinois Valley and, after passing the Lake Erie Lobe, the Lake Michigan Lobe spread southeastward. The Smithboro Till also may have been deposited by a glacier from the Green Bay Lobe, or some other lobe, that moved southward to the west of a Lake Michigan Lobe ice sheet that did not extend as far south. The western lobe ice then could have curved southeastward around the south side of the older lobe. No definite conclusion as to this exact history can be reached, for the areas where evidence for such deflections might exist are largely covered with thick tills of Wisconsin age, and the older tills are largely eroded.

Pebble orientations confirm that the Vandalia Till was deposited by an ice sheet moving to the southwest in the eastern part of the area and to the west in the western part of the area (Fig. 3). This is consistent with the mineralogy of the Vandalia Till.

ACKNOWLEDGMENTS

Studies of till stratigraphy in south-central Illinois were part of a cooperative study with A. M. Jacobs of the Illinois State Geological Survey. Grain-size analyses and X-ray mineralogy analyses were made in the Illinois State Geological Survey laboratories by W. A. White and H. D. Glass, respectively.

REFERENCES

Jacobs, A. M., and Lineback, J. A., 1969, Glacial geology of the Vandalia, Illinois, region: Illinois Geol. Survey Circ. 442, 24 p.

Willman, H. B., and Frye, J. C., 1970, Pleistocene stratigraphy of Illinois: Illinois Geol. Survey Bull. 94, 204 p.

Willman, H. B., Glass, H. D., and Frye, J. C., 1963, Mineralogy of glacial tills and their weathering profiles in Illinois. Part 1. Glacial tills: Illinois Geol. Survey Circ. 347, 55 p.

*Evidence for Reorientation
of a Till Fabric in the
Edmonton Area, Alberta*

John Ramsden and John A. Westgate

ABSTRACT

Two distinct Pleistocene till units exist in the Edmonton area. The north-easterly provenance of the upper till is indicated by well-developed sole markings and supported by measurements of the orientation of elongate pebbles. Pebbles in the lower till are preferentially oriented, at some places with a northwesterly trend and at other places with a northeasterly trend. The parallelism of the northeasterly preferred trend in the lower till to the basal groove molds and preferred trend of elongate pebbles in the upper till led to the hypothesis that the pebbles of the lower till had been locally reoriented by the northeast-southwest movement of the later glacier.

Structural observations support this hypothesis. At all locations where the preferred trend of the long axes of pebbles in the lower till is north-easterly or southwesterly, the till is highly fractured and contains numerous closely spaced shear surfaces, commonly with northeast-southwest linea-tions. A well-developed shear zone containing slickensides with north-easterly trends is present at the base of the lower till at many places.

It is suggested (1) that pebbles in the lower till, whose long axes orig-inally were preferentially aligned with a northwesterly or southeasterly trend, have in places been reoriented by the movement of the later glacier so that their long axes now have a preferred trend that parallels the north-east-southwest direction of movement of the later glacier, and (2) that parts of the lower till have been moved *en masse* by the later glacier. Reorientation appears to have resulted principally from shearing of the till along numerous closely spaced shear planes.

335

INTRODUCTION

Since 1967 the authors have been engaged in a program designed to determine the value of measurements of the orientation of till pebbles as an aid in identifying and distinguishing tills in the Edmonton, Alberta, area. The senior author has recently completed a comprehensive report on this work (Ramsden, 1970). During the course of these studies, it became apparent that, in parts of the area studied, the reorientation of pebbles in the lower of the two tills by later overriding ice was a possibility that required careful consideration. The purpose of the present paper is to document the evidence for such reorientation. Although data from throughout the study area are presented, detailed reference is made to an exposure in a gravel pit in the Clover Bar area just east of Edmonton where the relevant features are particularly well exhibited.

The term *fabric* was defined by Pettijohn (1957, p. 72) as "the orientation, or lack of it, of the elements of which a rock is composed." All the fabric information presented in this paper, however, is restricted to the orientations of the long axes of pebbles. Therefore, the term *fabric*, as used here, refers only to such pebble orientations.

QUATERNARY DEPOSITS IN THE EDMONTON AREA

The succession of rock-stratigraphic units in the Edmonton area is shown

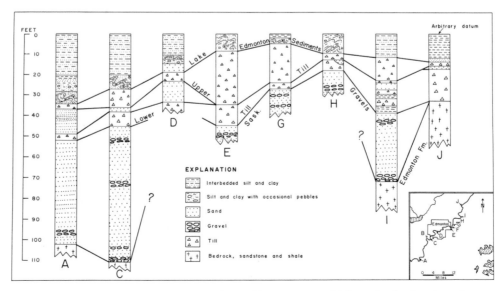

Figure 1. Stratigraphy of Pleistocene deposits in the Edmonton area, with inset showing locations of measured sections. (After Westgate, 1970.)

in Figure 1. These deposits have been described in some detail by Westgate (1969).

The oldest Quaternary deposits are fluviatile gravels and sands known as the Saskatchewan Gravels. Vertebrate fossils suggest that the youngest beds of this formation are of late Pleistocene age (Reimchen, 1968). A grayish-brown, dense, clay-loam till with some inclusions of stratified sand overlies the Saskatchewan Gravels or lies directly on bedrock. It varies in thickness from 0 to more than 20 feet, is highly jointed, and possesses some folded joint-surfaces. The direction of preferred stone orientation is north-west-southeast; several two-dimensional fabric diagrams for this till are shown in Figure 2. The exact age of this lower till is not known, but it is probably of early Wisconsin age. Stratified sediments, 40 feet thick in places, commonly separate this till from an overlying till.

The upper till, which is loamy and yellowish brown, has a pronounced columnar structure; the greatest observed thickness is 25 feet. Elongated till stones are preferentially oriented with a northeasterly trend; Figure 2 also includes some samples from this late Wisconsin till. Well-developed sole markings (Westgate, 1968) indicate ice movement from N 25°E (Plate 1-A). Up to 50 feet of well-bedded Lake Edmonton sands, silts, and clays rest on the compact upper till.

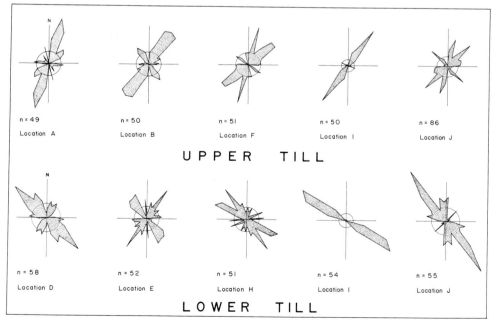

Figure 2. Disribution diagrams for the trends of the long axes of stones in till. Radius of circle represents 5% of the total of number of measurements (n). Locations are shown in Figure 1. (After Westgate, 1970.)

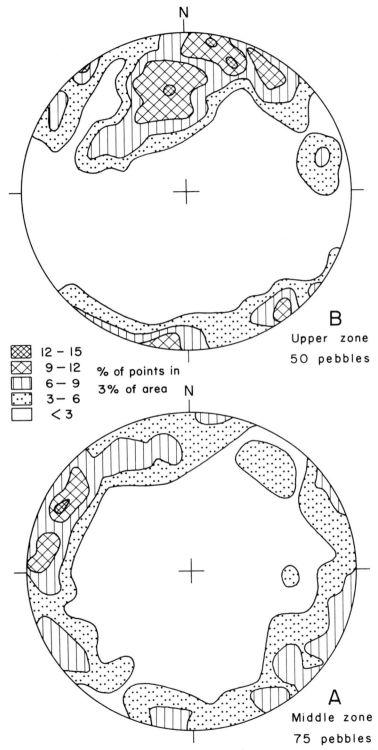

Figure 3. Equal-area lower-hemisphere projections showing the distribution of the long axes of pebbles in (a) the middle zone and (b) the upper zone of the lower till. Diagrams obtained by contouring a grid of density values produced by a computer using a program developed at the University of Alberta by G. K. Muecke (1965) under the direction of Dr. H. A. K. Charlesworth.

EVIDENCE FOR FABRIC ALTERATION

The gravel pit to which detailed reference is made is operated by Alberta Concrete Products and is situated approximately one quarter mile south of Highway 16, one and one-half miles east of the Edmonton city limits, in the area known as Clover Bar. It lies approximately one quarter mile west of location G in Figure 1, in the NW¼ of Sec. 8, T53, R23. Four stratigraphic units are present in the pit (Plate 1-E): the Saskatchewan Gravels (a), the lower till (b), the upper till (c), and the Lake Edmonton sediments (d). No sand beds separate the tills at this locality; however, the tills are readily distinguishable by their color difference, the contact being sharp and regular.

The lower till contains three structurally distinct vertical zones. At its base is a layer of intensely sheared, continuous, black clayey till one to two inches thick. Plate 1-B shows the basal portion of a till block collected from the base of the lower till at the gravel pit. The contact with the underlying Saskatchewan Gravels can be seen, and the basal sheared zone is clearly discernible. At one point within this layer, near-horizontal slickensides were observed, having a trend of N 15°E. At another pit half a mile to the southwest, a horizontal shear surface within this zone had slickensides with trends of between N 27°E and N 35°E.

The middle zone is dense and massive, and has a columnar structure (Plate 1-D, E). Its thickness varies from zero to about four feet. The two fabric analyses that were made at this level showed similar characteristics. One of these is illustrated in Figure 3-A. The distribution of the long axes of 75 pebbles shows two important features: (1) a near-horizontal girdle, and (2) a tendency for the density on this girdle to reach a maximum close to a line trending N 62°W and plunging 12 degrees, and a minimum 90 degrees away from this line. The distribution diagrams of the long-axis orientations of pebbles from till in areas of ground moraine usually show one of two patterns: (1) a near-horizontal girdle with a single maximum on this girdle, or (2) a near-horizontal girdle with two maxima 90 degrees apart on this girdle (Holmes, 1941, p. 1313; Harrison, 1957, p. 283; Kauranne, 1960, p. 87). The distribution diagram in Figure 3-A is of the first type; thus the distribution of the long axes of pebbles in the middle zone of the lower till is similar to that found in till of many ground-moraine areas.

The upper zone of the lower till is highly fractured and contains many shear surfaces (Plate 1-D, E). Lineations (Plate 1-C) and grooves (Plate 1-D) on near-horizontal shear surfaces have trends between N 20°E and N 25°E. The distribution of the long axes of 50 pebbles from this zone is shown in Figure 3-B. There is an apparent tendency for the axes to be grouped about two directions — one with a trend of N 20°E and a plunge of 16 degrees, the other with a trend of N 10°W and a plunge of 35

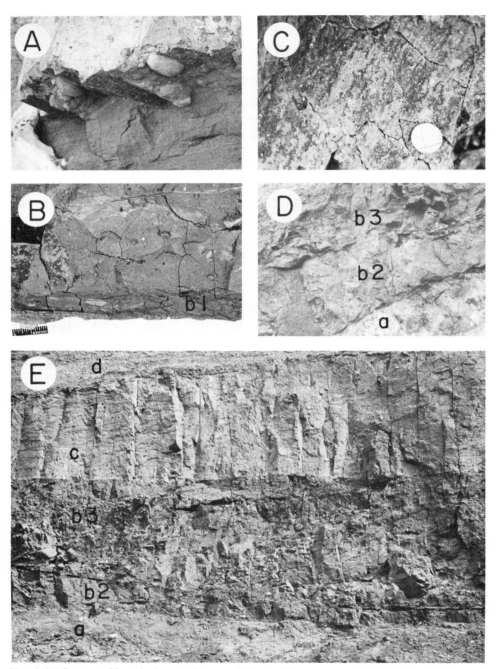

Plate 1. (A) Groove molds at the base of the upper till in the Edmonton area. The undercut is about two and a half feet deep. (B) Till block collected from the base of the lower till at the Clover Bar pit. The basal shear zone is clearly visible (b1). (C) Lineations on a near-horizontal shear surface in the upper zone of the lower till (see text). The coin is a dime. (D) Undulating shear surface separating the upper zone (b3) from the middle zone (b2) of the lower till. The undulations are approximately parallel to the direction of movement of the later glacier. Photo covers about 8 feet vertically at center. Letters and numbers have same meaning as in E. (E) The section exposed in Clover Bar pit: (a) Saskatchewan Gravels, (b2) Middle zone of lower till, (b3) Upper zone of lower till, (c) Upper till, and (d) Lake Edmonton sediments. The prominent shear surface separating the middle and upper zones of the lower till is indicated by dashed lines. The pick is 17 inches long.

degrees. This distribution appears to represent a modified fabric. It differs from the expected patterns described above in that it lacks the near-horizontal girdle. Also, the presence of two maxima whose trends differ by only 30 degrees, and the large plunge of 35 degrees of one of these maxima, are both anomalous in a ground-moraine fabric.

INTERPRETATION

Several lines of evidence suggest that the fabric of the middle zone of the lower till, with its near-horizontal northwesterly trending maximum (Fig. 3-A), is the original depositional fabric of the lower till. First, the majority of fabric observations on the lower till within the study area reveal a tendency for the long axes of pebbles to have a northwesterly trend (Fig. 2). Second, the fabric itself has the characteristic features of an undisturbed ground-moraine fabric. Third, the middle zone is much less sheared and fractured than is the upper zone; it is comparatively massive and dense, and its columnar structure resembles that of the undisturbed upper till. Thus structural observations support the fabric evidence, and both suggest that the middle zone of the lower till is relatively undisturbed, and that its fabric is the original depositional fabric of the lower till.

Both structural and fabric observations are important in interpreting the fabric of the upper zone of the lower till. The near-parallelism of the slickensides in the basal sheared zone of the lower till, with trends of between N 15°E and N 35°E, to the grooves and lineations in the upper zone (Plate 1-C, D), with trends of N 20°E to N 25°E, suggests that all these similarly oriented features were produced at about the same time by the same agent of deformation. Furthermore, the close parallelism of the trends of these features to the known direction of movement of the later glacier (Fig. 4, Plate 1-A) strongly suggests that the movement of this glacier was the cause of the deformation in both zones.

The grooves and lineations in the upper zone of the lower till are also nearly parallel to the N 20°E trend of the fabric (Fig. 3-B). This suggests that shearing altered the original fabric of this till, producing the present orientation. This theory is supported by observations at other locations in the study area, where the long axes of pebbles in the lower till tend to have a northeasterly trend; at all such locations the lower till is highly fractured and sheared. It is believed that the other maximum in the fabric of this upper zone represents a remnant of the northwesterly trending maximum of the original fabric Other fabrics possessing both northwesterly and northeasterly trending maxima have been observed in the lower till in the area studied. One of these is the fabric at location E shown in

Figure 2. The lower till at this location is intensely fractured, and it is possible that this also represents a partially altered fabric.

The nature of the basal sheared zone of the lower till, with its slickensides paralleling the movement direction of the overriding glacier, together with the undisturbed character of the middle zone, suggests that locally the lower till was moved *en masse* by the overriding ice by shearing of the basal clayey zone.

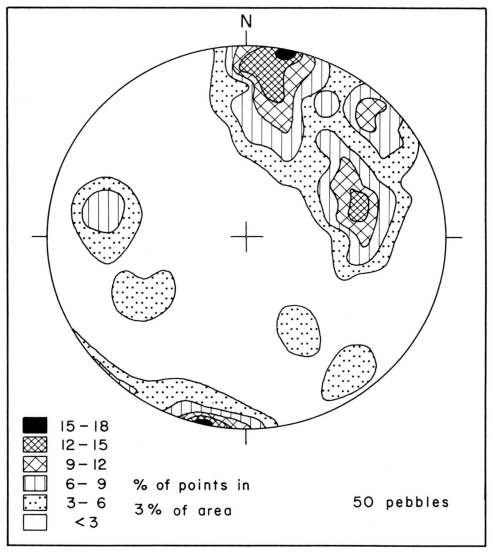

Figure 4. Equal-area lower-hemisphere projection showing the distribution of the long axes of pebbles in the upper till.

MECHANISM OF FABRIC ALTERATION

Alteration of the fabric in the upper zone of the lower till is thought to have resulted from reorientation of those elongate pebbles intersected by any of the numerous shear surfaces. As the shearing took place, these pebbles would tend to become re-aligned parallel to the direction of movement along the shear plane. Pebbles reoriented in this way are thought to be the cause of the northeasterly trending maximum in Figure 3-B. Elongate pebbles not intersected by any shear surface would retain their original orientation; the north-northwesterly trending maximum in Figure 3-B is believed to be due to such pebbles. The shift of the original maximum from a trend of N 62°W and a plunge of 12 degrees to a trend of N 10°W and a plunge of 35 degrees may be due to a certain amount of plastic deformation accompanying the shearing of the till.

MacClintock and Dreimanis (1964) described a till fabric in the St. Lawrence valley that had been modified by overriding ice. There, reoriented fabrics were found as deep as 35 feet below the top of the lower till, and effects of folding and faulting caused by the later advance extended to as deep as 65 to 70 feet below the top of the lower till. Mac-Clintock and Dreimanis suggested three possible mechanisms of reorientation. (1) The till was displaced along associated relatively incompetent lacustrine clay and silt layers. A similar kind of displacement appears to have taken place in the intensely sheared clayey layer at the base of the lower till in the Clover Bar area of Alberta. (2) The till itself was sheared. Although such shear planes were "seldom recognizable except at those places where lacustrine clay and silt was dragged in" (MacClintock and Dreimanis, 1964, p. 141), MacClintock and Dreimanis suggested that "It is possible that considerable parts of the Malone till were so thoroughly sheared along minute shearing planes, not now visible because of lack of clay partings, that the fabric was rearranged toward the southeast" (Mac-Clintock and Dreimanis, 1964, p. 141). This mechanism corresponds to that proposed by the present authors for reorientation of the pebbles of the lower till in the Clover Bar area, where numerous shear surfaces can be seen in the till. (3) The third possible mechanism suggested by Mac-Clintock and Dreimanis was "viscous flow," made possible by shear stresses "remoulding" the till and making it into a viscous mass. However, Mac-Clintock and Dreimanis found no evidence to confirm that this mechanism had in fact been operative in their St. Lawrence valley area, and no evidence for this kind of deformation was seen in the lower till in the Clover Bar area.

ACKNOWLEDGMENTS

The exposure described in this paper was examined with the permission of Alberta Concrete Products, Ltd. Financial support from the National Research Council of Canada is gratefully acknowledged. Thanks are due to Dr. H. A. K. Charlesworth, University of Alberta, who commented on the structural aspects of the paper.

REFERENCES

Harrison, P. W., 1957, A clay-till fabric: its character and origin: Jour. Geology, v. 65, p. 275-308.

Holmes, C. D., 1941, Till fabric: Geol. Soc. America Bull., v. 52, p. 1299-1354.

Kauranne, L. K., 1960, A statistical study of stone orientation in glacial till: Comm. Geol. de Finlande, Bull. No. 188, p. 87-97.

MacClintock, P., and Dreimanis, A., 1964, Reorientation of till fabric by overriding glacier in the St. Lawrence valley: Amer. Jour. Sci., v. 262, p. 133-42.

Muecke, G. K., 1965, Fracture analysis in the Canadian Rocky Mountains: M.S. thesis, University of Alberta, 117 p.

Pettijohn, F. J., 1957, Sedimentary rocks: New York, Harper and Bros., 718 p.

Ramsden, J., 1970, Till fabric studies in the Edmonton area, Alberta, with special emphasis on methodology: M. Sc. thesis, University of Alberta, 205 p.

Reimchen, T. H. F., 1968, Pleistocene mammals from the Saskatchewan gravels in Alberta, Canada: M. Sc. thesis, University of Alberta, 92 p.

Westgate, J. A., 1968, Linear sole markings in Pleistocene till: Geol. Mag., v. 105, p. 501-505.

———, 1969, The Quaternary geology of the Edmonton area, Alberta: *in* Pawluk, S., Editor, Proceedings of the symposium on pedology and Quaternary research: University of Alberta Press, Edmonton, p. 129-51.

The Relationship of
Macro- and Microfabric of Till
and the Genesis of Glacial Landforms
in Jefferson County, Wisconsin

Edward B. Evenson

ABSTRACT

Macrofabric analysis performed at fourteen exposures in a drumlin field and recessional moraine complex in Jefferson County, Wisconsin, yielded mean azimuthal directions corresponding closely to regional drumlin orientation. Long-axis plunge was oriented up-glacier at an average of 11°.

Microfabric analyses of thirty-five samples were performed, twenty-one from oriented cores and fourteen from the same exposures used in macrofabric analysis. Sand-size-particle orientation was measured in the horizontal plane and in the vertical plane parallel to the horizontal lineation. Average directions in both planes are expressed as vector means. Microfabrics, although slightly more variable, showed the same relationship to regional ice flow as macrofabrics.

Both micro- and macrofabrics have pronounced long-axis maxima parallel to past glacier flow. A transverse fabric normal to the main linear fabric is also present in most samples. The major azimuthal mode is attributed to shearing within the till due to the movement of the overriding ice. Transverse fabrics may be the result of protracted flow. Up-glacier imbrication is attributed to particle transport in upward-rising shear planes which develop in till as a function of: (1) transport of till from high-pressure to low-pressure zones beneath the ice, (2) extrusion due to the weight of overlying ice, and (3) thrusting resulting from the couple created by the overriding ice.

Based on fabric analyses and field relationships, drumlin development in the area is attributed to transport of till to drumlin crests as a result of flow from areas of high pressure to areas of low pressure under the ice. End moraines of the area appear to be overridden washed deposits capped with a thin cover of till and, like the ground moraine, have fabrics that are best explained as the result of shearing and microthrusting caused by the overriding ice.

INTRODUCTION

The works of Richter (1932), Holmes (1941), and Harrison (1957) have established till-fabric analysis as a technique by which the direction of flow of the former glacier can be determined. The early work demonstrated that the long axis of elongate pebbles tends to become aligned parallel to the direction of flow of the former ice, a concept documented by many workers: Harrison (1957); Glen, Donner, and West (1957); MacClintock and Dreimanis (1964); and Wright (1962, 1957). Harrison (1957) reported that, in addition to becoming aligned parallel to the direction of glacial flow, elongate elements have a preferred plunge in an up-glacier direction. He suggested that the plunge was the result of particles being transported in shear planes within the ice. Because shear planes usually dip up-glacier, it should be possible to determine the bearing of previous ice flow and the absolute direction of flow. Wright (1957, 1962) used these concepts to establish the absolute direction of ice flow in the Wadena drumlin field of Minnesota.

The "in field" portion of an average macrofabric analysis takes three to four hours and requires a suitable exposure and the presence of pebble-sized till stones. Where these requirements are not met, or where pebble-deficient tills occur, the problem can be overcome by using microfabric analysis, in which an oriented core is obtained in less than one hour and exposures are not needed.

Previous work on microfabric is limited. Harrison (1957) performed two and Gravenor and Meneley (1958) performed three microfabric analyses. They report that the alignment and plunge of sand-size particles have the same relationship to ice flow as found using macrofabric analysis. Ostry and Dean (1963) measured long-axis orientation in 32 sections and compared the results to macrofabric data. They report a high correlation between the two methods, but as they did not determine the plunge of the axes, their data cannot be applied to studies of absolute flow direction.

The current study attempts to show: (1) the relationships between the fabric revealed by macroanalysis and the fabric found by microanalysis; (2) that relative and absolute directions of glacial flow can be determined by fabric analysis; (3) how fabric elements in a till become oriented; and (4) that fabric analysis can be used to determine the mode of genesis of glacial landforms.

It should be made clear that true statistical analysis using hypothesis testing, correlation, and so forth, has not been used. The conclusions made are semi-quantitative, but are believed to be adequate for the purpose of this study. Likewise, operator variation, precision of measurement, and reproducibility have not been calculated. The mechanics of flow and fabric development have been inferred from the geological evidence rather than derived from calculations or experimentation.

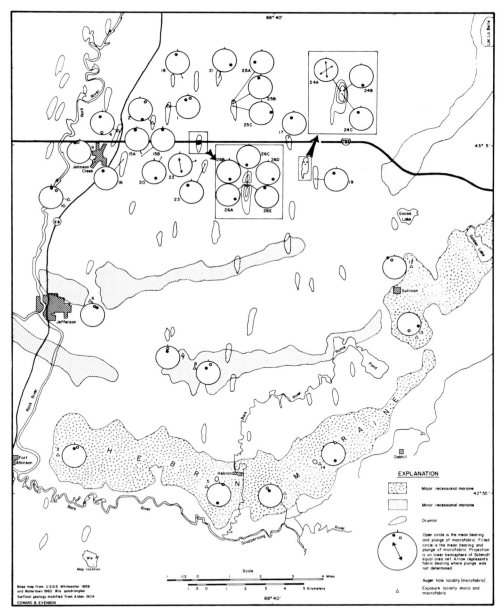

Figure 1. Relationships of macrofabric and microfabric in Jefferson area, Wisconsin.

Location and Geologic Setting

The study area is located in the southeast quarter of Jefferson County, Wisconsin, approximately 50 miles due west of Milwaukee (Fig. 1). The surficial geology here consists of a field of north-south-trending drumlins, and a set of three recessional moraines (Fig. 1). The Hebron Moraine (Alden, 1904) is the best-developed moraine, and in places obscures the drumlin topography which it overlies. Elsewhere the drumlin shapes are clearly visible through the morainal cover. The two other recessional moraines (Fig. 1) are poorly developed and are topographically indistinct, and are mappable only as arcuate washed deposits that parallel, but lie north of, the Hebron Moraine.

The till in Jefferson County is Woodfordian in age (Frye, Willman, and Black, 1965), is of similar composition throughout the area, and ranges in thickness from a few feet to over 100 feet. The only structure present in the till is a nearly horizontal jointing, which develops when the overlying till is removed.

The Jefferson County area was selected for this study because the orientation of the drumlins provides an absolute control of the direction of regional ice flow, and because the area provides an opportunity to examine the fabrics of drumlins, end moraines, and ground moraine.

PROCEDURES

Sample Locations

Thirty-five microfabric and fourteen macrofabric analyses were performed. At locations 1 through 14 (Fig. 1), a macrofabric analysis was performed and an oriented sample was collected for microfabric analysis. At locations 15A through 26E, an oriented sample was collected from a cleaned auger hole using a split spoon.

Samples 3, 5, 7, 8, 11, 12, 13, and 14 were collected from areas mapped by Alden (1904) as recessional moraines. Samples 18, 20, and 22 were obtained from ground moraine. The remaining samples were taken from drumlins or drumlinoid features.

Macrofabric Analysis

Macrofabric analysis was performed following the procedure of Wright (1962). Measurements were made in freshly exposed roadcuts and gravel pits. At each site the bearing and plunge of the long axis of fifty stones, ranging in size from one-half to three inches, were measured. Only stones with a 3:2 or greater long- to short-axis ratio were measured. After taking fifty measurements, a fresh horizontal platform was prepared and marked with a north orientation. A vertical trench was cut to free the oriented sample, which was then wrapped in aluminum foil and newsprint to prevent dessication. These samples (1-14) were analyzed using the micro-fabric technique described in the following section.

The values of bearing and plunge of the long axes of the pebbles were plotted using a contour interval of one percent, on a Schmidt equal-area net, lower-hemisphere projection. Mean bearing and plunge were determined by visual inspection of the contoured plots.

Microfabric Analysis

Sample Collection

Samples used for microfabric analysis were collected by two methods. Samples 1 through 14 were obtained as described above from exposures used to perform macrofabric analyses. Twenty-one samples (Nos. 15A through 26E) were obtained from augered holes. A truck-mounted auger was used to bore six-inch-diameter holes ranging in depth from six to eight feet. The holes were cleaned and a three-inch-diameter split-spoon sampler was lowered into the hole. The spoon was then set with three or four blows of a standard 140-pound driving hammer, and then driven a measured foot by successive blows. The drilling rod was checked to assure that all joints remained tight. The drill rod was oriented using masking tape marked with waterproof pen; bearings were determined by sighting with a Brunton compass. Using a rotary winch, the drill rod and spoon were recovered from the hole and laid out for orientation transfer. The orientation from the drill rod was transferred to the outer surface of the split spoon, which was then removed from the drilling rod. Sample collection was completed by opening the split spoon and transferring the orientation markings from the spoon exterior to the sample.

From the oriented core, two four- to six-inch sections were removed, wrapped, and returned to the laboratory for analysis. In all instances, the top few inches of the core were discarded, to make sure that any material which had fallen into the auger hole was not sampled.

Sample Preparation and Impregnation

In the laboratory, samples were examined for any disturbance that might have occurred in transport (most samples were undisturbed). Samples approximately one cubic inch in volume were oven-dried for 24 hours and then impregnated with a mixture of 60 percent Castolite plastic and 40 percent monomeric styrene, using the method of Boul and Fadness (1961). The mixture is a low-viscosity medium which reduces surface tension and allows impregnation without grain reorientation. At 90°C, the plastic sets in approximately 12 hours, depending on the amount of hardener used.

To achieve impregnation, dried samples are submerged in plastic and placed in a vacuum desiccator. The pressure is gradually lowered to 7 mm. of Hg and maintained for 10 to 20 minutes to allow the air contained within the sample to escape. Escaping air is indicated by a continuous stream of bubbles rising from the sample, but should not be confused with boiling which occurs below 7 mm. of Hg, or when excessive styrene is used. When the air has all been removed, bubbling ceases, and air is slowly released into the vacuum. At this time, one atmosphere of pressure forces the impregnating medium into the evacuated voids of the sample. Boul and Fadness (1961) report that the technique causes no observable grain reorientation, even when used on soils. The sample should remain submerged in plastic for 15 minutes to allow the plastic to flow into all voids. If the sample is removed prematurely from the plastic, the outermost surface becomes depleted of plastic by migration of the plastic toward the center.

If sectioning reveals small unimpregnated areas, the above procedure may be repeated, and the outer disturbed layer ground off, producing a new surface suitable for thin-section preparation. Standard procedure was followed in the preparation of thin sections.

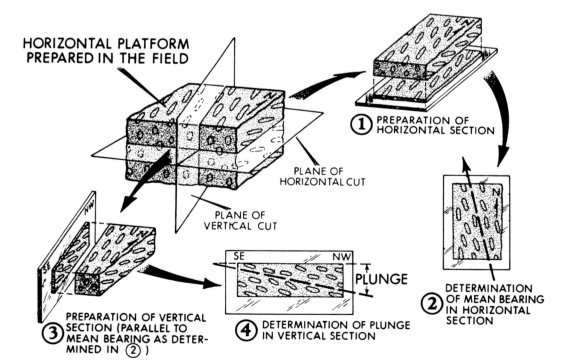

HORIZONTAL PLATFORM PREPARED IN THE FIELD

PLANE OF HORIZONTAL CUT

PLANE OF VERTICAL CUT

① PREPARATION OF HORIZONTAL SECTION

② DETERMINATION OF MEAN BEARING IN HORIZONTAL SECTION

③ PREPARATION OF VERTICAL SECTION (PARALLEL TO MEAN BEARING AS DETERMINED IN ②)

④ DETERMINATION OF PLUNGE IN VERTICAL SECTION

PLUNGE

Figure 2. Preparation of thin sections to obtain bearings and plunges of microfabrics.

Sectioning Technique and Fabric Analysis

Fabric analysis is performed using two mutually perpendicular thin sections (see Fig. 2). A horizontal thin section is used to determine the modal lineation in the horizontal plane. The vertical section is used to determine the direction and angle of plunge.

Prior to thin sectioning, the impregnated sample is trimmed so that one of its vertical edges is parallel to the north arrow, and the sample is then mounted so that the north orientation parallels the edge of the slide. At this time, the slide is scribed with sample number and north orientation. A thin section, parallel to the horizontal plane, is then prepared (Fig. 2, Step 1).

Using a microprojector, the grain images (enlarged 10×) are projected onto a sheet of graph paper oriented such that one of the coordinates is parallel to the north orientation of the slide. All sand grains with an axial ratio of 3:2 or greater are traced onto the graph paper. Using an ellipse template or the least-projection-elongation method of Dapples and Rominger (1945), the long axes of the grains are drawn and the bearings measured using a protractor. The bearings are then plotted on polar coordinate paper. A spoke diagram is prepared by grouping the data into 10-degree intervals. From the spoke diagram, the major modal orientation may be determined by visual inspection or by the vector-mean method (Pincus, 1961).

To determine plunge, a vertical section is prepared parallel to the determined horizontal lineation mode (Fig. 2, Step 3). The section is mounted so that the horizontal plane parallels one of the slide edges and the trend of the horizontal section is marked on the appropriate ends of the vertical section. Plunge directions and angles are measured using the same techniques as described for the horizontal section. The major modal orientations from the horizontal and vertical sections are then combined to give apparent bearing and plunge of the microfabric.

A number of internal checks were performed to determine the effects of different operators performing the analysis. Sample 19 was analyzed twice for elongation direction, once by myself and once by another operator. The two distribution maxima obtained differed by only 2°; both showed a marked cross fabric. The similarity of the two fabrics suggests that operator bias is minimal. Sample 16 was analyzed by a third operator with approximately the same result.

To establish the reproducibility of results over small distances, two samples, 15A and 15B, which were obtained at the same locality, but in auger holes approximately 10 feet apart, were analyzed. The bearings differed by only 3°, the plunge by 6°.

Sample 1 was analyzed twice. In the first analysis, 70 grains with axial ratios of 2:1 or higher were measured. In the second, 210 grains with ratios of 3:2 or higher were measured. The results of the two trials differed by less than 2°, which may indicate that, as the axial ratio is increased, the number of grains that need to be counted is decreased. The 2:1 axial ratio is near the limit of grain enlongation.

RESULTS

Comparison of Macrofabric and Microfabric Analysis through Matched Sampling at Locations 1 through 14

A summary of the relationship of the 14 analyses is presented in Table 1, and circles 1 through 14 (Fig. 1) contain plots of the macro- and micro-fabric bearings and plunge modes. The macrofabrics range in bearing from N 20°W to N 34°E, while the microfabrics, with the exception of Number 13, (which is discussed below), range from N 30°W to N 17°E. I consider the agreement close enough to assume that, on a regional scale, the direction of glacial flow indicated by macrofabric and microfabric analyses is equivalent.

Nearly all microfabrics display a secondary mode, hereafter referred to as "cross fabric," at approximately 90° to the major modal orientation. A cross fabric is also found, although not as markedly, in the macrofabric diagrams. The anomalous microfabric bearing of sample 13 may be an example of an instance in which the cross fabric has become the dominant lineation.

Comparison of the plunges of macro- and microfabrics presents a special problem — that of the choice of a zero line, or datum. Because most tills lack stratification, and the subglacial slope at the time of deposition is commonly unknown, an arbitrary datum, usually the horizontal plane, must be used as a reference with respect to which plunge angles are measured. Although the work with matched samples yields no clear-cut relationship

TABLE 1

MEAN BEARINGS AND PLUNGES OF MACROFABRICS AND MICROFABRICS

Sample No.	Bearing and Plunge Macrofabric	Bearing and Plunge Microfabric
1	N14°E 10°N	N10°W 27°N
2	N34°E 5°N	N 3°W 5°S
3	N17°E 8°N	N 30°N
4	N 3°E 10°S	N13°W 30°N
5	N18°E 0°	N 5°E 10°S
6	N 7°E 14°N	N17°E 17°N
7	N 10°N	N 1°W 42°N
8	N 8°E 8°N	N20°W 43°N
9	N 2°W 15°N	N10°W 6°N
10	N29°E 10°N	N 2°E 8°N
11	N12°W 10°N	N 5°W 35°N
12	N20°W 20°N	N30°W 0°
13	N 2°W 25°S	N80°W 3°E
14	N 5°E 0°	N 3°W 20°S

between the plunge values of macro- and microfabrics (Table 1), the results obtained using all the fabric analyses performed (see Fig. 4 and discussion in next section) suggest that plunge values as determined by macro- and microfabric methods are similar.

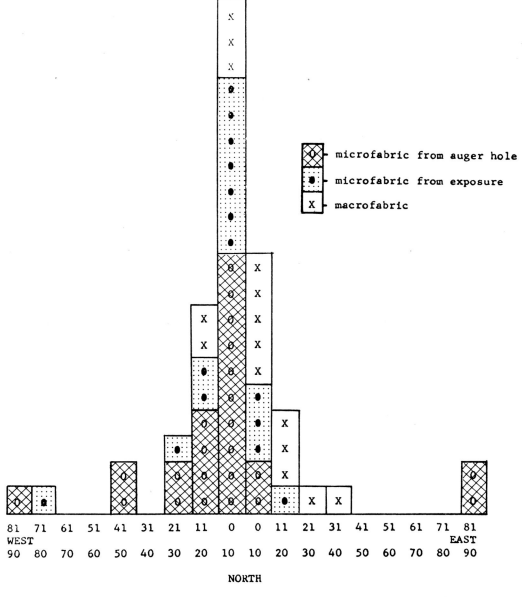

Figure 3. Bearings of long axes of fabrics. Samples falling on north-south azimuth plotted in 0-10° east and and 0-10° west columns; those falling on east-west azimuth plotted in 81-90° west column. Sample 24A not shown, as bearing could not be determined.

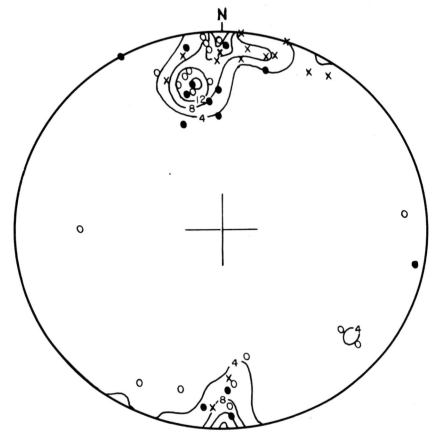

Figure 4. Composite diagram of all fabric analyses performed. Schmidt equal-area net, lower hemisphere projection. Note similarity of this diagram to that of Harrison (1957, p. 35).

The Relationship of Microfabrics Taken from Cores to the Matched Samples 1 through 14

A summary of mean bearing and plunge data obtained from auger holes 15A through 26E is presented in Table 2. Figure 3 shows the mean bearings, relative to north and plotted in 10-degree intervals, of all analyses performed. This figure reveals a strong concentration of analyses centered in the 0-10-degree-west interval and a secondary concentration normal to the maximum. The histogram indicates a strong agreement between the results obtained using any of the macro- or microfabric methods. The occurrence of two samples in the 41-50°-west interval in Figure 3 may be the result of ice moving around some obstacle or, in the case of 26A (which was taken from the lower west flank of a drumlin — Fig 1), may be the result of redeposition by mass wasting.

TABLE 2

MEAN BEARING AND PLUNGE OF MICROFABRICS OBTAINED FROM AUGERED HOLES

Sample No.	Bearing and Plunge		Sample No.	Bearing and Plunge	
15A	N 3°W	10°N	24A	NR†	
15B	N	4°N	24B	N 3°W	10°S
16	N22°W	18°N	24C	N16°W	25°N
17	N12°W	30°N	25A	N90°W	31°W
18	N 3°W	26°N	25B	N 4°W	5°N
19	N12°W	23°N	25C	N85°E	10°E
20	N10°W	34°S	26A	N50°W	12°S
			26B	N 3°W	21°S
22	NR*		26C	N13°E	17°S
23	N10°W	20°N	26D	N27°W	14°S
			26E	N 5°W	3°N

*Vertical section had no preferred plunge (Fig. 1, No. 22).
†No vertical section cut (see Fig. 1, No. 24A) due to two modes of equal magnitude.

A composite diagram of all fabrics for which bearing and plunge were determined is presented in Figure 4. Of the 21 samples collected from auger holes, bearing and plunge were determined for only 19; Sample 24A (Fig. 1, No. 24A) had two modes of equal strength and Sample 22 had no plunge maximum (Fig. 1, No. 22). Sample size prohibited cutting sections parallel to each of the modes in Sample 22.

The strong north-south-trending, north plunging maximum (Fig. 4) and the relatively low amount of scatter confirm the hypothesis that the fabrics obtained using macro- and microfabric analyses are similar and that microfabrics from augered split-spoon samples yield results similar to fabrics of samples taken from exposures.

THE GENESIS OF FABRIC

Relation of Long-Axis Orientation to Ice Flow

Several theories explain the preferred parallelism and transversity to glacial flow displayed by elongate elements in tills. All theories have one point in common, shear. Shear is produced in any flowing medium which has a velocity gradient perpendicular or transverse to the velocity. Where such a velocity gradient occurs, particles in adjacent flow surfaces are necessarily sheared with respect to each other and subsequently become aligned parallel to the velocity. Any theory which attempts to explain the generation of till fabric must therefore explain how shearing conditions are created, maintained, and distributed in glacier ice and in the material transported within its base.

Holmes (1941) first reported that till stones have: (1) a strong tendency to become aligned, such that their long axes coincide with the direction of glacial flow, and (2) a tendency toward orientation transverse to the direction of flow. Pebble alignment parallel to flow was attributed to the dragging of pebbles along shear planes in the basal ice and at the ice-till contact, because such an orientation presents the least cross-sectional area against ice thrust. Stones with an orientation transverse to flow were explained as having rolled into position by rotation about their longest axis while totally immersed in the ice.

Three-dimensional fabric analyses performed on tills by Harrison (1957) resulted in a dominant long-axis maximum plunging upstream against the direction of former glacial flow. On the basis of kinematic considerations, Harrison suggests that most till fabric is inherited directly from the transportational environment, as a result of material moving up along shear surfaces in the glacier. Subsequently, upon stagnation and melting, deposition of the material occurs without reorientation.

Parallel orientation of fabric elements have been suggested by Glen, Donner, and West (1957) to occur by viscous flow in shearing ice. They state that shearing probably occurs continuously within the tectonite, and is adequate to account for the orientation of fabric elements. To explain the behavior of fabric elements in a flowing medium, they applied Jeffrey's examination (1922) of the behavior of ellipsoidal particles suspended in a viscous liquid. Jeffrey's calculations showed that, in slow-flowing liquids, prolate spheroids which were introduced with a random orientation initially, developed an orientation parallel to flow, and with increasing time, would slowly develop a long-axis orientation transverse to flow. Taylor (1923) placed ellipses with an axial ratio of 2:1 in a shearing waterglass medium and found that development of the transverse long-axis orientation was very slow. Lindsay (1968) subjected Jeffrey's equations to computer cycling and reported that a recognizable fabric developed rapidly, and then degenerated with time.

MacClintock and Dreimanis (1964) have shown, through the use of fabric analysis, that ice of continental thickness can cause reorientation of the fabric elements in unfrozen, overridden deposits to a depth of 35 feet. Gravenor and Meneley (1958) have reported fabrics developed as the result of subglacial transport of tills (see below, p. 358). If it is assumed that the viscous flow suggested by Glen, Donner, and West (1957) to explain the orientation of till stones in ice can also occur within water-saturated tills, it is then possible to orient particles in tills by the action of overriding ice. The salient point is that final particle orientation need not necessarily occur within the ice, and in fact may develop within the till, under the ice in response to shear which is created by the action of the overriding ice If this assumption is correct, orientation parallel to glacial flow may develop in the basal till, the difficulties of depositing oriented materials from the ice proper without reorientation are eliminated.

Development of the Up-Glacier Plunge

The combined viscous flow theory of Glen, Donner, and West (1957) and of MacClintock and Dreimanis (1964) seems to account for the alignment of fabric elements, but does not explain the up-glacier plunge of fabric elements. Several theories have been suggested to account for this direction of plunge of fabric elements. The theories fall into three main groups: (1) ice-shear theory (for example, Harrison, 1957; Wright, 1957 and 1962); (2) slope theory (Glen, Donner, and West, 1957); and (3) till-shear theories (for example, Hoppe, 1963; MacClintock and Dreimanis, 1964). The first and third theories (ice-shear and till-shear) involve flow that produces the shear necessary to cause element parallelism. The second (slope theory) requires that the particles have parallelism developed from some previous flow regime.

Ice-Shear Theory

Harrison (1957) and Wright (1957, 1962) have suggested that ice shear causes fabric elements to become aligned as transport occurs along upward-curving shear planes developed within the ice. Harrison's evidence from analysis of englacial material taken from an active glacier strongly suggests that shear planes can cause both parallel alignment and up-glacier plunge. He describes debris-filled shear planes in ice dipping up-glacier at an angle of 37° (1957, p. 286). Fabric analysis performed on the englacial block produced a pattern very similar to the fabrics obtained in this study. Boulton (1968) reports debris-filled shear planes up to five meters thick at the edge of an active glacier. Wright (1957, 1962), in twenty-two fabric analyses performed on drumlins, attributed the up-glacier plunge, averaging 23°, to orientation obtained in shear planes during transportation in the glacial tectonite. Wright also presents a review of the literature documenting the existence of debris-filled shear planes in recent active glaciers. There is little double that shear planes in ice orient particles parallel to glacial flow and plunging up-glacier. The problem with the theory lies in an explanation of how the orientation of the particles can be maintained during the melting process.

Melting at the glacier margin has been reported by Boulton (1968) to produce slump and flow-till deposits and to almost completely destroy the orientation of the relic shear planes. Boulton does report one instance where englacial debris was released from the ice by melting without disturbance (p. 405), but his diagrams imply that this type of deposition is far less common than are flow and slump deposition. Until it is demonstrated that alignment developed within ice commonly persists through melting, the ice-shear theory should be considered suspect.

Slope Theory

Slope theory attributes the plunge of an element to the slope of the material over which the ice is moving. If this theory is correct, there should be a near one-to-one correlation between the slope of a particular glacial landform and the plunge of the fabric elements near the surface. Boulton (1968) reports that long-axis plunges are slope-conformable for the upper part of some flow tills formed at the margin of recent glaciers. Cowan (1968) reports that fabrics taken on the proximal sides of ribbed moraines showed plunges up-glacier, but at angles greater than those of the slope on which they occurred. Distal fabrics were reported to be slope-conformable, indicating that they were probably of flow-till origin. Gravenor and Meneley (1958) report an up-glacier plunge for fabrics occurring on ridges oriented parallel to glacial flow; therefore, the plunge is normal to the slope direction. In this study, amount and direction of plunge were compared to landform slope (measured along the bearings of determined plunge) for 26 fabrics. Assuming that the angles presented for slopes are accurate to 3°, only four samples were judged to have plunges that were slope-conformable.

Samples taken in moraines were not included because of the difficulty of obtaining meaningful slope angles in a moraine complex. Due to the lack of agreement between angle of plunge and angle of slope, it appears that plunge is not a function of the depositional slope here. Therefore, the slope theory is not adequate to explain the consistent up-glacier plunge of the fabric maxima in this study.

Till-Shear Theories

Till-shear theories are based on the premise that tills behave as viscous

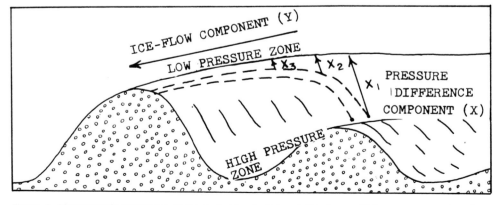

Figure 5. Diagrammatic illustration of the path of a particle moving from a high-pressure zone to a low-pressure zone. In this system, the ice-flow component (Y) is a constant and the pressure-difference component (X) is a decreasing function, $X_1 > X_2 > X_3$. The dashed line indicates the part of particle movement.

liquids, thereby flowing and shearing in response to stress. Each of the theories in this category account for long-axis alignment by shear and requires a different factor to explain the preferred direction of plunge.

Till is transported outward and upward from its place of deposition by the weight of overlying ice, according to Hoppe (1952). As the till is extruded from its initial location upward into crevasses in the basal ice, the pebbles are oriented, by flow shear, so that their long axes plunge toward the block of ice which displaced them. Hoppe demonstrated that this effect was applicable to dead-ice moraine systems, where large blocks of ice are isolated from the active ice by melting.

Alignment parallel to flow can be produced by subglacial transport of till in response to pressure gradients which develop under the ice, according to Gravenor and Meneley (1958). If vector analyses are made of the shear zones which transport materials from high-pressure troughs to lower pressure ridges, the up-glacier imbrication observed by Gravenor and Meneley in glacial flutings and the imbrication found in the drumlins of Jefferson County are satisfactorily explained.

The path traversed by a particle moving in a shear zone is diagrammatically illustrated in Figure 5. The component of ice flow (Y) is assumed to be constant, while the component due to pressure difference (X), which is normal to the ridge side, decreases as the particle moves into the area of low pressure. To simplify analysis, the surface XY can be approximated, over small areas, by a plane. Figure 6 illustrates the vectors X and Y in the

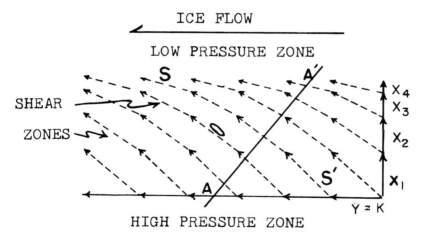

Figure 6. Tentative flow model illustrating development of shear zones resulting from flow of material from high- to low-pressure zones. The low-pressure zone is at the top of the diagram. The plan of analysis is the slide of the drumlin. The ice-flow component (Y) is assumed to be a constant $X_1 > X_2 > X_3$, where X is the component due to pressure difference. A-A^1 is orthogonal to the X_2Y-vector component of the flow zone S-S^1. The ellipse illustrates the orientation of elongate particles transported in the shear zones.

plane of XY. The dashed vectors diagrammatically illustrate the curved shear zones along which particles in the plane XY would move. Line A-A^1 is orthogonal to the X_2Y vector component of the flow zone S-S^1. As velocities are not constant in adjacent flow zones along A-A^1, shear results. The ellipse diagrammatically illustrates the orientation of a particle in transport in a shear zone.

Thus, Figures 5 and 6 illustrate that up-glacier imbrication (resulting from the shape of the shear zones) and orientation parallel, or nearly parallel, to the flow can occur in the transport of materials from low- to high-pressure zones.

Using microfabric analysis, Gravenor and Meneley (1958) examined fabrics in glacial flutings (which they consider transitional forms of drumlins) at depths of 4 to 5, and 10, 11, and 13 feet. The fabrics taken at the two shallowest depths had long-axis plunges up glacier and a bearing parallel to the ridge. Fabrics from depths of 11 and 13 feet had trends parallel to earlier ice advances rather than parallel to the ridge trends. The divergence in fabric orientation led Gravenor and Meneley to suggest the theory of erosion and redeposition for the upper 10 feet of till at this locality. The flow paths and vector considerations in Figure 5 and 6 are suggested to supplement their work.

The concept that till, under the pressure of overriding ice, reacts to stress as does any other viscous or plastic medium, was discussed earlier. If till does react to stress in this way, then it should react in the same manner as does debris-charged ice which is also a plastic medium. Many workers report the presence of upward-curving shear planes at the edges of glaciers (for example, Boulton, 1968; Nye, 1967). These same shear planes can therefore be expected in over-run tills. MacClintock and Dreimanis (1964) report the existence of upward-curving shear planes, filled with clays and sands which act as lubricants, in a sequence of till 70 feet thick. By applying the concepts of particle orientation as a response to flow and imbrication due to shear-zone shape, a strong case for the development of fabric in basal tills can be presented.

SUGGESTED ORIGIN OF GLACIAL LANDFORMS
IN JEFFERSON COUNTY

The Origin of the Drumlin Fabric

Two theories discussed in the previous section adequately account for the fabric observed in the drumlins: the till-shear theory and the pressure-differential theory. Figure 7 shows a south-facing exposure at location number 2. Recumbently folded gravels are in contact with the till. The gravel units appear dragged at both their upper and lower contacts. Folding and preservation of the bedding suggests that they were transported and injected, in a frozen but viscous state, into a crevasse or low-pressure area such as was described earlier in this paper. The relationship between the drumlin form and the fabric within is still far from clear. The only consistant relationship is the parallelism between the long axis fabric maximum and long axis orientation of the drumlin form. At present, plunge data are so variable that no meaningful relationships can be deciphered (see Fig. 1).

The Origin of End Moraines

Three morainal belts developed at the edge of the receding Woodfordian ice are delineated in the Jefferson County area (Fig. 1). None of these moraines are thick enough to completely obscure the drumlin complex which they overlie. The two unnamed recessional moraines, which lie three and six miles to the north of the large Hebron Moraine (Fig. 1), mark brief halts in the retreat of the ice. The moraines consist primarily of outwash sands and gravels which form a veneer over the existing topography.

The Hebron Moraine is the best developed of the recessional moraines.

Figure 7. Recumbently folded gravels exposed in a drumlin cut.

In places, the moraine consists of till up to 60 feet in thickness; in other locations, till is completely absent, yielding breaks in the moraine trend (Fig. 1). A quarry at location number 14 exposes 50 feet of gravels overlain by a two-foot layer of till; the same stratigraphic relationship is found at location number 13. The common occurrence of highly folded glaciofluvial deposits overlain by till (Fig. 8) indicates that, after deposition of the gravel, the ice probably readvanced, "bulldozing" the frozen gravels into their present configuration. Fabrics reported from Locations 12, 13, and 14 (Fig. 1) are therefore more likely developed as a result of readvance than as a result of moraine-building. It should be noted that both micro- and macrofabrics at location number 13 plunge to the south instead of to the north; this may be the result of slumping in the underlying sands and gravels. Fabrics at locations 11 and 5 were taken in the moraine complex from features definitely identifiable as drumlins and should not be considered in an analysis of end moraines. I feel that, in most of the exposures studied, the fabric appears to have developed in response to deposition during overriding or rapid retreat.

Figure 8. Folded sands and gravels overlain by till.

CONCLUSIONS

Assuming that the fabric observed in tills is directly related to the genesis of the landform in which the fabric occurs, and based on evidence presented in this paper, the following conclusions are reached:

1. When applied on a regional scale, the results obtained using micro- and macrofabric analysis are equivalent, except that microfabric gives slightly less consistent plunge data than does macrofabric analysis.

2. The fabric observed in the landforms of Jefferson County has a strong maximum parallel to the direction of past glacial flow, a secondary maximum (cross fabric) normal to that flow, and a long-axis plunge up-glacier, opposite to the direction of inferred flow.

3. The long-axis orientation of fabric elements is attributed to particle readjustment to velocity gradients (shear). It is suggested that fabric develops in till in response to flow induced in it by the action of the overriding ice.

4. The transverse maximum of the fabric diagrams is attributed to *protracted* flow and frequent stone collisions.

5. The up-glacier plunge found in most till fabrics (see Fig. 4) is attributed: (a) to the orientation obtained when materials are transported from areas of high pressure to areas of low pressure under the ice, or (b) to upcurving shear zones which develop in the till in response to the stress created by overriding ice.

6. The end moraines of Jefferson County are composed of thick sequences of washed material which were subsequently overrun by minor readvances of the ice. Tills ranging in thickness from 5 to 20 feet were deposited over the gravels during readvance.

ACKNOWLEDGMENTS

The writer wishes to thank Professors N. P. Lasca and H. J. Pincus for their helpful discussions and suggestions. This work was supported in part by the Wisconsin Geological and Natural History Survey, which provided a truck-mounted auger and drill crew. Typing and figure preparation was provided by Pan American Petroleum Corporation and is gratefully acknowledged.

REFERENCES

Alden, W. C., 1904, The Delevan Lobe of the Lake Michigan glacier of the Wisconsin Stage of glaciation and associated phenomena: U.S. Geol. Survey Prof. Paper No. 34, 106 p.

Boul, S. W., and Fadness, D. M., 1961, New method of impregnating fragile material for thin sectioning: Soil Sci. Soc. Amer. Proc., v. 25, p. 253.

Boulton, G. S., 1968, Flow tills and related deposits on some Vestspitsbergen glaciers: Jour. Glaciology, v. 7, no. 51, p. 391-412.

Cowan, W. R., 1968, Ribbed moraine: till fabric analysis and origin: Can. Jour. Earth Sci., v. 5, p. 1145-59.

Dapples, E. C., and Rominger, J. F., 1945, Orientation analysis of fine grained clastic sediments: Jour. Geology, v. 53, p. 246-61.

Frye, J. C., Willman, H. B., and Black, R. F., 1965, Outline of glacial geology of Illinois and Wisconsin: *in* Wright, H. E., Jr. and Frey, D. G. (eds.), The Quaternary of the United States, Princeton, N. J., Princeton University Press, p. 43-62.

Glen, J. W., Donner, J. J., and West, R. G., 1957, On the mechanism by which stones in till become oriented: Amer. Jour. Sci., v. 255, p. 195-205.

Gravenor, C. P., and Meneley, W. A., 1958, Glacial flutings in central and northern Alberta: Amer. Jour. Sci., v. 256, p. 715-28.

Harrison, P. W., 1957, A clay-till fabric: its character and origin: Jour. Geology, v. 65, no. 3, p. 275-308.

Holmes, C. D., 1941, Till Fabric: Geol. Soc. America Bull., v. 52, p. 1299-1354.

Hoppe, G,. 1952, Hummocky Moraine Regions: Geog. Annaler, v. 34, p. 1-72.

Jeffery, G. B., 1922, The motion of ellipsoidal particles immersed in a viscous fluid: Royal Soc. London Proc. ser. A., v. 102, no. 715, p. 161-79.

Lindsay, J. F., 1968, The Development of clast fabric in mudflows: Jour. Sed. Petrology, v. 38, no. 4, p. 1242-53.

MacClintock, P., and Dreimanis, A., 1964, Reorientation of till fabric by overriding glacier in the St. Lawrence Valley: Amer. Jour. Sci., v. 262, p. 133-42.

Nye, J. F., 1967, Plasticity solution for a glacier snout: Jour. Glaciology, v. 6, p. 695-715.

Ostry, R. C., and Deane, R. E., 1963, Microfabric analyses of till: Geol. Soc. America Bull., v. 74, p. 165-68.

Pincus, H. J., 1956, Some vector and arithmetic operations on two-dimensional orientation variates, with applications to geological data: Jour. Geology, v. 64, no. 6, p. 533-57.

Richter, K., 1932, Die Bewegungsrichtung des Inlandeis rekonstrueirt aus den Kirtzen und Langsachen des Geschiebe: Zeitschr. f. Geschiebeforschurg, v. 8, p. 62-66.

Taylor, G. I., 1923, The motion of ellipsoidal particles in a viscous fluid: Royal Soc. London Proc. ser. A., v. 103, no. 720, p. 58-61.

Wright, H. E., Jr., 1957, Stone orientation in the Wadena drumlin field, Minnesota, Geog. Annaler, v. 39, p. 19-31.

————, 1962, Role of the Wadena Lobe in the Wisconsin Glaciation of Minnesota: Geol. Soc. America Bull., v. 73, p. 73-100.

7

Other Phenomena

Pleistocene Mudflow along the Shelbyville Moraine Front, Macon County, Illinois

Norman C. Hester and Paul B. DuMontelle

ABSTRACT

A fan-shaped geomorphic feature covering an area of approximately five square miles lies immediately west of the Shelbyville Moraine in Macon County, Illinois. From the top downward, the stratigraphy consists of approximately four feet of loess, four to 12 feet of till-like material (diamicton), and sand and fine gravel. The similarity of the sand-silt-clay percentages and clay mineralogy of the diamicton to the till of the Shelbyville Moraine implies that both were formed from the same source material. The smoothness of the surface, low angle of repose, stratigraphy, and spatial relation to the Shelbyville Moraine indicate that the deposit is an ancient mudflow that occurred during the waning stage of Shelbyville ice. Features of this type, if properly interpreted, could contribute to more precise reconstructions of Pleistocene history and provide details useful in resource and engineering investigations.

INTRODUCTION

During exploration for sand and gravel in Macon County, Illinois, several fan-shaped geomorphic features were observed adjacent to and immediately west of the Shelbyville Moraine front. The more conspicuous of these features were found at the following locations:

Southwest of Macon, Macon County, T. 14 N., R. 1 E., secs. 1, 2, 11, and 12

North of Macon and west of Boody, Macon County, T. 15 N., R. 1 E., secs. 3, 4, 5, 9 and 10

West of Warrensburg, Macon County, T. 17 N., R. 1 E., secs. 7, 8, 9, 10, 16, 17, 18, 19, 20, 30, 31

South of Midland City, DeWitt County, T. 19 N., R. 1 E., secs. 5, 6, 7, 8, and T. 20 N., R 1 E., sec. 31

East of Emden, Logan County, T. 21 N., R. 3 W., secs. 3, 4, 5, 6, 7, 8, 9, 10, 11, 14, 15, 16, 17, 18

West of Delavan, Tazewell County, T. 23 N., R. 4 W., secs. 31, 32, and T. 22 N., R. 4 W., secs. 5, 6, 7, 8

North of Strasburg, Shelby County, T. 11 N., R. 5 E., secs. 19, 20, 21, 22, 23, 26, 27, 28, 29

Because of their position with respect to the moraine and their alluvial-fan shape, the features were considered as possible sources of sand and gravel. However, test augering of the one southwest of Macon and west of the Shelbyville Moraine (Fig. 1) showed surficial loess approximately four feet thick underlain by four to 12 feet of till-like material that rests with sharp contact on sand and fine gravel. A desire to determine the origins of both this till-like material and the fan-shaped feature instigated this study.

Previous Studies

Materials similar to those found in the fan-shaped feature have been described from much the same geologic setting in other areas of Wisconsinan and Recent glaciations. In southeastern Massachusetts, Hartshorn (1958, p. 477) described material overlying stratified sand and gravel as "flow-till," deposited as a result of supraglacial debris flowing laterally from the glacier to lower adjacent areas. Boulton (1968, p. 411) pointed out that the Pleistocene tills in lowland glacial deposits in Europe and North America are identical to many of the flowtills on the margins of Recent Vestspitsbergen glaciers and that their origin may be the same. During a study of the development of end moraines in east-central Baffin Island, Goldthwait (1951, p. 575 and 577) observed that mudflows and rock avalanches were common along the steep outer faces of ice-cored moraines.

Distinguishing mudflows from tills is difficult in ancient sediments. Deposits of possible mudflow origin have been recognized and described from several ages of the geologic past by Woodford (1925); Crowell (1957); Tracey, Oriel, and Rubey (1961); and Lindsay (1966). These

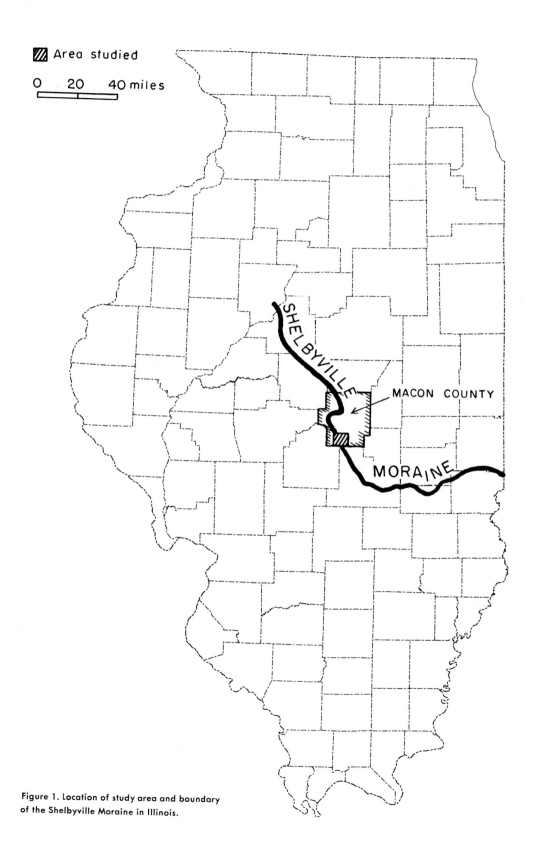

Figure 1. Location of study area and boundary
of the Shelbyville Moraine in Illinois.

deposits, which are made up of nonsorted to poorly sorted, nonlithified, terrigenous sediment, consisting of sand and/or larger particles in a muddy matrix, have been defined as *diamictons* (Flint, Sanders, and Rogers, 1960; Flint, 1960).

Descriptions of Recent mudflows, fanglomerates, or mud avalanches have

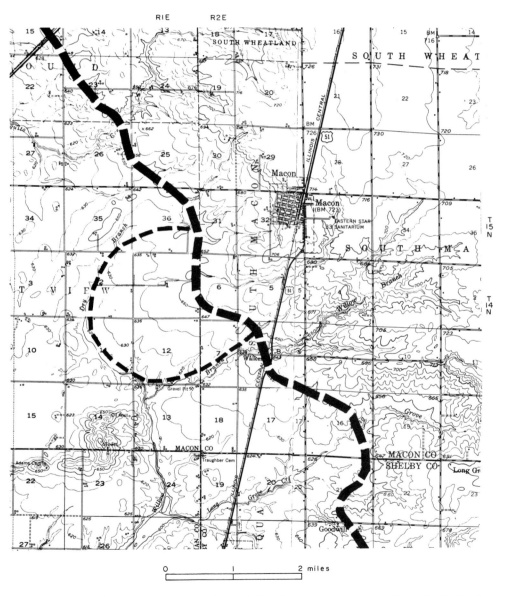

Figure 2. Topographic map of study area. Heavy dashed line represents the margin of the Shelbyville Moraine defined by the topographic break. Thinner dashed line encloses the fan-shaped geomorphic feature studied. (From U.S. Geological Survey 15-minute topographic quadrangle maps, Assumption and Dalton City.)

provided useful information concerning shape, volume, texture, lithology, geographic and geologic settings, and manner and mechanics of deposition (Dickson, 1892; Conway, 1907; Cross, 1909; Howe, 1909; Pack, 1923; Blackwelder, 1928; Crawford and Thackwell, 1931; Scrivenor, 1929; Chawner, 1935; Fryxell and Horberg, 1943; Sharp and Nobles, 1953; Bull, 1962; Landim and Frakes, 1968). Many of these workers have pointed out the similarity of the mudflow material to glacial till.

METHODS OF INVESTIGATION

In this study, topographic maps and aerial photographs were used to determine the topographic form, topographic texture of the land surface, slope, drainage patterns, and spatial relation of the major topographic features. Because outcrops are scarce in the area, a truck-mounted power auger was used to collect samples for determining the subsurface stratigraphy. Both continuous and specific-interval cores were taken with a soil-coring tube three inches in diameter. The hydrometer method was used to find sand-silt-clay percentages of the <2 mm fraction, and clay mineralogy was determined by X-ray diffraction (CuKα radiation) of oriented aggregates of the <2 micron fraction.

GEOMORPHOLOGY

The area of study, approximately five square miles, is in southwestern Macon County, Illinois (Fig. 1), west and southwest of the town of Macon, in secs. 35 and 36, T. 15 N., R. 1 E., secs. 1, 2, 11, and 12, T. 14 N., R. 1 E., secs. 5, 6, 7, and 8, T. 14 N., R. 2 E., and secs. 31, T. 15 N., R. 2 E. (Fig. 2). Physiographically it is within, and extends west from, the boundary of the Shelbyville Moraine, which here trends northwest-southeast. To the south and west of the fan-shaped feature are ridges of deeply leached sand and fine gravel.

A topographic map (Fig. 2) reveals that the moraine has two distinct levels, one at an elevation of approximately 660 feet and the other at approximately 720 feet. West of the lower level, the land surface is exceptionally smooth and dips very gently to the west, with a slope of approximately 20 feet to the mile. The principal drainage system is Willow Branch, which flows generally to the southwest, and its tributary from the north, Dry Branch. Dry Branch is a consequent stream that is arcuate, circumscribing the fan-shaped topographic feature.

Aerial photographs show that the surfaces on the Shelbyville Moraine, its front slope, and the Illinoian till plain to the west each have a distinct topographic texture (Fig. 3). The end moraine has a pockmarked surface, while the till plain has a surface less distinctly pockmarked. The transitional area between the moraine and till plain, however, has a very regular texture, suggesting that the surface, unlike those east and west of it, is exceptionally smooth.

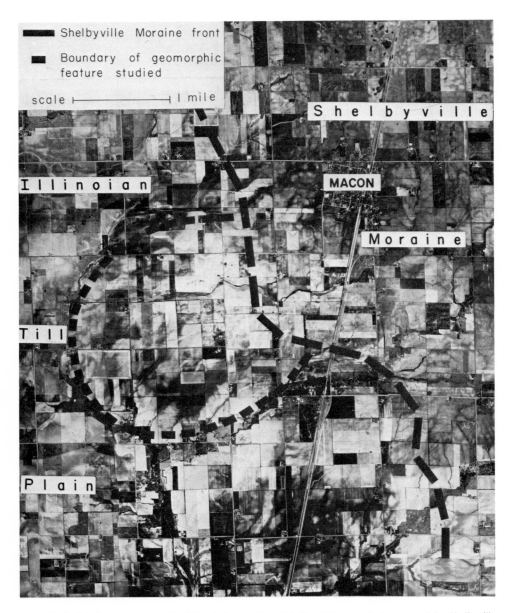

Figure 3. Aerial photograph mosaic of the study area in which the differences in textures of the Shelbyville Moraine, the fan-shaped area, and the Illinoian till plain are evident.

STRATIGRAPHY

Deposits from two Pleistocene stages have been recognized in Macon County. The southwestern margin of the physiographic end moraine called the Shelbyville Moraine marks the surface boundary between deposits of the Illinoian Stage to the west and of the Wisconsinan Stage to the east. A Sangamonian profile 14 feet thick was encountered in boring 15 (Fig. 4) on the elongate ridge west of the moraine, confirming the age of these deposits as Illinoian. Accretion-gley found underlying the thin loess cover in hole 11 also appears to be a result of Sangamonian-Altonian weathering. At the same location, peat found at a depth of 12 feet, just above the

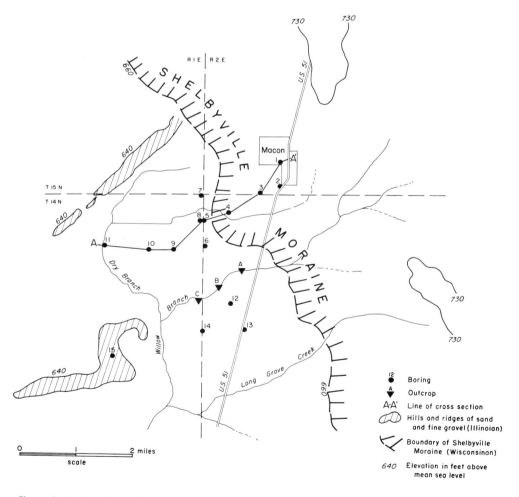

Figure 4. Location of test borings, outcrops, drainage, Wisconsinan Shelbyville Moraine front, and Illinoian sand and gravel ridges. Connected points A-A' appear as cross section in Figure 5.

accretion-gley, has been dated at 25,500±600 radiocarbon years B. P. (ISGS-21), identifying it as Farmdalian. The overlying materials, therefore, belong to the Woodfordian Substage.

Power auger holes 6, 7, 9, and 10, located on the fan-shaped topographic form (Fig. 4) encountered approximately four feet of loess overlying approximately four feet of brown, calcareous, till-like material, which in turn overlay sand and fine gravel (Fig. 5). Approximately 12 feet of till-like material was found in borings 5 and 8. In boring 5, the entire thickness of sand and gravel was penetrated and gray, calcareous till was encountered at a depth of 29 feet. The till-like material is thickest near the moraine and thins rapidly to approximately four feet in the remaining area of the fan (Figs. 4 and 5).

A similar stratigraphic sequence was encountered in holes 12 and 13, but its relation to that north of Willow Branch has not been determined.

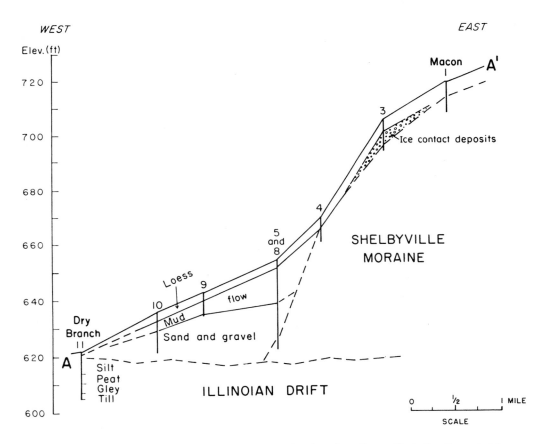

Figure 5. Geologic cross section A-A' (see Figure 4) from Dry Branch to Macon, illustrating the materials encountered by power-auger borings.

Hole 14, augered to a depth of 25 feet, encountered only silts and accretion-gley. Perhaps the till-like materials south and north of Willow Branch were at one time a continuous deposit, but they are now separated by sand and fine gravel along the stream. Till-like material, underlying as much as eight feet of sand and gravel, was observed in outcrops A, B, and C along Willow Branch (Figs. 2 and 4) at elevations ranging from 628 to 618 feet. Because of their elevation and stratigraphic position, this material is considered to be till of the Shelbyville Moraine and is not correlated with the till-like material overlying the sand and gravel of the fan-shaped feature.

CHARACTER OF THE TILL-LIKE DEPOSIT

The till-like material in the fan-shaped feature is a heterogeneous mixture of clay, silt, sand, and pebbles generally less than one inch in diameter. It fits Pettijohn's (1957) description of a paraconglomerate (conglomeratic mudstone), but is best described as a diamicton. The physical properties of the diamicton were compared with those of the oxidized till of the Shelbyville Moraine and were found to be similar. They are both calcareous and grayish orange (7.5 YR 7/3), their sand-silt-clay percentages (averaging 33-38-29) being similar, and their clay mineral ratios (3 parts chlorite to 2 parts illite) being the same. Other workers have also found that tills and mudflow deposits have almost identical physical characteristics. Hartshorn (1958, p. 477) said that the Pleistocene flowtills do not differ from ground moraine in physical description nor in grain-size distribution. In their study of the mudflow at Wrightwood, California, Sharp and Nobles (1953, p. 554) found that the mudflow material looked much like that composing some glacial tills. Blackwelder (1928, p. 465) described as mudflows fan deposits of some unsorted, unstratified, earthy material that resembled glacial till. Scrivenor (1929, p. 433) stated that mudstream deposits resemble glacial deposits.

ORIGIN

Two hypotheses for the origin of the diamicton were considered. It could have been deposited by the readvance of glacial ice or by a mudflow from the ablating ice.

Formation by Readvance of Ice

If the diamicton were deposited as a result of a local ice readvance (Fig. 6), a very irregular ground surface would have been produced by the ablating ice and the topography should appear much the same as that on the adjacent terminal moraine. However, the surface of the fan-shaped form, as compared with that of the terminal moraine, is exceptionally smooth (Fig. 3). The loess cannot be considered as a masking agent for the supposed irregularities on the surface of the underlying till-like material, because in 13 holes augered on and near the fan-shaped topographic form, it was found to be uniformly only three to five feet thick. This strongly suggests that the smoothness of the ground surface reflects a smooth surface on the material underlying the loess.

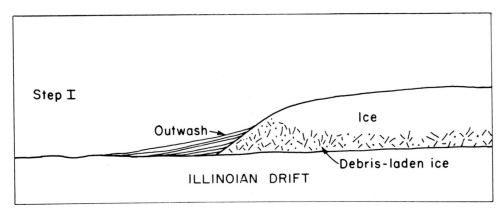

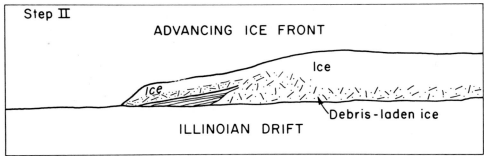

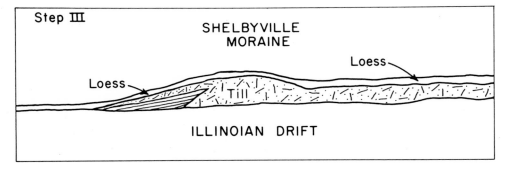

Figure 6. Schematic diagram of origin of diamicton during ice readvance. Step I: Ablating ice forming terminal moraine and outwash plain. Step II: Readvance of debris-laden ice over outwash plain. Step III: Complete ablation of ice leaving terminal moraine, till over outwash, and loess cover.

The thickness of till in terminal moraines characteristically varies a great deal from place to place. In contrast, the till-like material underlying the loess of the fan consistently ranges from 3.9 to 4.3 feet thick over most of the area. Such uniformity makes it most unlikely that the diamicton was deposited as a result of the readvance of ice.

A variety of materials is commonly encountered in augering a terminal moraine, suggesting that several mechanisms are responsible for the emplacement of the drift. Norris and White (1961) showed that till in end moraines is quite variable in texture, composition, permeability, and engineering properties. In a report on end moraines on east-central Baffin Island, Goldthwait (1951, p. 577) suggested that nearly all end moraines are composed of superglacial (ablation) load; thus the till is sandier because some of the finer fraction has been removed by meltwater.

In hole 3 (Fig. 4) on the Shelbyville Moraine, four feet of loess was encountered above 7 feet of clayey, pebbly silt with sand and pebbly sand layers, overlying till. In hole 1, three feet of pebbly, brown silt was found underlying six feet of loess and overlying till. However, at none of the 12 holes augered in the fan-shaped form were sediments of the above description encountered. Instead, the stratigraphy was strikingly consistent throughout — the loess directly overlying till-like material, which in turn was in sharp contact with the underlying sand and fine gravel. Moreover, on the Shelbyville Moraine, a thin, sandy, gravelly layer is commonly encountered between the loess and the till. This zone is considered to have been formed by surface run-off of water during ablation. No sign of this deposit was encountered, at the same stratigraphic position, above the till-like material in any of the power-augered holes. The consistent stratigraphy and the absence of a sandy, pebbly zone between the loess and the underlying till-like material strongly suggest that the mechanism responsible for deposition of the diamicton was not the same as that for the till of the Shelbyville Moraine.

Formation by Mudflow

Several lines of evidence were examined in considering the diamicton as a mudflow (Fig. 7). The surface expression of the deposit has the shape, in plan view, of an alluvial fan or mudflow. According to Rickmers (1913), when a mudflow is allowed to come to rest on a gentle incline, it will form a snout or tongue if it is not too liquid. The largest mudflow described by Fryxell and Horberg (1943, p. 465) was also a fan-shaped feature.

A uniform thickness is characteristic of mudflows. With the exception of holes 5 and 8 (Figs. 4 and 5), the thickness most commonly encountered for the diamicton was four feet. Similar thicknesses have been noted by others. In discussing Recent mudflows in semiarid areas, Pack (1923, p. 354) described a flow that covered a field to a depth of three to four feet, and terminated with very abrupt margins. Crawford and Thackwell (1931, p. 104), in describing mudflows north of Salt Lake City, reported flows that measured 200 to 1000 feet across and three to 12 feet thick. Flowtill thicknesses similar to mudflow thicknesses encountered in our study have been reported from areas of Wisconsinan and Recent glaciation. Hartshorn (1958,

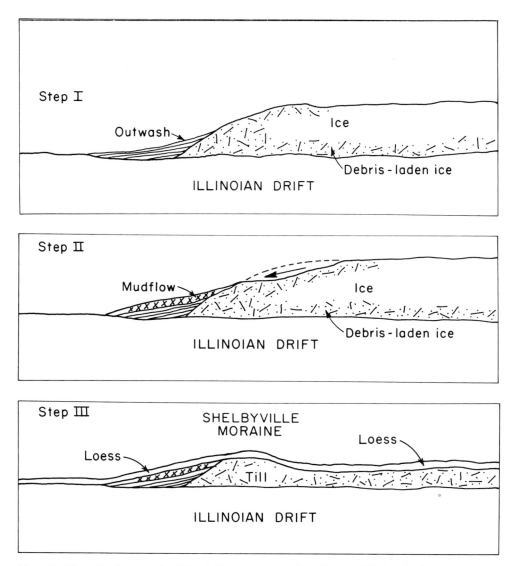

Figure 7. Schematic diagram of origin of diamicton as mudflow. Step I: Ablating ice forming terminal moraine and outwash plain with ablation material accumulating on ice surface. Step II: Unstable, water-saturated ablation material flows from surface of the ice out over the outwash fan. Step III: Complete melting of ice leaves terminal moraine with mudflow over outwash fan, subsequently covered by loess.

p. 478) reported a thickness of four to six feet for a Pleistocene flowtill over sand and gravel in southeastern Massachusetts. In the immediate proglacial zone of a Recent glacier, Boulton (1968, p. 399) found flowtill up to 0.75 m (29.5 inches) thick overlying gravel.

A gradient of less than 0.5 of a degree for the diamicton is not abnormally low for some mudflows. According to Boulton (1968, p. 399), some Recent glacial mudflows move very rapidly down slopes of 1 to 2 degrees. The gradient near the outer extremity of the Recent mudflow at Wrightwood, California, was less than 1 degree (Sharp and Nobles, 1953, p. 558).

Mudflows have been reported that are comparable in size to the one described here, which covers an area of approximately five square miles, extends out from the Shelbyville Moraine approximately two miles, and includes an estimated 23×10^6 cubic yards of sediment. Scrivenor (1929, p. 443) found that some Recent individual mudflows spread out laterally over more than four kilometers (2.5 miles), and Chawner (1935, p. 260) reported that approximately 7×10^5 cubic yards of sediment were deposited by a mudflow at Montrose, California. This represents approximately half a square mile covered by a three-foot thickness of mud.

No Recent mudflows associated with glaciers have been reported that are as large as the Macon County mudflow. In a study of an area of Recent glaciation on Stubendorfbreen, Boulton (1968, p. 410) recorded 0.5 kilometer (.31 mile) as the maximum distance traveled (in the direction of the flow) for flowtill overlying other sediments.

MANNER OF FAN DEVELOPMENT

The conditions that favor the development of mudflows were outlined by Blackwelder (1928, p. 478) as follows:

1. unconsolidated material that becomes slippery when wet,

2. slopes steep enough to induce flow in such viscous material,

3. abundant water, and

4. insufficient protection of the ground by forest.

All of the above conditions seem to be satisfied in the study area.

1. The unconsolidated till-like material has sand-silt-clay percentages of 33-38-29. Tills of the Shelbyville Moraine with similar sand-silt-clay percentages have the following Atterberg limits:

Liquid limit	— 20 to 25
Plastic limit	— 10 to 15
Plasticity index	— 8 to 10

The numerical value of the above liquid limit (20 to 25) is the water content, expressed as a percentage of the dry weight of the soil solids, required to cause a material to flow. Also important, according to Sharpe (1938, p. 56), was the presence of sufficient clay or silt for lubrication.

2. Slopes steep enough to induce the water-saturated ablation material to flow were present on the ice front of the glacier that deposited the Shelbyville Moraine. Orvig's (1953, p. 245, Fig. 2, traverse A-B) slope measurements on the present glacier at Balfur Island show a slope of approximately 160 feet per mile. Slope reconstruction of the Pleistocene ice cap by Harrison (1958, p. 86, Fig. 7) was conservatively estimated at approximately 120 feet per mile on the ice front. These slope gradients are considerably steeper than those observed on the Shelbyville Moraine front in Macon County, which generally average 40 feet per mile, with occasional slopes as steep as 80 feet per mile.

3. Boulton (1968, p. 398) reported that ablation releases considerable amounts of water and the till can become water-logged if drainage is impeded, as happens in hollows, on flats, and on low-angle slopes. The presence of abundant water in such a situation is attested by Richardson (1968), who stated that the principal source of the glacial outburst floods in the Pacific Northwest was the large volume of water that is stored at times within and beneath a glacier.

4. No evidence for a forest or vegetated cover on the terminal moraine at the time the till-like material was deposited has been found.

The manner of emplacement of the diamicton is interpreted to be similar to that described by Hartshorn (1958, p. 481). He believed flowtills were formed when ablation material, which accumulated on the ice as superglacial debris, became so water-saturated and unstable that the debris moved down the ice slope as a mudflow. The spatial relation of this fan-shaped topographic feature to the Shelbyville Moraine and the thinning of the diamicton from east to west indicate that the material was derived from the east. The similarity of the sand-silt-clay percentages and clay mineralogy of the mudflow material to the till of the Shelbyville Moraine implies that both were formed of material from the same source.

There is no topographic expression which suggests that the mudflow deposit originated on the present Shelbyville Moraine. It follows, therefore, that ice must have stood in the immediate area at the time of the mudflow implacement. The topographic low (Figs. 2 and 4) southeast of Macon, aligned with the headwater drainage of Willow Branch creek on the present Shelbyville Moraine, is probably consequent to a topographic sag that existed on the ice. This sag probably concentrated the waters of a drainage system developed on the ice, which debouched onto the Illinoian till plain and deposited the sand and fine gravel encountered in the borings on the fan. During the waning stage of the Shelbyville ice, water-saturated ablation material could have been accumulated in the drainage system by sliding and flowing, thereby forming dams. With continued impoundment of water, the point would finally have been reached where breaching of one or more of these dams took place, triggering the break-up of the system of dams along the main drainage way. The event was probably cataclysmic, much like that in the Pacific Northwest (Richardson, 1968), and the entire mud-flow could have been built in a matter of hours, possibly minutes.

ACKNOWLEDGMENTS

Thanks are extended to the Illinois Division of Highways (District 5) for drilling two deep holes in the Shelbyville Moraine front. The writers gratefully acknowledge the discussion and constructive criticism provided by J. E. Lamar of the Illinois State Geological Survey.

REFERENCES

Blackwelder, Eliot, 1928, Mudflow as a geologic agent in semi-arid mountains: Geol. Soc. America Bull., v. 39, p. 465-80.

Boulton, G. S., 1968, Flow tills and related deposits on some Vestspitsbergen glaciers: Jour. Glaciology, v. 7, no. 51, p. 391-412.

Bull, W. B., 1962, Relation of textural (CM) patterns to depositional environment of alluvial-fan deposits: Jour. Sed. Petrology, v. 32, p. 211-16.

Chawner, W. D., 1935, Alluvial fan flooding, the Montrose, California flood of 1934: Geographic Rev., v. 25, p. 255-63.

Conway, W. M., 1907, Comment on "The Fan Mountains in the Duab of Turkestan by W. R. Rickmers,": Geog. Jour., v. 30, p. 501-2.

Crawford, A. L., and Thackwell, F. E., 1931, Some aspects of the mudflows north of Salt Lake City: Utah Acad. Sci. Proc., v. 88, p. 97-105.

Cross, C. W., 1909, The Slumgullion mudflow (abs.): Science, new ser., v. 30, p. 126-27.

Crowell, J. C., 1957, Origin of pebbly mudstones: Geol. Soc. America Bull., v. 68, p. 993-1010.

Dickson, E., 1892, Mud avalanches: Liverpool Geol. Soc. Proc., v. 6, pt. 4, p. 387-95.

Flint, R. F., 1960, Diamictite, a substitute term for symmictite: Geol. Soc. America Bull., v. 71, p. 1809.

Flint, R. F., Sanders, J. E., Rodgers, John, 1960, Symmictite: A name for nonsorted terrigenous sedimentary rocks that contain a wide range of particle sizes: Geol. Soc. America Bull., v. 71, p. 507-9.

Fryxell, F. M., and Horberg, C. L., 1943, Alpine mud flows in Grand Teton National Park, Wyoming: Geol. Soc. America Bull., v. 54, p. 457-72.

Goldthwait, R. P., 1951, Development of end moraines in east-central Baffin Island: Jour. Geology, v. 59, p. 567-77.

Harrison, W., 1958, Marginal zones of vanished glaciers reconstructed from the preconsolidation-pressure values of overridden silts: Jour. Geology, v. 66, p. 72-95.

Hartshorn, J. H., 1958, Flowtill in southeast Massachusetts: Geol. Soc. America Bull., v. 69, p. 477-82.

Howe, Ernest, 1909, Landslides in the San Juan Mountains, Colorado: Including a consideration of their causes and their classification: U.S. Geol. Survey Prof. Paper 67, 58 p.

Landim, P. M. B., and Frakes, L. A., 1968, Distinction between tills and other diamictons based on textural characteristics: Jour. Sed. Petrology, v. 38, p. 1213-23.

Lindsay, J. F., 1966, Carboniferous subaqueous mass movement in the Manning-Macleay Basin, Kempsey, New South Wales: Jour. Sed. Petrology, v. 36, p. 719-32.

Norris, S. E., and White, G. W., 1961, Hydrologic significance of buried valleys in glacial drift: Art. 17 *in* U. S. Geol. Survey Prof. Paper 424-B, p. B34-B35.

Orvig, S., 1953, The glaciological studies of the Baffin Island Expedition, 1950, Pt. V. On the variation of the shear stress on the bed of an icecap: Jour. Glaciology, v. 2, p. 242-47.

Pack, F. J., 1923, The torrential potential of desert waters (Utah): Pan-American Geol., v. 4, (40), p. 349-56.

Pettijohn, F. J., 1957, Sedimentary rocks: 2nd ed., New York, Harper and Bros., 718 p.

Richardson, D., 1968, Glacier outburst floods in the Pacific northwest: U.S. Geol. Survey Prof. Paper 600-D, p. D79-86.

Rickmers, W. R., 1913, The Duab of Turkestan: Cambridge University Press (England), 563 p.

Scrivenor, J. B., 1929, The mudstreams ("Lahar") of Gunong Keloet in Java: Geol. Mag., v. 66, p. 433-34.

Sharp, R. P., and Nobles, L. H., 1953, Mudflow of 1941 at Wrightwood, southern California: Geol. Soc. America Bull., v. 64, p. 547-60.

Sharpe, C. F. S., 1938, Landslides and related phenomena: New York, Columbia Univ. Press, 137 p.

Tracey, J. I., Jr., Oriel, S. S., and Rubey, W. W., 1961, Diamictite facies of the Wasatch Formation in the Fossil Basin, southwestern Wyoming: U.S. Geol. Survey Prof. Paper 424-B, p. B149-B150.

Woodford, A. O., 1925, The San Onofre Breccia; its nature and origin: Univ. Calif. Publ., Dept. Geol. Sci. Bull., v. 15, p. 159-280.

Till / a Symposium

Unarmored Till Balls
in Unusual Abundance
near Minot, North Dakota

Wayne A. Pettyjohn and Richard W. Lemke

ABSTRACT

Unarmored till balls are exposed in unusual abundance in a gravel pit near Minot, North Dakota, occurring in crudely stratified layers in a channel fill deposited in late Wisconsin time. They lack distinctive internal structure and appear to have been formed by abrasion and rounding of chunks of till (perhaps in a frozen state) that fell into a meltwater stream. It is believed that the till balls were transported by the stream only a short distance from their source area.

These till balls are compared to clay, mud, and till balls — reported from various other parts of the world — which have been found in alluvial deposits, on the shores of lakes and oceans, in stratified glacial deposits, and much more rarely in bedrock, and which range in size from 20 inches to less than one-half inch, are armored or unarmored, and which may or may not have a distinctive internal structure depending on whether they are formed by abrasion or by accretion.

INTRODUCTION

Unarmored till balls in glacial outwash near Minot, North Dakota, were exposed in great abundance when examined in May, 1967. The purpose of this paper is to compare these till balls with clay, mud, and till balls found elsewhere and to discuss their origin and the environmental conditions that permitted their development. To achieve this purpose and also to update a review of the literature on the subject, a fairly extensive literature summary is included.

The terms clay balls and mud balls are commonly used interchangeably in the literature. On the other hand, it might be more appropriate to consider mud balls as having been formed, in some manner, by accretion of wet or damp clay. Clay balls, as considered herein, are mainly those formed by abrasion of chunks of dry or nearly dry clay or shale that commonly sloughs from banks or cliffs. Till balls also are formed by abrasion. They differ from clay balls only in that they consist entirely of till.

Summary of Previous Investigations

More or less spherical masses, variously described as clay, mud, or till balls, have been reported from many parts of the world. They have been found in alluvial deposits mainly along ephemeral streams, on the shores of lakes and oceans beneath shoreline cliffs, in stratified glacial deposits, and much more rarely in bedrock formations. They are said to range in diameter from 20 inches to less than one-half inch. Some have surface coatings of pebbles or other particles and are classed as "armored"; others that do not have the surface coating are classed as "unarmored." Most appear to have been formed by stream abrasion and rounding of larger, more rectangular chunks of material; some, however, apparently are formed by accretion of material as they are rolled along the floor of a stream. Those formed entirely by abrasion lack an internal structure related to their formation, whereas those apparently formed by accretion may have a concentric structure.

As early as 1875, armored clay balls, found along beaches, were described by Jones and King (see Bell, 1940, p. 3). They attributed the forming of the balls to waves rolling and abrading chunks of clay that had fallen from nearby cliffs. As the chunks of clay were rounded by rolling and softening by the water, pebbles and other particles adhered to the balls so that they became "armored."

Gardner (1908, p. 455) attributed the formation of mud balls found in ephemeral streams in New Mexico to the process of accretion — either by accretion of clay around a nucleus or by accretion without a nucleus. He suggested that soft clay particles are pressed together much as are finely disseminated particles of butter during churning; the resulting soft nodule continues to grow in a manner similar to that of forming a large snowball by the process of rolling and rotating a small ball. Growth continues until the stream is no longer able to move the ball.

For nearly 20 years, Gardner's hypothesis for the formation of mud balls went unchallenged. Then Haas (1927), as a result of studies in the Mesa Verde region of southwestern Colorado, and also along a stretch of Lake Michigan beach, expressed doubt that large balls could be formed in that manner. His studies in the Mesa Verde region showed that, where floodwaters flow through deep arroyos, large blocks of shale slump from the undercut arroyo walls into the stream. According to Haas, the blocks are abraded and become progressively more spherical as they are rolled along the stream channel. The surfaces of the balls are soft sticky masses to which adhere particles of shale and other material, but the major part of the interior of the larger balls is dry. In his studies along Lake Michigan north of Evanston, Illinois, Haas further described irregularly rounded balls of clay that were formed of material slumped from nearby high cliffs, and rolled back and forth by waves until rounded into ball-shaped masses. The balls thus formed ranged from one-half inch to 14 inches in diameter. Of the scores of balls Haas broke apart in his studies, he reports that none showed a concentric internal structure.

That mud balls can form by accretion, although probably not quite in the manner suggested by Gardner (1908, p. 455), and that they can have a concentric structure have been demonstrated by several workers (Twenhofel, 1932, p. 694; Cartwright, 1928, p. 241; and Nordin and Curtis, 1962). Photographs by Nordin and Curtis especially are convincing as to the concentric nature of at least some of the mud balls.

The mud balls described by Nordin and Curtis (1962) were formed in a reach of the Rio Pueblo in New Mexico. Here the erosional scarp of an arroyo had moved upstream, cutting through a sand bed the upper 6 inches of which was impregnated with clay. Lumps of this clay formed mud balls as large as 16 inches in diameter, which were both armored and unarmored. Those that had moved only over the clay-impregnated bed were not armored and contained negligible amounts of sand and pebbles. On the other hand, those that subsequently were carried over exposures of the sand bed had become armored with sand and pebbles — the armor constituting 15 to 75 percent of the weight of the balls. Some of the balls had no perceivable internal structure; some had a concentric structure (a manifestation of the accretion process) and some had a laminated structure, which constituted the original bedding. Presence of these laminations indicated that the central part of the ball had remained dry during formation and transportation of the ball, for otherwise this structure would have been distorted or destroyed by plastic deformation. Nordin and Curtis also were able to demonstrate, from comparison of size curves of bank material and of material constituting the balls, that the mud balls had not been derived primarily from bank sloughing or from headcutting through the clay, but from accretion on clay masses rolling along on the clay layer. Furthermore, they showed that the mud balls selectively accreted the coarse part of the bed material "probably because angular coarse particles penetrate farther into the ball and adhere to the clay more firmly than do the smaller, well-rounded sand grains" (Nordin and Curtis, 1962, p. B38).

Comprehensive field and laboratory studies were made by Bell (1940) concerning the origin, properties, and role in sedimentation of armored clay balls. Following a major flood in 1938, tens of thousands of clay balls were available for study along the steep-walled channel of the Las Posas barranca between El Rio and Camarillo, California. The balls had formed as a result of clay from a landslide mass, several acres in size, sloughing off into the stream channel. Bell noted that the irregularly shaped clay balls near the slide area were unarmored. However, within a distance downstream of half a mile, they were lightly armored with sand, gravel, shells, and even a bottle cap and a nail, and they progressively increased in armor and sphericity with increasing distance (of 2¾ miles). The largest ball measured was 20 inches in diameter, but most were between 2 and 12 inches.

Some of the more important conclusions reached by Bell from his studies are paraphrased below:

1. Floodwater conditions that lead to extensive undercutting of banks or to rapid headward erosion in areas where clay is present are ideal for producing clay balls.

2. The maximum-sized balls that can be formed is a function not so much of transportation (the stream energy necessary to overturn it) as of the structural strength of the ball. Cohesion within the ball is called upon to resist the forces of impact that tend to destroy it.

3. The rolling process is essential to forming clay balls, and the balls formed are large in comparison with the size of the other particles in the stream.

4. The sphericity of a ball is a measure of the distance it has traveled.

5. Between 17.1 and 44.0 percent of the total weight of a ball can be attributed to armor. The general tendency is for the heavier balls to have a higher percentage of armor than the lighter ones. Once the ball is covered with armor, further growth is effectively stopped.

6. Perhaps most of the balls, whether armored or not, have dry interiors at the time they are forming. This is due to the rapid rate of formation of the ball (commonly less than 15 minutes), coupled with its general impermeability. Thus, the balls retain some of their original buoyancy. The resulting buoyancy combines with the high sphericity of the balls and their ability to acquire armor to make clay balls powerful and efficient vehicles for the transportation of erosional debris as bed load. Not only can larger loads be moved in this manner, but they can be moved farther.

7. Further careful study of clay balls composed of a considerable number of different clays under a variety of conditions may furnish geologists with a means of estimating maximum velocities of ancient streams, the nature of their bed material, approximate distances to source material, and the kind of water body that formed them.

Armored clay balls, similar in some respects to those studied by Bell, were reported from Trinidad, West Indies, and from Venezuela, and Ecuador by Kugler and Saunders (1959). They stated that the balls can be found in Trinidad along streambeds, but are much more abundant on some of the present-day beaches below unstable cliffs of marl and clay. Sliding of this material during the wet season produces a large volume of debris subject to wave action. The debris is rolled across the beach to form clay and marl balls coated with limonitic sand, which are subsequently further armored with bits of less-disintegrated clay and marl.

Fossil-armored mud balls, although fairly rare, do exist in sedimentary rocks. Kugler and Saunders (1959, p. 564) described one such mud ball from the Cruse clay, of Oligocene and Miocene age, in Trinidad. This most interesting find was ellipsoidal in shape, was approximately six inches long, and had a core consisting of a very hard mudstone. The pebbles forming its surface armor included chert, sandstone, and quartz. Kugler and Saunders attribute the presence of the mud ball in the Cruse clay to slumping of debris from an ancient shoreline and forming in a manner similar to that of balls that are forming today in the area.

Till balls *per se* have not been described to any great extent in the literature. This is somewhat surprising in view of the fact that such balls are not uncommon — being formed both at the present time and during Pleistocene time. Grabau (1932, p. 711) probably was one of the first to describe armored till balls. He observed fragments of "glacial clay" (till?) being broken loose from sea cliffs in Scotland, rolled about by the waves, and fashioned into "pebbles and boulders of elongate but well-rounded outline." Where these rounded masses were rolled over a pebble beach, the pebbles were pressed into the mass, which then assumed the appearance of a worn conglomerate. Grabau also noted that there were reports of similar boulders, armored with shell fragments, being found in Spitsbergen.

Leney and Leney (1957) described armored till balls from Pleistocene outwash in southeastern Michigan. The balls, which range in diameter from one-half to six inches (average two inches), were exposed in a sand pit on the edge of the Defiance Moraine near Ann Arbor. The interiors of the balls were identical to the till that caps the moraine (mostly clay and silt-size particles with some sand and a few pebbles). The armor was described as thin, generally a single layer of sand and pebbles pressed into the till matrix. Where a number of balls were nested together, they commonly were pressed together, as shown by the preserved indentations in the surfaces of the balls. This indicates that, when they were made, the outer parts of the balls were soft and fairly plastic. Leney and Leney are of the opinion that the till balls were formed in a manner similar to that forming modern mud balls, but at a time after retreat of the ice and before encroachment of vegetation. Thus, torrential rains and meltwaters easily eroded the glacial till and reworked the underlying gravel.

In 1952 the junior author of this paper had the opportunity to examine newly formed till balls in the Portage quadrangle east of Great Falls, Montana. During a torrential rainstorm one night, a flooding intermittent stream in a steep-walled tributary of the Missouri River undercut a 50-foot-high vertical cliff of very compact till. The till, which consisted chiefly of well-compacted silt and clay-size particles, exhibited prismatic jointing, as well as indistinct horizontal parting planes. Thus, when the bank was undercut and sloughing occurred, more or less cubic-to-rectangular chunks of till dropped into the stream. As these chunks were rolled along the stream bed, their corners were abraded and nearly spherical till balls were formed. The balls were deposited as part of a newly formed sand and gravel bar only a few hundred feet from their point of origin. Most were three to four inches in diameter and were unarmored. When examined the following morning, they were found to be hard and completely dry except for a very thin damp outer rind.

Similar clay and till balls, both armored and unarmored, are forming along streams in parts of North and South Dakota, as well as in Ohio. The major prerequisites for their formation are an oversteepened bank, a source of till or clay, and generally a stream subject to flash flooding. In most cases, modern clay or till balls rapidly disintegrate upon drying, unless they are buried.

Till balls also have been studied in deposits of Wisconsin age. Unarmored till balls were described by Lemke (1960, p. 57, 73, 77, and 87) from various parts of the Souris River area in north-central North Dakota. They were found in what he described as "overridden ice-contact deposits" consisting chiefly of poorly sorted sand and gravel, in poorly sorted parts of "river terrace deposits" along the Souris and Des Lacs Rivers, in "kame and esker deposits," and in "ice-marginal outwash-channel deposits" where, locally, till balls constitute as much as 25 percent of the deposit. The till balls, all of which were unarmored, ranged from one to eight inches in diameter, but more commonly were two to four inches across. Most were well-rounded and consisted of till that appeared to be identical to the till constituting the ground moraine of the area. In explaining the origin of the till balls in the river-terrace deposits, Lemke concluded that the balls originated from "chunks of till that were dropped into a glacial meltwater stream by the undercutting of till walls of the stream channel and became rounded during transportation." He further stated, "It does not seem possible they could have survived transportation more than a short distance" (Lemke, 1960, p. 73).

UNARMORED TILL BALLS IN THE VICINITY OF MINOT

Till balls were exposed in unusual abundance, in May 1967, in a gravel pit about 1 mile east of Minot, in north-central North Dakota. Although Lemke (1960) had found till balls previously in glacial deposits in this general area, nowhere were they exposed, in 1967, as abundantly as they were in this gravel pit. This, combined with the rather obvious mode of formation of the balls, their unarmored nature, and their depositional characteristics, prompted us to briefly examine and describe the deposits.

The pit exposing the till balls was excavated into the top of the northeast wall of the Souris River valley and was bounded on three sides by gravel representing glacial diversion channel deposits (see map, Lemke, 1960). The other side (southwest) of the pit was bounded by the floodplain of the Souris River. Figure 1 is a sketch showing the stratigraphic relations in the pit and the locations shown in the photographs (Figs. 2, 3, 4, and 5).

The north and northeast faces of the pit revealed till overlying a buried glacial outwash (Fig. 2). The east face of the pit exposed two phases of channel cutting and filling: (1) an earlier channel fill cut into the buried glacial outwash (Fig. 3), and a slightly later channel fill truncating the earlier channel fill and also cut into the buried glacial outwash (Figs. 4 and 5). This later channel deposit was about 100 feet wide and as much as 15 feet thick.

The till overlying the buried outwash was as much as 15 feet thick in the north face of the pit and represented ground moraine deposited by the Souris River ice lobe during late Wisconsinan time (Lemke, 1960; Lemke,

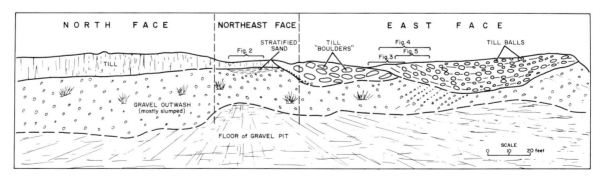

Figure 1. Sketch showing stratigraphic relations in the north, northeast, and east faces of gravel pit one mile east of Minot. Approximate locations shown in Figures 2, 3, 4, and 5 are indicated by brackets.

Figure 2. View of the northeast face of gravel pit showing about three feet of till overlying glacial outwash.

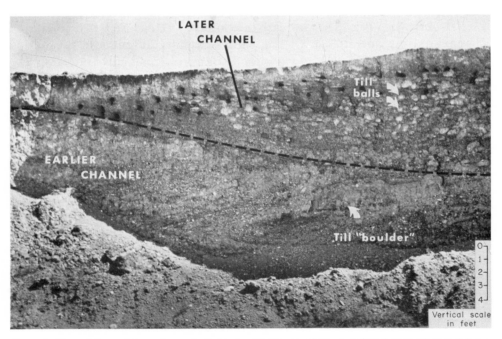

Figure 3. View of the east face of gravel pit showing till "boulders" in an earlier channel fill underlying a later channel fill containing till balls. Note that the "boulders" in the channel fill have been deposited in a crudely stratified fashion.

Figure 4. View of the east face of gravel pit (slightly south of Figure 3) showing till balls in a later channel fill. Note that the till balls have been deposited in a crudely stratified fashion.

Figure 5. Closer view of till balls shown in Figure 4.

Laird, Tipton, and Lindvall, 1965). According to size analyses of material collected only a few miles away from the pit (Lemke, 1960, p. 48), the till consisted of about 25 percent clay-size particles, 40 percent silt, 30 percent sand, and 5 percent gravel; a few cobbles and boulders were intermixed. In the exposure, the till was oxidized to a light buff color, exhibited a vertical prismatic structure, and stood nearly vertically. It was nearly impermeable and very hard when dry.

On the basis of studies done in a nearby area to the southwest (Pettyjohn, 1967), the buried outwash deposits may be considerably older than the overlying till, and may represent deposits of an earlier stage of Wisconsinan time. The outwash was obscured in the pit, except for the upper one to three feet, by talus. It appeared to consist, however, of moderately to well-sorted and well-bedded sand and gravel. No till balls were seen in these deposits.

The two channel-fill deposits consisted of sand and gravel, probably mostly reworked from the buried outwash deposits, and contained till "boulders" and till balls in crudely stratified layers that apparently were derived from a nearby till source.

The oldest of these channel deposits contained till "boulders," which were larger, much more irregularly shaped, and more angular than the till balls in the younger channel deposit. A few were five to seven inches in longest dimension, but most were more than one foot long. Many were roughly rectangular to tabular in outline, and a few were ovoid to subrounded; none were armored and all were identical in appearance to the till overlying the buried outwash. They commonly were concentrated along crudely stratified depositional planes, where they were locally so abundant as to actually touch each other. However, those touching were not indented, and therefore, suggest that they were not in a plastic state at the time of deposition.

The later channel deposits were essentially the same as the earlier channel deposits except that the incorporated till masses were somewhat smaller and were sufficiently spherical to be called till balls. It appeared that the till balls either had been reworked from the "boulders" of the earlier channel fill or that they had been transported from a slightly greater distance upstream. They rarely were more than a foot in diameter and more commonly were four to eight inches across. Most ranged in shape from subrounded to ovoid, a few were nearly spherical, and some were roughly rectangular. The till balls were arranged in crudely stratified layers, similar to the "boulders" in the earlier channel fill. Along these layers they constituted as much as 80 to 90 percent of that part of the deposit, and as much as 50 percent or more of the entire channel deposit. They were also identical in appearance to the till overlying the outwash; they were unarmored, and lacked a concentric structure.

The till "boulders" and the till balls apparently were formed in a manner similar to that of the unarmored till balls described previously from other localities. Presumably, a relatively steep till face was undercut by stream action, causing chunks of till to slough off into a glacial meltwater stream. On the basis of the reconstructed history of the area, it is likely that the meltwater stream was flowing between a mass of till (represented by the till overlying the outwash in the pit) and a tongue of ice that filled the valley of the Souris River. The chunks of till first were somewhat rounded into the "boulders" deposited in the earlier channel fill deposits. Later, the "boulders" were abraded into smaller and more spherical balls, either because the channel-fill deposits were reworked by glacial meltwaters slightly upstream, or because the till constituting the "boulders" originated farther upstream and was subjected to a greater distance of transport. In any event, the till balls probably were transported only a very short distance, perhaps a few hundred feet or less. This is indicated because, if they had been transported a considerable distance, they probably would have become sufficiently wetted and plastic on their outer surfaces to have become armored. However, even with an implied short distance of transport, it still seems likely that the till balls would have disintegrated if the till from which they were derived had not been in a dry or frozen state. Inasmuch as the balls were formed in Pleistocene time and presumably at the edge of the ice sheet, the idea of their being frozen is the more likely of the two alternatives. With an indicated short time of transport, they well may have remained frozen even on their outer surfaces, thus preventing the development of armor.

There is no doubt but that there are many other localities of unarmored and armored till balls throughout many parts of the glaciated region of North America and elsewhere. It is hoped that this paper, in some small measure, will encourage other investigators to describe such interesting and unusual localities.

ACKNOWLEDGMENTS

Publication of this paper has been approved by the Director, U.S. Geological Survey.

REFERENCES

Bell, H. S., 1940, Armored mud balls—their origin, properties, and role in sedimentation: Jour. Geology, v. 68, no. 1, p. 1-31.

Cartwright, L. D., Jr., 1928, Sedimentation of the Pico formation in the Ventura quadrangle, California: Am. Assoc. Petroleum Geologists Bull., v. 12, no. 3, p. 235-369.

Gardner, J. H., 1908, The physical origin of certain concretions: Jour. Geology, v. 16, p. 452-58.

Grabau, A. W., 1932, Principles of stratigraphy (3d ed.): New York, A. G. Seiler, 1185 p.

Haas, W. H., 1927, Formation of clay balls: Jour. Geology, v. 35, no. 2, p. 150-57.

Kugler, H. G., and Saunders, J. B., 1959, Occurrence of armored mud balls in Trinidad, West Indies: Jour. Geology, v. 67, no. 5, p. 563-65.

Lemke, R. W., 1960, Geology of the Souris River area, North Dakota: U.S. Geol. Survey Prof. Paper 325, 138 p.

Lemke, R. W., Laird, W. M., Tipton, M. J., and Lindvall, R. M., 1965, Quaternary geology of northern Great Plains, in The Quaternary of the United States: Princeton, N. J., Princeton Univ. Press, p. 15-27.

Leney, G. W., and Leney, A. T., 1957, Armored till balls in the Pleistocene outwash of southeastern Michigan: Jour. Geology, v. 65, no. 1, p. 105-6.

Nordin, C. F., Jr., and Curtis, W. F., 1962, Formation and deposition of clay balls, Rio Puerco, New Mexico, in Short papers in geology, hydrogeology, and topography: U.S. Geol. Survey Prof. Paper 450-B, p. B37-B40.

Pettyjohn, W. A., 1967, Multiple drift sheets in southwestern Ward County, North Dakota, in Clayton, Lee, and Freers, T. F., Glacial Geology of the Missouri Coteau and adjacent areas: North Dakota Geol. Survey Misc., Ser. 30, p. 123-29.

Twenhofel, W. H., 1932, Treatise on sedimentation (2d ed.): Baltimore, Williams and Wilkins Co., 926 p.

J. T. Andrews is associated with the Institute of Arctic and Alpine Research and the Department of Geological Science at the University of Colorado.

G. S. Boulton is professor of geology at the University of East Anglia, Norwich, England.

E. A. Christiansen is geology division director with the Saskatchewan Research Council in Saskatoon.

L. D. Drake is assistant professor of geology at the University of Iowa, Iowa City.

L. R. Drees is a technical assistant in the Department of Agronomy at the Ohio State University, Columbus.

A. Dreimanis is professor of geology at the University of Western Ontario, London, Ontario.

P. B. DuMontelle is assistant geologist with the Illinois State Geological Survey, Natural Resources Building, Urbana.

E. B. Evenson is NSF trainee in the Department of Geology at the University of Michigan, Ann Arbor.

Jane L. Forsyth is associate professor of geology at Bowling Green State University.

H. D. Glass is geologist with the Illinois State Geological Survey, Natural Resources Building, Urbana.

R. P. Goldthwait is professor of geology at the Ohio State University, Columbus.

D. L. Gross is assistant geologist with the Illinois State Geological Survey, Natural Resources Building, Urbana.

G. F. Hall is associate professor of agronomy at the Ohio State University, Columbus.

N. C. Hester is geologist with the Illinois State Geological Survey, Natural Resources Building, Urbana.

N. Holowaychuk is professor of agronomy at the Ohio State University, Columbus.

W. H. Johnson is associate professor of geology at the University of Illinois, Urbana.

J. P. Kempton is geologist with the Illinois State Geological Survey, Natural Resources Building, Urbana.

R. W. Lemke is geologist with the U.S. Geological Survey, Denver, Colorado.

J. A. Lineback is geologist with the Illinois State Geological Survey, Natural Resources Building, Urbana.

P. MacClintock (now deceased) was emeritus professor of geology at Princeton University.

S. R. Moran is geologist with the North Dakota Geological Survey at Grand Forks.

L. H. Nobles is professor of geology at Northwestern University.

F. Pessl, Jr., is project geologist with the U.S. Geological Survey in Middletown, Connecticut.

W. A. Pettyjohn is professor of Geology at the Ohio State University, Columbus.

J. Ramsden is assistant professor of geology at the University of Alberta, Edmonton.

N. E. Smeck is assistant professor of agronomy at the University of North Dakota, Grand Forks.

J. R. Steiger is soils scientist with the USDA Soil Conservation Service at Bucyrus, Ohio.

D. P. Stewart is associate professor of geology at Miami University, Oxford, Ohio.

U. J. Vagners is lecturer in geology at Acadia University, Wolfville, Nova Scotia.

J. H. Weertman is professor of materials science at Northwestern University.

J. A. Westgate is associate professor of geology at the University of Alberta, Edmonton.

G. W. White is emeritus research professor of geology at the University of Illinois, Urbana.

L. P. Wilding is professor of agronomy at the Ohio State University, Columbus.